INSTRUMENT PROCEDURES HANDBOOK

INSTRUMENT PROCEDURES HANDBOOK

FAA-H-8261-1A

U.S. Department of Transportation
FEDERAL AVIATION ADMINISTRATION
Flight Standards Service

SKYHORSE PUBLISHING

Skyhorse Publishing books may be purchased in bulk at special discounts for sales promotion, corporate gifts, fund-raising, or educational purposes. Special editions can also be created to specifications. For details, contact the Special Sales Department, Skyhorse Publishing, 307 West 36th Street, 11th Floor, New York, NY 10018 or info@skyhorsepublishing.com.

Skyhorse® and Skyhorse Publishing® are registered trademarks of Skyhorse Publishing, Inc.®, a Delaware corporation.

www.skyhorsepublishing.com

10 9 8 7 6 5 4 3 2 1

Library of Congress Cataloging-in-Publication Data is available on file.
ISBN: 978-1-61608-271-0

Printed in China

PREFACE

This handbook supercedes FAA-H-8261-1, *Instrument Procedures Handbook*, dated 2004. It is designed as a technical reference for professional pilots who operate under instrument flight rules (IFR) in the National Airspace System (NAS). It expands on information contained in the FAA-H-8083-15, *Instrument Flying Handbook*, and introduces advanced information for IFR operations. Instrument flight instructors, instrument pilots, and instrument students will also find this handbook a valuable resource since it is used as a reference for the Airline Transport Pilot and Instrument Knowledge Tests and for the Practical Test Standards. It also provides detailed coverage of instrument charts and procedures including IFR takeoff, departure, en route, arrival, approach, and landing. Safety information covering relevant subjects such as runway incursion, land and hold short operations, controlled flight into terrain, and human factors issues also are included.

This handbook conforms to pilot training and certification concepts established by the FAA. Where a term is defined in the text, it is shown in blue. Terms and definitions are also located in Appendix C. The discussion and explanations reflect the most commonly used instrument procedures. Occasionally, the word "must" or similar language is used where the desired action is deemed critical. The use of such language is not intended to add to, interpret, or relieve pilots of their responsibility imposed by Title 14 of the Code of Federal Regulations (14 CFR).

It is essential for persons using this handbook to also become familiar with and apply the pertinent parts of 14 CFR and the *Aeronautical Information Manual* (AIM). The CFR, AIM, this handbook, AC 00.2-15, Advisory Circular Checklist, which transmits the current status of FAA advisory circulars, and other FAA technical references are available via the Internet at the FAA Home Page http://www.faa.gov. Information regarding the purchase of FAA subscription products such as charts, Airport/Facility Directory, and other publications can be accessed at http://www.naco.faa.gov/.

Comments regarding this handbook should be sent to AFS420.IPH@FAA.Gov or U.S. Department of Transportation, Federal Aviation Administration, Flight Procedure Standards Branch, AFS-420, P.O. Box 25082, Oklahoma City, OK 73125.

ACKNOWLEDGEMENTS

The following individuals and their organizations are gratefully acknowledged for their valuable contribution and commitment to the publication of this handbook:

FAA: Project Manager–Steve Winter; Assistant Project Manager–Lt Col Paul McCarver (USAF); Program Analyst–Alan Brown; Chief Editor–Fran Chaffin; Legal Review–Mike Webster; Subject Matter Experts–Dean Alexander, John Bickerstaff, Barry Billmann, Larry Buehler, Dan Burdette, Jack Corman, Dave Eckles, Gary Harkness, Hooper Harris, Harry Hodges, John Holman, Bob Hlubin, Gerry Holtorf, Steve Jackson, Scott Jerdan, Alan Jones, Norm Le Fevre, Barry Miller, John Moore, T. J. Nichols, Jim Nixon, Dave Olsen, Don Pate, Gary Powell, Phil Prasse, Larry Ramirez, Mark Reisweber, Dave Reuter, Jim Seabright, Eric Secretan, Ralph Sexton, Tom Schneider, Lou Volchansky, Mike Webb, and Mike Werner.

Jeppesen: Project Managers–Pat Willits, James W. Powell, and Dick Snyder; Technical Editors–Dave Schoeman and Chuck Stout; Media Manager–Rich Hahn; Artists–Paul Gallaway and Pat Brogan.

NOTICE

The U.S. Govt. does not endorse products or manufacturers. Trade or manufacturers' names appear herein solely because they are considered essential to the objective of this handbook.

CONTENTS

Chapter 3 — En Route Operations

Chapter 6 — System Improvement Plans

Chapter 7 — Helicopter Instrument Procedures

Appendix A — Airborne Navigation Databases

CHAPTER 1
IFR OPERATIONS IN THE NATIONAL AIRSPACE SYSTEM

Today's National Airspace System (NAS) consists of a complex collection of facilities, systems, equipment, procedures, and airports operated by thousands of people to provide a safe and efficient flying environment. The NAS includes:

- More than 690 air traffic control (ATC) facilities with associated systems and equipment to provide radar and communication service.

- Volumes of procedural and safety information necessary for users to operate in the system and for Federal Aviation Administration (FAA) employees to effectively provide essential services.

- More than 19,800 airports capable of accommodating an array of aircraft operations, many of which support instrument flight rules (IFR) departures and arrivals.

- Approximately 11,120 air navigation facilities.

- Approximately 45,800 FAA employees who provide air traffic control, flight service, security, field maintenance, certification, systems acquisitions, and a variety of other services.

- Approximately 13,000 instrument flight procedures as of September 2005, including 1,159 instrument landing system (ILS), 121 ILS Category (CAT) II, 87 ILS CAT III, 7 ILS with precision runway monitoring (PRM), 3 microwave landing system (MLS), 1,261 nondirectional beacon (NDB), 2,638 VHF omnidirectional range (VOR), and 3,530 area navigation (RNAV), 30 localizer type directional aid (LDA), 1,337 localizer (LOC), 17 simplified directional facility (SDF), 607 standard instrument departure (SID), and 356 standard terminal arrival (STAR).

- Approximately 48,200,000 instrument operations logged by FAA towers annually, of which 30 percent are air carrier, 27 percent air taxi, 37 percent general aviation, and 6 percent military.

America's aviation industry is projecting continued increases in business, recreation, and personal travel. The FAA expects airlines in the United States (U.S.) to carry about 45 percent more passengers by the year 2015 than they do today. [Figure 1-1]

Figure 1-1. IFR Operations in the NAS.

BRIEF HISTORY OF THE NATIONAL AIRSPACE SYSTEM

About two decades after the introduction of powered flight, aviation industry leaders believed that the airplane would not reach its full commercial potential without federal action to improve and maintain safety standards. In response to their concerns, the U.S. Congress passed the Air Commerce Act of May 20, 1926, marking the onset of the government's hand in regulating civil aviation. The act charged the Secretary of Commerce with fostering air commerce, issuing and enforcing air traffic rules, licensing pilots, certifying aircraft, establishing airways, and operating and maintaining aids to air navigation. As commercial flying increased, the Bureau of Air Commerce — a division of the Department of Commerce—encouraged a group of airlines to establish the first three centers for providing ATC along the airways. In 1936, the bureau took over the centers and began to expand the ATC system. [Figure 1-2] The pioneer air traffic controllers used maps, blackboards, and mental calculations to ensure the safe separation of aircraft traveling along designated routes between cities.

helped controllers to keep abreast of the postwar boom in commercial air transportation.

Following World War II, air travel increased, but with the industry's growth came new problems. In 1956 a midair collision over the Grand Canyon killed 128 people. The skies were getting too crowded for the existing systems of aircraft separation, and with the introduction of jet airliners in 1958 Congress responded by passing the Federal Aviation Act of 1958, which transferred CAA functions to the FAA (then the Federal Aviation Agency). The act entrusted safety rulemaking to the FAA, which also held the sole responsibility for developing and maintaining a common civil-military system of air navigation and air traffic control. In 1967, the new Department of Transportation (DOT) combined major federal transportation responsibilities, including the FAA (now the Federal Aviation Administration) and a new National Transportation Safety Board (NTSB).

By the mid-1970s, the FAA had achieved a semi-automated ATC system based on a marriage of radar and computer technology. By automating certain routine tasks, the system allowed controllers to concentrate more efficiently on the task of providing aircraft separation. Data appearing directly on the controllers' scopes provided the identity, altitude, and groundspeed of aircraft carrying radar beacons. Despite its effectiveness, this system required continuous enhancement to keep pace with the increased air traffic of the late 1970s, due in part to the competitive environment created by airline deregulation.

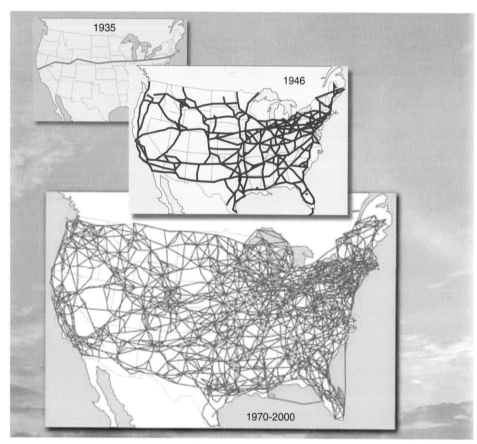

Figure 1-2. ATC System Expansion.

On the eve of America's entry into World War II, the Civil Aeronautics Administration (CAA)—charged with the responsibility for ATC, airman and aircraft certification, safety enforcement, and airway development—expanded its role to cover takeoff and landing operations at airports. Later, the addition of radar

To meet the challenge of traffic growth, the FAA unveiled the NAS Plan in January 1982. The new plan called for more advanced systems for en route and terminal ATC, modernized flight service stations, and improvements in ground-to-air surveillance and communication. Continued ATC modernization under the NAS Plan included such steps as the implementation of Host Computer Systems (completed in 1988) that were able to accommodate new programs needed for the future. [Figure 1-3]

Figure 1-3. National Airspace System Plan.

- Airport weather conditions.
- En route severe weather.

The goal of the OEP is to expand capacity, decrease delays, and improve efficiency while maintaining safety and security. With reliance on the strategic support of the aviation community, the OEP is limited in scope, and only contains programs to be accomplished over a ten-year period. Programs may move faster, but the OEP sets the minimum schedule. Considered a living document that matures over time, the OEP is continually updated as decisions are made, risks are identified and mitigated, or new solutions to operational problems are discovered through research.

An important contributor to FAA plans is the Performance-Based Operations Aviation Rulemaking Committee (PARC). The objectives and scope of PARC are to provide a forum for the U.S. aviation community to discuss and resolve issues, provide direction for U.S. flight operations criteria, and produce U.S. consensus positions for global harmonization.

In February 1991, the FAA replaced the NAS Plan with the more comprehensive Capital Investment Plan (CIP), which outlined a program for further enhancement of the ATC system, including higher levels of automation as well as new radar, communications, and weather forecasting systems. One of the CIP's programs currently underway is the installation and upgrading of airport surface radars to reduce runway incursions and prevent accidents on airport runways and taxiways. The FAA is also placing a high priority on speeding the application of the GPS satellite technology to civil aeronautics. Another notable ongoing program is encouraging progress toward the implementation of Free Flight, a concept aimed at increasing the efficiency of high-altitude operations.

The general goal of the committee is to develop a means to implement improvements in operations that address safety, capacity, and efficiency objectives, as tasked, that are consistent with international implementation. This committee provides a forum for the FAA, other government entities, and affected members of the aviation community to discuss issues and to develop resolutions and processes to facilitate the evolution of safe and efficient operations.

NATIONAL AIRSPACE SYSTEM PLANS

FAA planners' efforts to devise a broad strategy to address capacity issues resulted in the Operational Evolution Plan (OEP)—the FAA's commitment to meet the air transportation needs of the U.S. for the next ten years.

To wage a coordinated strategy, OEP executives met with representatives from the entire aviation community—including airlines, airports, aircraft manufacturers, service providers, pilots, controllers, and passengers. They agreed on four core problem areas:

- Arrival and departure rates.
- En route congestion.

Current efforts associated with NAS modernization come with the realization that all phases must be integrated. The evolution to an updated NAS must be well orchestrated and balanced with the resources available. Current plans for NAS modernization focus on three key categories:

- Upgrading the infrastructure.
- Providing new safety features.
- Introducing new efficiency-oriented capabilities into the existing system.

It is crucial that our NAS equipment is protected, as lost radar, navigation signals, or communications capabilities can slow the flow of aircraft to a busy city, which in turn, could cause delays throughout the entire region, and possibly, the whole country.

The second category for modernization activities focuses on upgrades concerning safety. Although we cannot control the weather, it has a big impact on the NAS. Fog in San Francisco, snow in Denver, thunderstorms in Kansas, wind in Chicago; all of these reduce the safety and capacity of the NAS. Nevertheless, great strides are being made in our ability to predict the weather. Controllers are receiving better information about winds and storms, and pilots are receiving better information both before they take off and in flight—all of which makes flying safer. [Figure 1-4]

Another cornerstone of the FAA's future is improved navigational information available in the cockpit. The Wide Area Augmentation System (WAAS) initially became operational for aviation use on July 10, 2003. It improves conventional GPS signal accuracy by an order of magnitude, from about 20 meters to 2 meters or less.

Moreover, the local area augmentation system (LAAS) is being developed to provide even better accuracy than GPS with WAAS. LAAS will provide localized service for final approaches in poor weather conditions at major airports. This additional navigational accuracy will be available in the cockpit and will be used for other system enhancements. More information about WAAS and LAAS is contained in Chapters 5 and 6.

The Automatic Dependent Surveillance (ADS) system, currently being developed by the FAA and several airlines, enables the aircraft to automatically transmit its location to various receivers. This broadcast mode, commonly referred to as ADS-B, is a signal that can be received by other properly equipped aircraft and ground based transceivers, which in turn feed the automation system accurate aircraft position information. This more accurate information will be used to improve the efficiency of the system—the third category of modernization goals.

Other key efficiency improvements are found in the deployment of new tools designed to assist the controller. For example, most commercial aircraft already have equipment to send their GPS positions automatically to receiver stations over the ocean. This key enhancement is necessary for all aircraft operating in oceanic airspace and allows more efficient use of airspace. Another move is toward improving text and graphical message exchange, which is the ultimate goal of the Controller Pilot Data Link Communications (CPDLC) Program.

In the en route domain, the Display System Replacement (DSR), along with the Host/Oceanic Computer System Replacement (HOCSR) and Eunomia projects, are the platforms and infrastructure for the future. These provide new displays to the controllers, upgrade the computers to accept future tools, and provide modern surveillance and flight data processing capabilities. For CPDLC to work effectively, it must be integrated with the en route controller's workstation.

RNAV PLANS

Designing routes and airspace to reduce conflicts between arrival and departure flows can be as simple as adding extra routes or as comprehensive as a full redesign in which multiple airports are jointly optimized. New strategies are in place for taking advantage of existing structures to departing aircraft through congested transition airspace. In other cases, RNAV procedures are used to develop new routes that reduce flow complexity by permitting aircraft to fly optimum routes with minimal controller intervention. These new routes spread the flow across the terminal and transition airspace so aircraft can be separated with optimal lateral distances and altitudes in and around the terminal area. In some cases, the addition

Figure 1-4. Modernization Activities Provide Improved Weather Information.

of new routes alone is not sufficient, and redesign of existing routes and flows are required. Benefits are multiplied when airspace surrounding more than one airport (e.g., in a metropolitan area) can be jointly optimized.

SYSTEM SAFETY

Although hoping to decrease delays, improve system capacity, and modernize facilities, the ultimate goal of the NAS Plan is to improve system safety. If statistics are any indication, the beneficial effect of the implementation of the plan may already be underway as aviation safety seems to have increased in recent years. The FAA has made particular emphasis to not only reduce the number of accidents in general, but also to make strides in curtailing controlled flight into terrain (CFIT) and runway incursions as well as continue approach and landing accident reduction (ALAR).

The term CFIT defines an accident in which a fully qualified and certificated crew flies a properly working airplane into the ground, water, or obstacles with no apparent awareness by the pilots. A runway incursion is defined as any occurrence at an airport involving an aircraft, vehicle, person, or object on the ground that creates a collision hazard or results in a loss of separation with an aircraft taking off, attempting to take off, landing, or attempting to land. The term ALAR applies to an accident that occurs during a visual approach, during an instrument approach after passing the initial approach fix (IAF), or during the landing maneuver. This term also applies to accidents occurring when circling or when beginning a missed approach procedure.

ACCIDENT RATES

The NTSB released airline accident statistics for 2004 that showed a decline from the previous year. Twenty-nine accidents on large U.S. air carriers were recorded in 2004, which is a decrease from the 54 accidents in 2003.

Accident rates for both general aviation airplanes and helicopters also decreased in 2004. General aviation airplane accidents dropped from 1,742 to 1,595, while helicopter accidents declined from 213 to 176. The number of accidents for commuter air services went up somewhat, from 2 accidents in 2003 to 5 in 2004. Air taxi operations went from 76 accidents in 2003 to 68 accidents in 2004. These numbers do not tell the whole story. Because the number of flights and flight hours increased in 2004, accident rates per 100,000 departures or per 100,000 flight hours will likely be even lower.

Among the top priorities for accident prevention are CFIT and ALAR. Pilots can decrease exposure to a CFIT accident by identifying risk factors and remedies prior to flight. [Figure 1-5] Additional actions on the CFIT reduction front include equipping aircraft with state-of-the art terrain awareness and warning systems (TAWS), sometimes referred to as enhanced ground proximity warning systems (EGPWS). This measure alone is expected to reduce CFIT accidents by at least 90

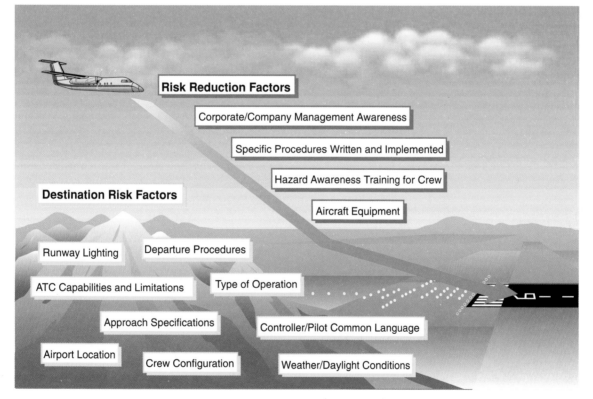

Figure 1-5. CFIT Reduction.

percent. With very few exceptions, all U.S. turbine powered airplanes with more than six passenger seats were required to be equipped with TAWS by March 29, 2005.

Added training for aircrews and controllers is part of the campaign to safeguard against CFIT, as well as making greater use of approaches with vertical guidance that use a constant angle descent path to the runway. This measure offers nearly a 70 percent potential reduction. Another CFIT action plan involves a check of ground-based radars to ensure that the minimum safe altitude warning (MSAW) feature functions correctly.

Like CFIT, the ALAR campaign features a menu of actions, three of which involve crew training, altitude awareness policies checklists, and smart alerting technology. These three alone offer a potential 20 to 25 percent reduction in approach and landing accidents. Officials representing Safer Skies—a ten-year collaborative effort between the FAA and the airline industry—believe that the combination of CFIT and ALAR interventions will offer more than a 45 percent reduction in accidents.

RUNWAY INCURSION STATISTICS

While it is difficult to eliminate runway incursions, technology offers the means for both controllers and flight crews to create situational awareness of runway incursions in sufficient time to prevent accidents. Consequently, the FAA is taking actions that will identify and implement technology solutions, in conjunction with training and procedural evaluation and changes, to reduce runway accidents. Recently established programs that address runway incursions center on identifying the potential severity of an incursion and reducing the likelihood of incursions through training, technology, communications, procedures, airport signs/marking/lighting, data analysis, and developing local solutions. The FAA's initiatives include:

- Promoting aviation community participation in runway safety activities and solutions.

- Appointing nine regional Runway Safety Program Managers.

- Providing training, education, and awareness for pilots, controllers, and vehicle operators.

- Publishing an advisory circular for airport surface operations.

- Increasing the visibility of runway hold line markings.

- Reviewing pilot-controller phraseology.

- Providing foreign air carrier pilot training, education, and awareness.

- Requiring all pilot checks, certifications, and flight reviews to incorporate performance evaluations of ground operations and test for knowledge.

- Increasing runway incursion action team site visits.

- Deploying high-technology operational systems such as the Airport Surface Detection Equipment-3 (ASDE-3) and Airport Surface Detection Equipment-X (ASDE-X).

- Evaluating cockpit display avionics to provide direct warning capability to flight crew(s) of both large and small aircraft operators.

Statistics compiled for 2004 show that there were 310 runway incursions, down from 332 in 2003. The number of Category A and Category B runway incursions, in which there is significant potential for collision, declined steadily from 2000 through 2003. There were less than half as many such events in 2003 as in 2000. The number of Category A incursions, in which separation decreases and participants take extreme action to narrowly avoid a collision, or in which a collision occurs, dropped to 10 per year.

SYSTEM CAPACITY

On the user side, there are more than 740,000 active pilots operating over 319,000 commercial, regional, general aviation, and military aircraft. This results in more than 49,500 flights per day. Figure 1-6 depicts over 5,000 aircraft operating at the same time in the U.S. shown on this Air Traffic Control System Command Center (ATCSCC) screen.

TAKEOFFS AND LANDINGS

According to the FAA Administrator's Fact Book for March 2005, there were 46,873,000 operations at airports with FAA control towers, an average of more than 128,000 aircraft operations per day. These figures do not include the tens of millions of operations at airports that do not have a control tower. User demands on the NAS are quickly exceeding the ability of current resources to fulfill them. Delays in the NAS for 2004 were slightly higher than in 2000, with a total of 455,786 delays of at least 15 minutes in 2004, compared to 450,289 in 2000. These illustrations of the increasing demands on the NAS indicate that current FAA modernization efforts are well justified. Nothing short of the integrated, systematic, cooperative, and comprehensive approach spelled out by the OEP can bring the NAS to the safety and efficiency standards that the flying public demands.

AIR TRAFFIC CONTROL SYSTEM COMMAND CENTER

The task of managing the flow of air traffic within the NAS is assigned to the Air Traffic Control System

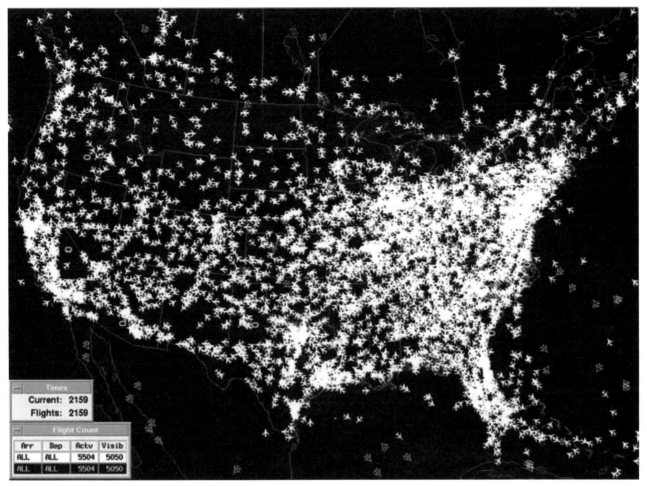

Figure 1-6. Approximately 5,000 Aircraft in ATC System at One Time.

Command Center (ATCSCC). Headquartered in Herndon, Virginia, the ATCSCC has been operational since 1994 and is located in one of the largest and most sophisticated facilities of its kind. The ATCSCC regulates air traffic at a national level when weather, equipment, runway closures, or other conditions place stress on the NAS. In these instances, traffic management specialists at the ATCSCC take action to modify traffic demands in order to remain within system capacity. They accomplish this in cooperation with:

- Airline personnel.

- Traffic management specialists at affected facilities.

- Air traffic controllers at affected facilities.

Efforts of the ATCSCC help minimize delays and congestion and maximize the overall use of the NAS, thereby ensuring safe and efficient air travel within the U.S. For example, if severe weather, military operations, runway closures, special events, or other factors affect air traffic for a particular region or airport, the ATCSCC mobilizes its resources and various agency personnel to analyze, coordinate, and reroute (if necessary) traffic to foster maximum efficiency and utilization of the NAS.

The ATCSCC directs the operation of the traffic management (TM) system to provide a safe, orderly, and expeditious flow of traffic while minimizing delays. TM is apportioned into traffic management units (TMUs), which monitor and balance traffic flows within their areas of responsibility in accordance with TM directives. TMUs help to ensure system efficiency and effectiveness without compromising safety, by providing the ATCSCC with advance notice of planned outages and runway closures that will impact the air traffic system, such as NAVAID and radar shutdowns, runway closures, equipment and computer malfunctions, and procedural changes. [Figure 1-7 on page 1-8]

HOW THE SYSTEM COMPONENTS WORK TOGETHER

The NAS comprises the common network of U.S. airspace, air navigation facilities, equipment, services, airports and landing areas, aeronautical charts, information and services, rules and regulations, procedures, technical information, manpower, and material. Included are system components shared jointly with the military. The underlying demand for air commerce is people's desire to travel for business and pleasure and to ship cargo by air. This demand grows with the economy independent of the capacity or performance of the NAS. As the economy grows, more and more people want to fly, whether the system can handle it or

Figure 1-7. A real-time Airport Status page displayed on the ATCSCC Web site (www.fly.faa.gov/flyfaa/usmap.jsp) provides general airport condition status. Though not flight specific, it portrays current general airport trouble spots. Green indicates less than five-minute delays. Yellow means departures and arrivals are experiencing delays of 16 to 45 minutes. Traffic destined to orange locations is being delayed at the departure point. Red airports are experiencing taxi or airborne holding delays greater than 45 minutes. Blue indicates closed airports.

not. Realized demand refers to flight plans filed by the airlines and other airspace users to access the system. It is moderated by the airline's understanding of the number of flights that can be accommodated without encountering unacceptable delay, and is limited by the capacity for the system.

USERS

Despite a drop in air traffic after the September 11 terrorist attacks, air travel returned to 2000 levels within three years and exceeded them in 2004. Industry forecasts predict growth in airline passenger traffic of around 4.3 percent per year. Commercial aviation is expected to exceed one billion passengers by 2015. The system is nearing the point of saturation, with limited ability to grow unless major changes are brought about.

Adding to the growth challenge, users of the NAS cover a wide spectrum in pilot skill and experience, aircraft types, and air traffic service demands, creating a challenge to the NAS to provide a variety of services that accommodate all types of traffic. NAS users range from professional airline, commuter, and corporate pilots to single-engine piston pilots, as well as owner-operators of personal jets to military jet fighter trainees.

AIRLINES

Though commercial air carrier aircraft traditionally make up less than 5 percent of the civil aviation fleet, they account for about 30 percent of the instrument operations flown in civil aviation. Commercial air carriers are the most homogenous category of airspace users, although there are some differences between U.S. trunk

carriers (major airlines) and regional airlines (commuters) in terms of demand for ATC services. Generally, U.S. carriers operate large, high performance airplanes that cruise at altitudes above 18,000 feet. Conducted exclusively under IFR, airline flights follow established schedules and operate in and out of larger and better-equipped airports. In terminal areas, however, they share airspace and facilities with all types of traffic and must compete for airport access with other users. Airline pilots are highly proficient and thoroughly familiar with the rules and procedures under which they must operate.

Some airlines are looking toward the use of larger aircraft, with the potential to reduce airway and terminal congestion by transporting more people in fewer aircraft. This is especially valuable at major hub airports, where the number of operations exceeds capacity at certain times of day. On the other hand, the proliferation of larger aircraft also requires changes to terminals (e.g., double-decker jetways and better passenger throughput), rethinking of rescue and fire-fighting strategies, taxiway fillet changes, and perhaps stronger runways and taxiways.

Commuter airlines also follow established schedules and are flown by professional pilots. Commuters characteristically operate smaller and lower performance aircraft in airspace that must often be shared by general aviation (GA) aircraft, including visual flight rules (VFR) traffic. As commuter operations have grown in volume, they have created extra demands on the airport and ATC systems. At one end, they use hub airports along with other commercial carriers, which contributes to growing congestion at major air traffic hubs. IFR-equipped and operating under IFR like other air carriers, commuter aircraft cannot be used to full advantage unless the airport at the other end of the flight, typically a small community airport, also is capable of IFR operation. Thus, the growth of commuter air service has created pressure for additional instrument approach procedures and control facilities at smaller airports. A growing trend among the major airlines is the proliferation of regional jets (RJs). RJs are replacing turboprop aircraft and they are welcomed by some observers as saviors of high-quality jet aircraft service to small communities. RJs are likely to be a regular feature of the airline industry for a long time because passengers and airlines overwhelmingly prefer RJs to turboprop service. From the passengers' perspective, they are far more comfortable; and from the airlines' point of view, they are more profitable. Thus, within a few years, most regional air traffic in the continental U.S. will be by jet, with turboprops filling a smaller role.

FAA and industry studies have investigated the underlying operational and economic environments of RJs on the ATC system. They have revealed two distinct trends: (1) growing airspace and airport congestion is exacerbated by the rapid growth of RJ traffic, and (2) potential airport infrastructure limitations may constrain airline business. The FAA, the Center for Advanced Aviation System Development (CAASD), major airlines, and others are working to find mitigating strategies to address airline congestion. With nearly 2,000 RJs already in use—and double that expected over the next few years—the success of these efforts is critical if growth in the regional airline industry is to be sustained. [Figure 1-8]

CORPORATE AND FRACTIONAL OWNERSHIPS
Though technically considered under the GA umbrella, the increasing use of sophisticated, IFR-equipped aircraft by businesses and corporations has created a niche of its own. By using larger high performance airplanes and

Figure 1-8. Increasing use of regional jets is expected to have a significant impact on traffic.

equipping them with the latest avionics, the business portion of the GA fleet has created demands for ATC services that more closely resemble commercial operators than the predominately VFR general aviation fleet.

GENERAL AVIATION
The tendency of GA aircraft owners to upgrade the performance and avionics of their aircraft increases the demand for IFR services and for terminal airspace at airports. In response, the FAA has increased the extent of controlled airspace and improved ATC facilities at major airports. The safety of mixing IFR and VFR traffic is a major concern, but the imposition of measures to separate and control both types of traffic creates more restrictions on airspace use and raises the level of aircraft equipage and pilot qualification necessary for access.

MILITARY
From an operational point of view, military flight activities comprise a subsystem that must be fully integrated within NAS. However, military aviation has unique requirements that often are different from civil aviation users. The military's need for designated training areas and low-level routes located near their bases sometimes conflicts with civilian users who need to detour around these areas. In coordinating the development of ATC systems and services for the armed forces, the FAA is challenged to achieve a maximum degree of compatibility between civil and military aviation objectives.

ATC FACILITIES
FAA figures show that the NAS includes more than 18,300 airports, 21 ARTCCs, 197 TRACON facilities, over 460 air traffic control towers (ATCTs), 58 flight service stations and automated flight service stations (FSSs/AFSSs), and approximately 4,500 air navigation facilities. Several thousand pieces of maintainable equipment including radar, communications switches, ground-based navigation aids, computer displays, and radios are used in NAS operations, and NAS components represent billions of dollars in investments by the government. Additionally, the aviation industry has invested significantly in ground facilities and avionics systems designed to use the NAS. Approximately 47,000 FAA employees provide air traffic control, flight service, security, field maintenance, certification, system acquisition, and other essential services.

Differing levels of ATC facilities vary in their structure and purpose. Traffic management at the national level is led by the Command Center, which essentially "owns" all airspace. Regional Centers, in turn, sign Letters of Agreement (LOAs) with various approach control facilities, delegating those facilities chunks of airspace in which that approach control facility has jurisdiction. The approach control facilities, in turn, sign LOAs with various towers that are within that airspace, further delegating airspace and responsibility. This ambiguity has created difficulties in communication between the local facilities and the Command Center. However, a decentralized structure enables local flexibility and a tailoring of services to meet the needs of users at the local level. Improved communications between the Command Center and local facilities could support enhanced safety and efficiency while maintaining both centralized and decentralized aspects to the ATC system.

AIR ROUTE TRAFFIC CONTROL CENTER

A Center's primary function is to control and separate air traffic within a designated airspace, which may cover more than 100,000 square miles, may span several states, and extends from the base of the underlying controlled airspace up to Flight Level (FL) 600. There are 21 Centers located throughout the U.S., each of which is divided into sectors. Controllers assigned to these sectors, which range from 50 to over 200 miles wide, guide aircraft toward their intended destination by way of vectors and/or airway assignment, routing aircraft around weather and other traffic. Centers employ 300 to 700 controllers, with more than 150 on duty during peak hours at the busier facilities. A typical flight by a commercial airliner is handled mostly by the Centers.

TERMINAL RADAR APPROACH CONTROL

Terminal Radar Approach Control (TRACON) controllers work in dimly lit radar rooms located within the control tower complex or in a separate building located on or near the airport it serves. [Figure 1-9] Using radarscopes, these controllers typically work an area of airspace with a 50-mile radius and up to an altitude of 17,000 feet. This airspace is configured to provide service to a primary airport, but may include other airports that are within 50 miles of the radar service area. Aircraft within this area are provided vectors to airports, around terrain, and weather, as well as separation from other aircraft. Controllers in TRACONs determine the arrival sequence for the control tower's designated airspace.

CONTROL TOWER

Controllers in this type of facility manage aircraft operations on the ground and within specified airspace around an airport. The number of controllers in the tower varies with the size of the airport. Small general aviation airports typically have three or four controllers, while larger international airports can have up to fifteen controllers talking to aircraft, processing flight plans, and coordinating air traffic flow. Tower controllers manage the ground movement of aircraft around the airport and ensure appropriate spacing between aircraft taking off and landing. In addition, it is the responsibility of the control tower to determine the landing sequence between aircraft under its control. Tower controllers issue a variety of instructions to

Figure 1-9. Terminal Radar Approach Control.

pilots, from how to enter a pattern for landing to how to depart the airport for their destination.

FLIGHT SERVICE STATIONS

Flight Service Stations (FSSs) and Automated Flight Service Stations (AFSSs) are air traffic facilities which provide pilot briefings, en route communications and VFR search and rescue services, assist lost aircraft and aircraft in emergency situations, relay ATC clearances, originate Notices to Airmen, broadcast aviation weather and NAS information, receive and process IFR flight plans, and monitor navigational aids (NAVAIDs). In addition, at selected locations, FSSs/AFSSs provide En route Flight Advisory Service (Flight Watch), take weather observations, issue airport advisories, and advise Customs and Immigration of transborder flights.

Pilot Briefers at flight service stations render preflight, in-flight, and emergency assistance to all pilots on request. They give information about actual weather conditions and forecasts for airports and flight paths, relay air traffic control instructions between controllers and pilots, assist pilots in emergency situations, and initiate searches for missing or overdue aircraft. FSSs/AFSSs provide information to all airspace users, including the military. In October 2005, operation of all FSSs/AFSSs, except those in Alaska, was turned over to the Lockheed Martin Corporation. In the months after the transition, 38 existing AFSSs are slated to close, leaving 17 "Legacy" stations and 3 "Hub" stations. Services to pilots are expected to be equal to or better than prior to the change, and the contract is expected to save the government about $2.2 billion over ten years.

FLIGHT PLANS

Prior to flying in controlled airspace under IFR conditions or in Class A airspace, pilots are required to file a flight plan. IFR (as well as VFR) flight plans provide air traffic center computers with accurate and precise routes required for flight data processing (FDP[1]). The computer knows every route (published and unpub-

---

[1] FDP maintains a model of the route and other details for each aircraft.

lished) and NAVAID, most intersections, and all airports, and can only process a flight plan if the proposed routes and fixes connect properly. Center computers also recognize preferred routes and know that forecast or real-time weather may change arrival routes. Centers and TRACONs now have a computer graphic that can show every aircraft on a flight plan in the U.S. as to its flight plan information and present position. Despite their sophistication, center computers do not overlap in coverage or information with other Centers, so that flight requests not honored in one must be repeated in the next.

RELEASE TIME

ATC uses an IFR release time[2] in conjunction with traffic management procedures to separate departing aircraft from other traffic. For example, when controlling departures from an airport without a tower, the controller limits the departure release to one aircraft at any given time. Once that aircraft is airborne and radar identified, then the following aircraft may be released for departure, provided they meet the approved radar separation (3 miles laterally or 1,000 feet vertically) when the second aircraft comes airborne. Controllers must take aircraft performances into account when releasing successive departures, so that a B-747 HEAVY aircraft is not released immediately after a departing Cessna 172. Besides releasing fast aircraft before slow ones, another technique commonly used for successive departures is to have the first aircraft turn 30 to 40 degrees from runway heading after departure, and then have the second aircraft depart on a SID or runway heading. Use of these techniques is common practice when maximizing airport traffic capacity.

EXPECT DEPARTURE CLEARANCE TIME

Another tool that the FAA is implementing to increase efficiency is the reduction of the standard expect departure clearance time[3] (EDCT) requirement. The FAA has drafted changes to augment and modify procedures contained in Ground Delay Programs (GDPs). Airlines may now update their departure times by arranging their flights' priorities to meet the controlled time of arrival. In order to evaluate the effectiveness of the new software and the airline-supplied data, the actual departure time parameter in relation to the EDCT has been reduced. This change impacts all flights (commercial and GA) operating to the nation's busiest airports. Instead of the previous 25-minute EDCT window (5 minutes prior and 20 minutes after the EDCT), the new requirement for GDP implementation is a 10-minute window, and aircraft are required to depart within 5 minutes before or after their assigned EDCT. Using reduced EDCT and other measures included in GDPs, ATC aims at reducing the number of arrival slots issued to accommodate degraded arrival capacity at an airport affected by weather. The creation of departure or ground delays is less costly and safer than airborne holding delays in the airspace at the arrival airport.

MANAGING SAFETY AND CAPACITY

SYSTEM DESIGN

The CAASD is aiding in the evolution towards free flight with its work in developing new procedures necessary for changing traffic patterns and aircraft with enhanced capabilities, and also in identifying traffic flow constraints that can be eliminated. This work supports the FAA's Operational Evolution Plan in the near-term. Rapid changes in technology in the area of navigation performance, including the change from ground-based area navigation systems, provide the foundation for aviation's global evolution. This progress will be marked by combining all elements of communication, navigation, and surveillance (CNS) with air traffic management (ATM) into tomorrow's CNS/ATM based systems. The future CNS/ATM operating environment will be based on navigation defined by geographic waypoints expressed in latitude and longitude since instrument procedures and flight routes will not require aircraft to overfly ground-based navigation aids defining specific points.

APPLICATION OF AREA NAVIGATION

RNAV airways provide more direct routings than the current VOR-based airway system, giving pilots easier access through terminal areas, while avoiding the circuitous routings now common in many busy Class B areas. RNAV airways are a critical component to the transition from ground-based navigation systems to GPS navigation. RNAV routes help maintain the aircraft flow through busy terminals by segregating arrival or departure traffic away from possibly interfering traffic flows.

Further, RNAV provides the potential for increasing airspace capacity both en route and in the terminal area in several important ways.

Strategic use of RNAV airways nationwide is reducing the cost of flying and providing aircraft owners more benefits from their IFR-certified GPS receivers. Several scenarios have been identified where RNAV routes provide a substantial benefit to users.

- Controllers are assigning routes that do not require overflying ground-based NAVAIDs such as VORs.

- The lateral separation between aircraft tracks is being reduced.

- RNAV routes lower altitude minimums on existing Victor airways where ground-based NAVAID performance (minimum reception altitude) required higher minimums.

---

[2] A release time is a departure restriction issued to a pilot by ATC, specifying the earliest and latest time an aircraft may depart.
[3] The runway release time assigned to an aircraft in a controlled departure time program and shown on the flight progress strip as an EDCT.

- RNAV routes may allow continued use of existing airways where the ground-based NAVAID has been decommissioned or where the signal is no longer suitable for en route navigation.

- The route structure can be modified quickly and easily to meet the changing requirements of the user community.

- Shorter, simpler routes can be designed to minimize environmental impact.

Dozens of new RNAV routes have been designated, and new ones are being added continuously. In order to designate RNAV airways, the FAA developed criteria, en route procedures, procedures for airway flight checks, and created new charting specifications. Some of the considerations include:

- Navigation infrastructure (i.e., the ground-based and space-based navigation positioning systems) provides adequate coverage for the proposed route/procedure.

- Navigation coordinate data meets International Civil Aviation Organization (ICAO) accuracy and integrity requirements. This means that all of the coordinates published in the Aeronautical Information Publication (AIP) and used in the aircraft navigation databases must be referenced to WGS 84, and the user must have the necessary assurance that this data has not been corrupted or inadvertently modified.

- Airborne systems meet airworthiness performance for use on the RNAV routes and procedures.

- Flight crews have the necessary approval to operate on the RNAV routes and procedures.

In the future, as aircraft achieve higher levels of navigation accuracy and integrity, closely spaced parallel routes may be introduced, effectively multiplying the number of available routes between terminal areas. RNAV can be used in all phases of flight and, when implemented correctly, results in:

- Improved situational awareness for the pilot.

- Reduced workloads for both controller and pilot.

- Reduced environmental impact from improved route and procedure designs.

- Reduced fuel consumption from shorter, more direct routes.

For example, take the situation at Philadelphia International Airport, located in the middle of some highly popular north-south traffic lanes carrying New York and Boston traffic to or from Washington, Atlanta, and Miami. Philadelphia's position is right underneath

these flows. Chokepoints resulted from traffic departing Philadelphia, needing to wait for a "hole" in the traffic above into which they could merge. The CAASD helped US Airways and Philadelphia airport officials establish a set of RNAV departure routes that do not interfere with the prevailing established traffic. Traffic heading north or south can join the established flows at a point further ahead when higher altitudes and speeds have been attained. Aircraft properly equipped to execute RNAV procedural routes can exit the terminal area faster — a powerful inducement for aircraft operators to upgrade their navigation equipment.

Another example of an RNAV departure is the PRYME TWO DEPARTURE from Washington Dulles International. Notice in Figure 1-10 the RNAV waypoints not associated with VORs help free up the flow of IFR traffic out of the airport by not funneling them to one point through a common NAVAID.

RNAV IFR TERMINAL TRANSITION ROUTES
The FAA is moving forward with an initiative to chart RNAV terminal transition routes through busy airspace. In 2001, some specific RNAV routes were implemented through Charlotte's Class B airspace, allowing RNAV-capable aircraft to cross through the airspace instead of

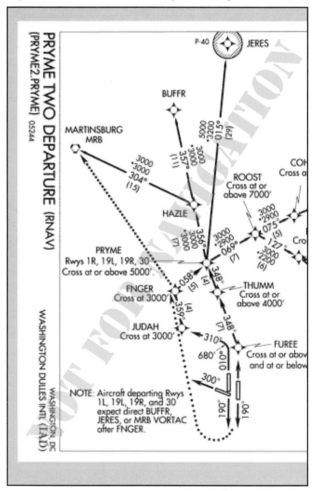

Figure 1-10. RNAV Departure Routes.

using costly and time-consuming routing around the Class B area. The original RNAV terminal transition routes have evolved into RNAV IFR terminal transition routes, or simply RITTRs.

Beginning in March 2005, with the publication of the notice of proposed rulemaking (NPRM) for the Charlotte, North Carolina, RITTRs, the FAA advanced the process of establishing and charting the first RITTRs on IFR en route low altitude charts. The five new RITTRs through Charlotte's Class B airspace took effect on September 1, 2005, making them available for pilots to file on their IFR flight plans. Additional RITTRs are planned for Cincinnati, Ohio, and Jacksonville, Florida.

The RITTRs allow IFR overflights through the Class B airspace for RNAV-capable aircraft. Without the RITTRS, these aircraft would be routinely routed around the Class B by as much as 50 miles.

REQUIRED NAVIGATION PERFORMANCE

The continuing growth of aviation places increasing demands on airspace capacity and emphasizes the need for the best use of the available airspace. These factors, along with the accuracy of modern aviation navigation systems and the requirement for increased operational efficiency in terms of direct routings and track-keeping accuracy, have resulted in the concept of required navigation performance—a statement of the navigation performance accuracy necessary for operation within a defined airspace. Required Navigation Performance (RNP) is a statement of the navigation performance necessary for operation within a defined airspace. RNP includes both performance and functional requirements, and is indicated by the RNP value. The RNP value designates the lateral performance requirement associated with a procedure. [Figure 1-11]

RNP includes a navigation specification including requirements for on-board performance monitoring and alerting. These functional and performance standards allow the flight paths of participating aircraft to be both predictable and repeatable to the declared levels of accuracy. More information on RNP is contained in subsequent chapters.

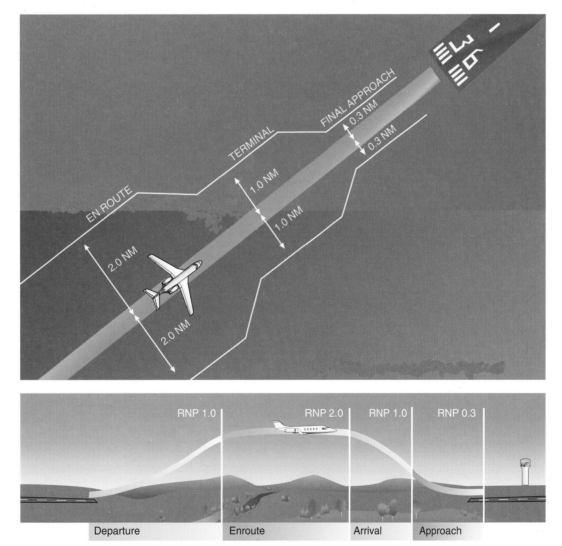

Figure 1-11. Required Navigation Performance.

The term RNP is also applied as a descriptor for airspace, routes, and procedures — including departures, arrivals, and instrument approach procedures (IAPs). The descriptor can apply to a unique approach procedure or to a large region of airspace. RNP applies to navigation performance within a designated airspace, and includes the capability of both the available infrastructure (navigation aids) and the aircraft. Washington National Airport (KDCA) introduced the first RNP approach procedure in September 2005. An example of an RNP approach chart is shown in Figure 1-12.

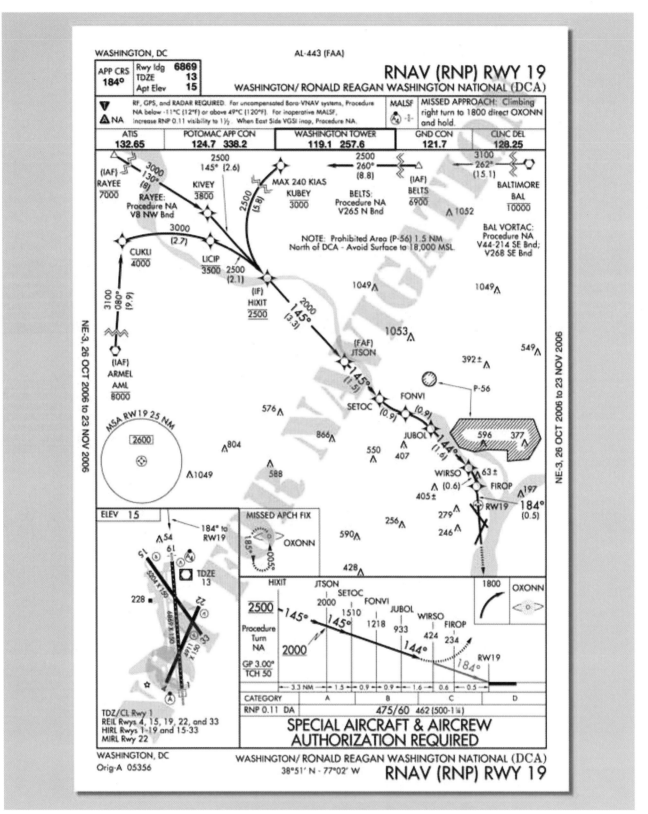

Figure 1-12. RNP Approach Chart.

The RNP value designates the lateral performance requirement associated with a procedure. The required performance is obtained through a combination of aircraft capability and the level of service provided by the corresponding navigation infrastructure. From a broad perspective:

Aircraft Capability + Level of Service = Access

In this context, aircraft capability refers to the airworthiness certification and operational approval elements (including avionics, maintenance, database, human factors, pilot procedures, training, and other issues). The level of service element refers to the NAS infrastructure, including published routes, signal-in-space performance and availability, and air traffic management. When considered collectively, these elements result in providing access. Access provides the desired benefit (airspace, procedures, routes of flight, etc.).

A key feature of RNP is the concept of on-board monitoring and alerting. This means the navigation equipment is accurate enough to keep the aircraft in a specific volume of airspace, which moves along with the aircraft. The aircraft is expected to remain within this block of airspace for at least 95 percent of the flight time. Additional airspace outside the 95 percent area is provided for continuity and integrity, so that the combined areas ensure aircraft containment 99.9 percent of the time. RNP levels are actual distances from the centerline of the flight path, which must be maintained for aircraft and obstacle separation. Although additional FAA-recognized RNP levels may be used for specific operations, the United States currently supports three standard RNP levels:

- RNP 0.3 – Approach

- RNP 1.0 – Terminal

- RNP 2.0 – Terminal and En Route

RNP 0.3 represents a distance of 0.3 nautical miles (NM) either side of a specified flight path centerline. The specific performance required on the final approach segment of an instrument approach is an example of this RNP level.

For international operations, the FAA and ICAO member states have led initiatives to apply RNP concepts to oceanic routes. Here are the ICAO RNP levels supported for international operations:

- RNP-1 – European Precision RNAV (P-RNAV)

- RNP-4 – Projected for oceanic/remote areas where 30 NM horizontal separation is applied

- RNP-5 – European Basic RNAV (B-RNAV)

- RNP-10 – Oceanic/remote areas where 50 NM lateral separation is applied

NOTE: Specific operational and equipment performance requirements apply for P-RNAV and B-RNAV.

GLOBAL POSITIONING SYSTEM

The FAA's implementation activities of the Global Positioning System (GPS) are dedicated to the adaptation of the NAS infrastructure to accept satellite navigation (SATNAV) technology through the management and coordination of a variety of overlapping NAS implementation projects. These projects fall under the project areas listed below and represent different elements of the NAS infrastructure:

- Avionics Development – includes engineering support and guidance in the development of current and future GPS avionics minimum operational performance standards (MOPS), as well as FAA Technical Standard Orders (TSOs) and establishes certification standards for avionics installations.

- Flight Standards – includes activities related to instrument procedure criteria research, design, testing, and standards publication. The shift from ground-based to space-based navigation sources has markedly shifted the paradigms used in obstacle clearance determination and standards development. New GPS-based Terminal Procedures (TERPS) manuals are in use today as a result of this effort.

- Air Traffic – includes initiatives related to the development of GPS routes, phraseology, procedures, controller GPS training and GPS outage simulations studies. GPS-based routes, developed along the East Coast to help congestion in the Northeast Corridor, direct GPS-based Caribbean routes, and expansion of RNAV activities are all results of SATNAV sponsored implementation projects.

- Procedure Development – includes the provision of instrument procedure development and flight inspection of GPS-based routes and instrument procedures. Today over 3,500 GPS-based IAPs have been developed.

- Interference Identification and Mitigation – includes the development and fielding of airborne, ground, and portable interference detection systems. These efforts are ongoing and critical to ensuring the safe use of GPS in the NAS.

To use GPS, WAAS, and/or LAAS in the NAS, equipment suitable for aviation use (such as a GPS receiver, WAAS receiver, LAAS receiver, or multi-modal receiver) must be designed, developed, and certified for use. To ensure standardization and safety of this equipment, the FAA plays a key role in the development and works closely with industry in this process. The avionics development process results in safe, standardized SATNAV avionics, developed in concurrence with industry. Due to the growing popularity of SATNAV and potential new aviation applications, there are several types of GPS-based receivers on the market, but only those that pass through this certification process can be used as approved navigation equipment under IFR conditions. Detailed information on GPS approach procedures is provided in Chapter 5–Approach.

GPS-BASED HELICOPTER OPERATIONS

The synergy between industry and the FAA created during the development of the Gulf of Mexico GPS grid system and approaches is an excellent example of what can be accomplished to establish the future of helicopter IFR SATNAV. The Helicopter Safety Advisory Council (HSAC), National Air Traffic Controllers Association (NATCA), helicopter operators, and FAA Flight Standards Divisions all worked together to develop this infrastructure. The system provides both the operational and cost-saving features of flying direct to a destination when offshore weather conditions deteriorate below VFR and an instant and accurate aircraft location capability that is invaluable for rescue operations.

The expansion of helicopter IFR service for emergency medical services (EMS) is another success story. The FAA worked with EMS operators to develop helicopter GPS nonprecision instrument approach procedures and en route criteria. As a result of this collaborative effort, EMS operators have been provided with hundreds of EMS helicopter procedures to medical facilities. Before the GPS IFR network, EMS helicopter pilots had been compelled to miss 30 percent of their missions due to weather. With the new procedures, only about 11 percent of missions are missed due to weather.

The success of these operations can be attributed in large part to the collaborative efforts between the helicopter industry and the FAA. There are currently 289 special use helicopter procedures, with more being added. There are also 37 public use helicopter approaches. Of these, 18 are to runways and 19 are to heliports or points-in-space (PinS).

REDUCED VERTICAL SEPARATION MINIMUMS

The U.S. domestic reduced vertical separation minimums (DRVSM) program has reduced the vertical separation from the traditional 2,000-foot minimum to a 1,000-foot minimum above FL 290, which allows aircraft to fly a more optimal profile, thereby saving fuel while increasing airspace capacity. The FAA has implemented DRVSM between FL 290 and FL 410 (inclusive) in the airspace of the contiguous 48 states, Alaska, and in Gulf of Mexico airspace where the FAA provides air traffic services. DRVSM is expected to result in fuel savings for the airlines of as much as $5 billion by 2016. Full DRVSM adds six additional usable altitudes above FL 290 to those available using the former vertical separation minimums. DRVSM users experience increased benefits nationwide, similar to those already achieved in oceanic areas where RVSM is operational. In domestic airspace, however, operational differences create unique challenges. Domestic U.S. airspace contains a wider variety of aircraft types, higher-density traffic, and an increased percentage of climbing and descending traffic. This, in conjunction with an intricate route structure with numerous major crossing points, creates a more demanding environment for the implementation of DRVSM than that experienced in applying RVSM on international oceanic routes. As more flights increase demands on our finite domestic airspace, DRVSM helps to reduce fuel burn and departure delays and increases flight level availability, airspace capacity, and controller flexibility.

FAA RADAR SYSTEMS

The FAA operates two basic radar systems; airport surveillance radar (ASR) and air route surveillance radar (ARSR). Both of these surveillance systems use primary and secondary radar returns, as well as sophisticated computers and software programs designed to give the controller additional information, such as aircraft speed and altitude.

AIRPORT SURVEILLANCE RADAR

The direction and coordination of IFR traffic within specific terminal areas is delegated to airport surveillance radar (ASR) facilities. Approach and departure control manage traffic at airports with ASR. This radar system is designed to provide relatively short-range coverage in the airport vicinity and to serve as an expeditious means of handling terminal area traffic. The ASR also can be used as an instrument approach aid. Terminal radar approach control facilities (TRACONs) provide radar and nonradar services at major airports. The primary responsibility of each TRACON is to ensure safe separation of aircraft transitioning from departure to cruise flight or from cruise to a landing approach.

Most ASR facilities throughout the country use a form of automated radar terminal system (ARTS). This system has several different configurations that depend on the computer equipment and software programs used. Usually the busiest terminals in the country have the most sophisticated computers and programs. The type of system installed is designated by a suffix of numbers and letters. For example, an ARTS-IIIA installation can detect, track, and predict primary, as well as secondary, radar returns. [Figure 1-13]

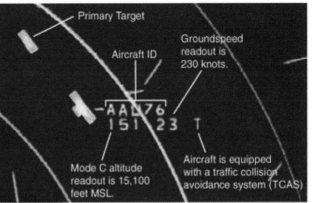

Figure 1-13. ARTS-III Radar Display.

On a controller's radar screen, ARTS equipment automatically provides a continuous display of an aircraft's position, altitude, groundspeed, and other pertinent information. This information is updated continuously as the aircraft progresses through the terminal area. To gain maximum benefit from the system, each aircraft in the area must be equipped with a Mode C altitude encoding transponder, although this is not an operational requirement. Direct altitude readouts eliminate the need for time consuming verbal communication between controllers and pilots to verify altitude. This helps to increase the number of aircraft that may be handled by one controller at a given time.

The FAA has begun replacing the ARTS systems with newer equipment in some areas. The new system is called STARS, for Standard Terminal Automation Replacement System. STARS is discussed in more detail later in this chapter.

AIR ROUTE SURVEILLANCE RADAR

The long-range radar equipment used in controlled airspace to manage traffic is the air route surveillance radar (ARSR) system. There are approximately 100 ARSR facilities to relay traffic information to radar controllers throughout the country. Some of these facilities can detect only transponder-equipped aircraft and are referred to as beacon-only sites. Each air route surveillance radar site can monitor aircraft flying within a 200-mile radius of the antenna, although some stations can monitor aircraft as far away as 600 miles through the use of remote sites.

The direction and coordination of IFR traffic in the U.S. is assigned to air route traffic control centers (ARTCCs). These centers are the authority for issuing IFR clearances and managing IFR traffic; however, they also provide services to VFR pilots. Workload permitting, controllers will provide traffic advisories and course guidance, or vectors, if requested.

PRECISION RUNWAY MONITORING

Precision runway monitor (PRM) is a high-update-rate radar surveillance system that is being introduced at selected capacity-constrained U.S. airports. Certified to provide simultaneous independent approaches to closely spaced parallel runways, PRM has been operational at Minneapolis since 1997. ILS/PRM approaches are conducted at Philadelphia International Airport. Simultaneous Offset Instrument Approach (SOIA)/PRM operations are conducted at San Francisco International and Cleveland Hopkins International Airports. Since the number of PRM sites is increasing, the likelihood is increasing that you may soon be operating at an airport conducting closely spaced parallel approaches using PRM. Furthermore, St. Louis Lambert International Airport began SOIA/PRM operations in 2005, and Atlanta Hartsfield International Airport will begin PRM operations in 2006. PRM enables ATC to improve the airport arrival rate on IFR days to one that more closely approximates VFR days, which means fewer flight cancellations, less holding, and decreased diversions.

PRM not only maintains the current level of safety, but also increases it by offering air traffic controllers a much more accurate picture of the aircraft's location on final approach. Whereas current airport surveillance radar used in a busy terminal area provides an update to the controller every 4.8 seconds, PRM updates every second, giving the controller significantly more time to react to potential aircraft separation problems. The controller also sees target trails that provide very accurate trend information. With PRM, it is immediately apparent when an aircraft starts to drift off the runway centerline and toward the non-transgression zone. PRM also predicts the aircraft track and provides aural and visual alarms when an aircraft is within 10 seconds of penetrating the non-transgression zone. The additional controller staffing that comes along with PRM is another major safety improvement. During PRM sessions, there is a separate controller monitoring each final approach course and a coordinator managing the overall situation.

PRM is an especially attractive technical solution for the airlines and business aircraft because it does not require any additional aircraft equipment, only special training and qualifications. However, all aircraft in the approach streams must be qualified to participate in PRM or the benefits are quickly lost and controller workload increases significantly. The delay-reduction benefits of PRM can only be fully realized if everyone participates. Operators that choose not to participate in PRM operations when arriving at an airport where PRM operations are underway can expect to be held until they can be accommodated without disrupting the PRM arrival streams.

EQUIPMENT AND AVIONICS
By virtue of distance and time savings, minimizing traffic congestion, and increasing airport and airway capacity, the implementation of RNAV routes, direct routing, RSVM, PRM, and other technological innovations would be advantageous for the current NAS. Some key components that are integral to the future development and improvement of the NAS are described below. However, equipment upgrades require capital outlays, which take time to penetrate the existing fleet of aircraft and ATC facilities. In the upcoming years while the equipment upgrade is taking place, ATC will have to continue to accommodate the wide range of avionics used by pilots in the nation's fleet.

ATC RADAR EQUIPMENT
All ARTCC radars in the conterminous U.S., as well as most airport surveillance radars, have the capability to interrogate Mode C and display altitude information to the controller. However, there are a small number of airport surveillance radars that are still two-dimensional (range and azimuth only); consequently, altitude information must be obtained from the pilot.

At some locations within the ATC environment, secondary only (no primary radar) gap filler radar systems are used to give lower altitude radar coverage between two larger radar systems, each of which provides both primary and secondary radar coverage. In the geographical areas serviced by secondary radar only, aircraft without transponders cannot be provided with radar service. Additionally, transponder-equipped aircraft cannot be provided with radar advisories concerning primary targets and weather.

An integral part of the air traffic control radar beacon system (ATCRBS) ground equipment is the decoder, which enables the controller to assign discrete transponder codes to each aircraft under his/her control. Assignments are made by the ARTCC computer on the basis of the National Beacon Code Allocation Plan (NBCAP). There are 4,096 aircraft transponder codes that can be assigned. An aircraft must be equipped with Civilian Mode A (or Military Mode 3) capabilities to be assigned a transponder code. Another function of the decoder is that it is also designed to receive Mode C altitude information from an aircraft so equipped. This system converts aircraft altitude in 100-foot increments to coded digital information that is transmitted together with Mode C framing pulses to the interrogating ground radar facility. The ident feature of the transponder causes the transponder return to "blossom" for a few seconds on the controller's radarscope.

AUTOMATED RADAR TERMINAL SYSTEM
Most medium-to-large radar facilities in the U.S. use some form of automated radar terminal system (ARTS), which is the generic term for the functional capability afforded by several automated systems that differ in functional capabilities and equipment. "ARTS" followed by a suffix Roman numeral denotes a specific system, with a subsequent letter that indicates a major modification to that particular system. In general, the terminal controller depends on ARTS to display aircraft identification, flight plan data, and other information in conjunction with the radar presentation. In addition to enhancing visualization of the air traffic situation, ARTS facilitates intra- and inter-facility transfers and the coordination of flight information. Each ARTS level has the capabilities of communicating with other ARTS types as well as with ARTCCs.

As the primary system used for terminal ATC in the U.S., ARTS had its origin in the mid-1960's as ARTS I, or Atlanta ARTS and evolved to the ARTS II and ARTS III configurations in the early to mid-1970's. Later in the decade, the ARTS II and ARTS III configurations were expanded and enhanced and renamed ARTS IIA and ARTS IIIA respectively. The vast majority of the terminal automation sites today remain either IIA or IIIA configurations, except for about nine of the largest IIIA sites, which are ARTS IIIE candidate systems. Selected ARTS IIIA/IIIE and ARTS IIA sites are scheduled to receive commercial off the shelf (COTS) hardware upgrades, which replace portions of the proprietary data processing system with standard off-the-shelf hardware.

STANDARD TERMINAL
AUTOMATION REPLACEMENT SYSTEM
The FAA has begun modernizing the computer equipment in the busiest terminal airspace areas. The newer equipment is called STARS, for Standard Terminal Automation Replacement System. The system's improvements will enhance safety while reducing

delays by increasing system reliability and lowering life-cycle operating and maintenance costs. STARS also will accommodate the projected growth in air traffic and provide a platform for new functions to support FAA initiatives such as Free Flight. STARS offers many advantages, including an open architecture and expansion capability that allow new software and capacity to be added as needed to stay ahead of the growth in air traffic. Under the first phase of terminal modernization, STARS is being deployed to 47 air traffic control facilities. As of July 2005, 37 FAA and 22 Department of Defense sites were fully operational with STARS. The first phase is expected to be complete in fiscal year 2007. By then, STARS will be operational at 18 of the FAA's 35 most critical, high-volume airports, which together handle approximately 50 percent of air traffic. STARS consists of new digital, color displays and computer software and processors that can track 435 aircraft at one time, integrating six levels of weather information and 16 radar feeds.

For the terminal area and many of the towers, STARS is the key to the future, providing a solid foundation for new capabilities. STARS was designed to provide the software and hardware platform necessary to support future air traffic control enhancements.

PRECISION APPROACH RADAR

While ASR provides pilots with horizontal guidance for instrument approaches via a ground-based radar, Precision Approach Radar (PAR) provides both horizontal and vertical guidance for a ground controlled approach (GCA). In the U.S., PAR is mostly used by the military. Radar equipment in some ATC facilities operated by the FAA and/or the military services at joint-use locations and military installations are used to detect and display azimuth, elevation, and range of aircraft on the final approach course to a runway. This equipment may be used to monitor certain non-radar approaches, but it is primarily used to conduct a precision instrument approach.

BRIGHT RADAR INDICATOR TERMINAL EQUIPMENT

Bright Radar Indicator Terminal Equipment (BRITE) provides radar capabilities to towers, a system with tremendous benefits for both pilots and controllers. Unlike traditional radar systems, BRITE is similar to a television screen in that it can be seen in daylight. BRITE was so successful that the FAA has installed the new systems in towers, and even in some TRACONs. In fact, the invention of BRITE was so revolutionary that it launched a new type of air traffic facility — the TRACAB, which is a radar approach control facility located in the tower cab of the primary airport, as opposed to a separate room.

In the many facilities without BRITE, the controllers use strictly visual means to find and sequence traffic. Towers that do have BRITE may have one of several different types. Some have only a very crude display that gives a fuzzy picture of blips on a field of green, perhaps with the capability of displaying an extra slash on transponder-equipped targets and a larger slash when a pilot hits the ident button. Next in sophistication are BRITEs that have alphanumeric displays of various types, ranging from transponder codes and altitude to the newest version, the DBRITE (digital BRITE). A computer takes all the data from the primary radar, the secondary radar (transponder information), and generates the alphanumeric data. DBRITE digitizes the image, and then sends it all, in TV format, to a square display in the tower that provides an excellent presentation, regardless of how bright the ambient light.

One of the most limiting factors in the use of the BRITE is in the basic idea behind the use of radar in the tower. The radar service provided by a tower controller is not, nor was it ever intended to be, the same thing as radar service provided by an approach control or Center. The primary duty of tower controllers is to separate airplanes operating on runways, which means controllers spend most of their time looking out the window, not staring at a radar scope.

RADAR COVERAGE

A full approach is a staple of instrument flying, yet some pilots rarely, if ever, have to fly one other than during initial or recurrency or proficiency training, because a full approach usually is required only when radar service is not available, and radar is available at most larger and busier instrument airports. Pilots come to expect radar vectors to final approach courses and that ATC will keep an electronic eye on them all the way to a successful conclusion of every approach. In addition, most en route flights are tracked by radar along their entire route in the 48 contiguous states, with essentially total radar coverage of all instrument flight routes except in the mountainous West. Lack of radar coverage may be due to terrain, cost, or physical limitations.

New developing technologies, like ADS-B, may offer ATC a method of accurately tracking aircraft in non-radar environments. ADS-B is a satellite-based air traffic tracking system enabling pilots and air traffic controllers to share and display the same information. ADS-B relies on the Global Positioning System (GPS) to determine an aircraft's position. The aircraft's precise location, along with other data such as airspeed, altitude, and aircraft identification, then is instantly relayed via digital datalink to ground stations and other equipped aircraft. Depending on the location of the ground based transmitters (GBT), ADS-B has the potential to work well at low altitudes, in remote locations, and mountainous terrain where little or no radar coverage exists.

COMMUNICATIONS

Most air traffic control communications between pilots and controllers today are conducted via voice. Each air traffic controller uses a radio frequency different from the ones used by surrounding controllers to communicate with the aircraft under his or her jurisdiction. With the increased traffic, more and more controllers have been added to maintain safe separation between aircraft. While this has not diminished safety, there is a limit to the number of control sectors created in any given region to handle the traffic. The availability of radio frequencies for controller-pilot communications is one limiting factor. Some busy portions of the U.S., such as the Boston-Chicago-Washington triangle are reaching toward the limit. Frequencies are congested and new frequencies are not available, which limits traffic growth to those aircraft that can be safely handled.

DATA LINK

The CAASD is working with the FAA and the airlines to define and test a controller-pilot data link communication (CPDLC), which provides the capability to exchange information between air traffic controllers and flight crews through digital text instead of voice messages. With CPDLC, communications between the ground and the air would take less time, and would convey more information (and more complex information) than by voice alone. Communications would become more accurate as up-linked information would be collected, its accuracy established, and then displayed for the pilot in a consistent fashion.

By using digital data messages to replace conventional voice communications (except during landing and departure phases and in emergencies) CPDLC is forecast to increase airspace capacity and reduce delays. Today the average pilot/controller voice exchange takes around 20 seconds, compared to one or two seconds with CPDLC. In FAA simulations, air traffic controllers indicated that CPDLC could increase their productivity by 40 percent without increasing workload. Airline cost/benefit studies indicate average annual savings that are significant in the terminal and en route phases, due to CPDLC-related delay reductions.

CPDLC for routine ATC messages, initially offered in Miami Center, will be implemented via satellite at all oceanic sectors. Communications between aircraft and FAA oceanic facilities will be available through satellite data link, high frequency data link (HFDL), or other subnetworks, with voice via HF and satellite communications remaining as backup. Eventually, the service will be expanded to include clearances for altitude, speed, heading, and route, with pilot initiated downlink capability added later.

MODE S

The first comprehensive proposal and design for the Mode S system was delivered to the FAA in 1975. However, due to design and manufacturing setbacks, few Mode S ground sensors and no commercial Mode S transponders were made available before 1980. Then, a tragic mid-air collision over California in 1986 prompted a dramatic change. The accident that claimed the lives of 67 passengers aboard the two planes and fifteen people on the ground was blamed on inadequate automatic conflict alert systems and surveillance equipment. A law enacted by Congress in 1987 required all air carrier airplanes operating within U.S. airspace with more than 30 passenger seats to be equipped with Traffic Alert and Collision Avoidance System (TCAS II) by December 1993. Airplanes with 10 to 30 seats were required to employ TCAS I by December 1995.

Due to the congressional mandate, TCAS II became a pervasive system for air traffic control centers around the world. Because TCAS II uses Mode S as the standard air-ground communication datalink, the widespread international use of TCAS II has helped Mode S become an integral part of air traffic control systems all over the world. The datalink capacity of Mode S has spawned the development of a number of different services that take advantage of the two-way link between air and ground. By relying on the Mode S datalink, these services can be inexpensively deployed to serve both the commercial transport aircraft and general aviation communities. Using Mode S makes not only TCAS II, but also other services available to the general aviation community that were previously accessible only to commercial aircraft. These Mode S-based technologies are described below.

TRAFFIC ALERT AND COLLISION AVOIDANCE SYSTEM

The traffic alert and collision avoidance system (TCAS) is designed to provide a set of electronic eyes so the pilot can maintain awareness of the traffic situation in the vicinity of the aircraft. The TCAS system uses three separate systems to plot the positions of nearby aircraft. First, directional antennae that receive Mode S transponder signals are used to provide a bearing to neighboring aircraft — accurate to a few degrees of bearing. Next, Mode C altitude broadcasts are used to plot the altitude of nearby aircraft. Finally, the timing of the Mode S interrogation/response protocol is measured to ascertain the distance of an aircraft from the TCAS aircraft.

TCAS I allows the pilot to see the relative position and velocity of other transponder-equipped aircraft within a 10 to 20-mile range. [Figure 1-14] More importantly, TCAS I provides a warning when an aircraft in the vicinity gets too close. TCAS I does not provide instructions on how to maneuver in order to avoid the aircraft, but

Figure 1-14. Traffic Alert and Collision Avoidance System.

does supply important data with which the pilot uses to evade intruding aircraft.

TCAS II provides pilots with airspace surveillance, intruder tracking, threat detection, and avoidance maneuver generations. TCAS II is able to determine whether each aircraft is climbing, descending, or flying straight and level, and commands an evasive maneuver to either climb or descend to avoid conflicting traffic. If both planes in conflict are equipped with TCAS II, then the evasive maneuvers are well coordinated via air-to-air transmissions over the Mode S datalink, and the commanded maneuvers do not cancel each other out.

TCAS and similar traffic avoidance systems provide safety independent of ATC and supplement and enhance ATC's ability to prevent air-to-air collisions. Pilots currently use TCAS displays for collision avoidance and oceanic station keeping (maintaining miles-in-trail separation). Recent TCAS technology improvements enable aircraft to accommodate reduced vertical separation above FL 290 and the ability to track multiple targets at longer ranges. The Airborne Collision Avoidance System (ACAS) is an international ICAO standard that is the same as the latest TCAS II, which is sometimes called "Change 7" or "Version 7" in the United States. ACAS has been mandated, based on varying criteria, throughout much of the world.

TRAFFIC INFORMATION SERVICE
Traffic Information Service (TIS) provides many of the functions available in TCAS; but unlike TCAS, TIS is a ground-based service available to all aircraft equipped with Mode S transponders. TIS takes advantage of the Mode S data link to communicate collision avoidance information to aircraft. Information is pre-

sented to a pilot in a cockpit display that shows traffic within 5 nautical miles and a 1,200-foot altitude of other Mode S-equipped aircraft. The TIS system uses track reports provided by ground-based Mode S surveillance systems to retrieve traffic information. Because it is available to all Mode S transponders, TIS offers an inexpensive alternative to TCAS. The increasing availability of TIS makes collision avoidance technology more accessible to the general aviation community. Beginning in 2005, the use of Mode S TIS is being discontinued at some sites as the ground radar systems are upgraded. In all, 23 sites are expected to lose TIS capability by 2012.

TERRAIN AWARENESS AND WARNING SYSTEM
The Terrain Awareness and Warning System (TAWS) is an enhanced ground proximity warning capability being installed in many aircraft. TAWS uses position data from a navigation system, like GPS, and a digital terrain database to display surrounding terrain. TAWS equipment is mandatory for all U.S registered turbine powered airplanes with six or more passenger seats. FAA and NTSB studies have shown that a large majority of CFIT accidents could likely have been avoided had the aircraft been equipped with enhanced ground proximity warning systems.

GRAPHICAL WEATHER SERVICE
The Graphical Weather Service provides a graphical representation of weather information that is transmitted to aircraft and displayed on the cockpit display unit. The service is derived from ground-based Mode S sensors and offers information to all types of aircraft, regardless of the presence of on-board weather avoidance equipment. The general aviation community has been very pro-active in evaluating this technology, as they have already participated in field evaluations in Mode S stations across the U.S. The service is provided through one of two types of flight information services (FIS) systems. Broadcast only systems, called FIS-B, include a ground- or space-based transmitter, an aircraft receiver, and a portable or installed cockpit display device. They allow pilots to passively collect weather and other operational data and to display that data at the appropriate time. They can display graphical weather products such as radar composite/mosaic images, temporary flight restricted airspace and other NOTAMs. In addition to graphical weather products, they can also show textual information, such as Aviation Routine Weather Reports (METARs)/Aviation Selected Special Weather Reports (SPECIs) and Terminal Area Forecasts (TAFs).

Two-way FIS systems are request/reply systems, that is, they permit the pilot to make specific requests for weather and other operational information. An FIS service provider will then prepare a reply in response to that specific request and transmit the product to that specific aircraft for display in the cockpit.

AVIONICS AND INSTRUMENTATION

The proliferation of advanced avionics and instrumentation has substantially increased the capabilities of aircraft in the IFR environment.

FLIGHT MANAGEMENT SYSTEM

A flight management system (FMS) is a flight computer system that uses a large database to allow routes to be preprogrammed and fed into the system by means of a data loader. The system is constantly updated with respect to position accuracy by reference to conventional navigation aids, inertial reference system technology, or the satellite global positioning system. The sophisticated program and its associated database ensures that the most appropriate navigation aids or inputs are automatically selected during the information update cycle. A typical FMS provides information for continuous automatic navigation, guidance, and aircraft performance management, and includes a control display unit (CDU). [Figure 1-15]

Courtesy of Universal Avionics Systems Corporation

Figure 1-15. FMS Control Display Unit. This depicts an aircraft established on the Atlantic City, NJ, RNAV (GPS) Rwy 13 instrument approach procedure at the Atlantic City International Airport, KACY. The aircraft is positioned at the intermediate fix UNAYY inbound on the 128 degree magnetic course, 5.5 nautical miles from PVIGY, the final approach fix.

ELECTRONIC FLIGHT INFORMATION SYSTEM

The electronic flight information system (EFIS) found in advanced aircraft cockpits offer pilots a tremendous amount of information on a colorful, easy-to-read display. Glass cockpits are a vast improvement over the earlier generation of instrumentation. [Figure 1-16]

Primary flight, navigation, and engine information are presented on large display screens in front of the flight crew. Flight management CDUs are located on the center console. They provide data display and entry capabilities for flight management functions. The display units generate less heat, save space, weigh less, and require less power than traditional navigation systems. From a pilot's point of view, the information display system is not only more reliable than previous systems, but also uses advanced liquid-crystal technology that allows displayed information to remain clearly visible in all conditions, including direct sunlight.

NAVIGATION SYSTEMS

Navigation systems are the basis for pilots to get from one place to another and know where they are and what course to follow. Since the 1930s, aircraft have navigated by means of a set of ground-based NAVAIDs. Today, pilots have access to over 2,000 such NAVAIDs within the continental U.S., but the system has its limitations:

- Constrained to fly from one NAVAID to the next, aircraft route planners need to identify a beacon-based path that closely resembles the path the aircraft needs to take to get from origin to destination. Such a path will always be greater in distance than a great circle route between the two points.

- Because the NAVAIDs are ground-based, navigation across the ocean is problematic, as is navigation in some mountainous regions.

- NAVAIDs are also expensive to maintain.

Since the 1980s, aircraft systems have evolved towards the use of SATNAV. Based on the GPS satellite constellation, SATNAV may provide better position information than a ground-based navigation system. GPS is universal so there are no areas without satellite signals. Moreover, a space-based system allows "off airway" navigation so that the efficiencies in aircraft route determination can be exacted. SATNAV is revolutionizing navigation for airlines and other aircraft owners and operators. A drawback of the satellite system, though, is the integrity and availability of the signal, especially during electromagnetic and other events that distort the Earth's atmosphere. In addition, the signal from space needs to be augmented, especially in traffic-dense terminal areas, to guarantee the necessary levels of accuracy and availability.

The CAASD is helping the navigation system of the U.S. to evolve toward a satellite-based system. The CAASD analysts are providing the modeling necessary to understand the effects of atmospheric phenomena on the GPS signal from space, while the CAASD is providing the architecture of the future navigation system and writing the requirements (and computer algorithms) to ensure the navigation system's integrity. Moving toward a satellite-based navigation system allows aircraft to divorce themselves from the constraints of ground-based NAVAIDs and formulate and fly those routes that aircraft route planners deem most in line with their own cost objectives.

With the advent of SATNAV, there are a number of applications that can be piggybacked to increase capacity in the NAS. Enhanced navigation systems will be capable of "random navigation," that is, capable of

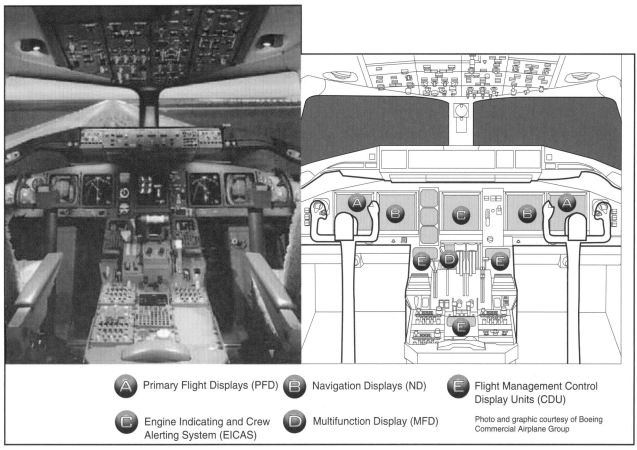

Figure 1-16. Airline Flight Deck Instrument Displays

A   Primary Flight Displays (PFD)   B   Navigation Displays (ND)   E   Flight Management Control Display Units (CDU)

C   Engine Indicating and Crew Alerting System (EICAS)   D   Multifunction Display (MFD)

Photo and graphic courtesy of Boeing Commercial Airplane Group

treating any latitude-longitude point as a radio navigation fix, and being able to fly toward it with the accuracy we see today, or better. New routes into and out of the terminal areas are being implemented that are navigable by on-board systems. Properly equipped aircraft are being segregated from other aircraft streams with the potential to increase volume at the nation's busy airports by keeping the arrival and departure queues full and fully operating.

The CAASD is working with the FAA to define the nation's future navigation system architecture. By itself, the GPS satellite constellation is inadequate to serve all the system's needs. Augmentation of the GPS signal via WAAS and LAAS is a necessary part of that new architecture. The CAASD is developing the requirements based on the results of sophisticated models to ensure the system's integrity, security, and availability.

SURVEILLANCE SYSTEMS

Surveillance systems are set up to enable the ATC system to know the location of an aircraft and where it is heading. Position information from the surveillance system supports many different ATC functions. Aircraft positions are displayed for controllers as they watch over the traffic to ensure that aircraft do not violate separation criteria. In the current NAS, surveillance is achieved through the use of long-range and terminal radars. Scanning the skies, these radars return azimuth

and slant range for each aircraft that, when combined with the altitude of the aircraft broadcast to the ground via a transceiver, is transformed mathematically into a position. The system maintains a list of these positions for each aircraft over time, and this time history is used to establish short-term intent and short-term conflict detection. Radars are expensive to maintain, and position information interpolated from radars is not as good as what the aircraft can obtain with SATNAV. ADS-B technology may provide the way to reduce the costs of surveillance for air traffic management purposes and to get the better position information to the ground.

New aircraft systems dependent on ADS-B could be used to enhance the capacity and throughput of the nation's airports. Electronic flight following is one example: An aircraft equipped with ADS-B could be instructed to follow another aircraft in the landing pattern, and the pilot could use the on-board displays or computer applications to do exactly that. This means that visual rules for landing at airports might be used in periods where today the airport must shift to instrument rules due to diminishing visibility. Visual capacities at airports are usually higher than instrument ones, and if the airport can operate longer under visual rules (and separation distances), then the capacity of the airport is maintained at a higher level longer. The CAASD is working with the Cargo Airline Association and the

FAA to investigate these and other applications of the ADS-B technology.

OPERATIONAL TOOLS

Airports are one of the main bottlenecks in the NAS, responsible for one third of the flight delays. It is widely accepted that the unconstrained increase in the number of airports or runways may not wholly alleviate the congestion problem and, in fact, may create more problems than it solves. The aim of the FAA is to integrate appropriate technologies, in support of the OEP vision, with the aim of increasing airport throughput.

The airport is a complex system of systems and any approach to increasing capacity must take this into account. Numerous recent developments contribute to the overall solution, but their integration into a system that focuses on maintaining or increasing safety while increasing capacity remains a major challenge. The supporting technologies include new capabilities for the aircraft and ATC, as well as new strategies for improving communication between pilots and ATC.

IFR SLOTS

During peak traffic, ATC uses IFR slots to promote a smooth flow of traffic. This practice began during the late 1960s, when five of the major airports (LaGuardia Airport, Ronald Reagan National Airport, John F. Kennedy International Airport, Newark International Airport, and Chicago O'Hare International Airport) were on the verge of saturation due to substantial flight delays and airport congestion. To combat this, the FAA in 1968 proposed special air traffic rules to these five high-density airports (the "high density rule") that restricted the number of IFR takeoffs and landings at each airport during certain hours of the day and provided for the allocation of "slots" to carriers for each IFR landing or takeoff during a specific 30 or 60-minute period. A more recent FAA proposal offers an overhaul of the slot-reservation process for JFK, LaGuardia, and Reagan National Airport that includes a move to a 72-hour reservation window and an online slot-reservation system.

The high density rule has been the focus of much examination over the last decade since under the restrictions, new entrants attempting to gain access to high density airports face difficulties entering the market. Because slots are necessary at high density airports, the modification or elimination of the high density rule could subsequently have an effect on the value of slots. Scarce slots hold a greater economic value than slots that are easier to come by.

The current slot restrictions imposed by the high density rule has kept flight operations well below capacity, especially with the improvements in air traffic control technology. However, easing the restrictions imposed by the high density rule is likely to affect airport oper-
ations. Travel delay time might be affected not only at the airport that has had the high density restrictions lifted, but also at surrounding airports that share the same airspace. On the other hand, easing the restrictions on slots at high density airports should help facilitate international air travel and help increase the number of passengers that travel internationally.

Slot controls have become a way of limiting noise, since it caps the number of takeoffs and landings at an airport. Easing the restrictions on slots could be politically difficult since local delegations at the affected airports might not support such a move. Ways other than imposing restrictions on slots exist that could diminish the environmental impacts at airports and their surrounding areas. Safeguards, such as requiring the quietest technology available of aircraft using slots and frequent consultations with local residents, have been provided to ensure that the environmental concerns are addressed and solved.

GROUND DELAY PROGRAM

Bad weather often forces the reconfiguration of runways at an airport or mandates the use of IFR arrival and departure procedures, reducing the number of flights per hour that are able to takeoff or land at the affected airport. To accommodate the degraded arrival capacity at the affected airport, the ATCSCC imposes a ground delay program (GDP), which allocates a reduced number of arrival slots to airlines at airports during time periods when demand exceeds capacity. The GDP suite of tools is used to keep congestion at an arrival airport at acceptable levels by issuing ground delays to aircraft before departure, as ground delays are less expensive and safer than in-flight holding delays. The FAA started GDP prototype operations in January 1998 at two airports and expanded the program to all commercial airports in the U.S. within nine months.

Ground Delay Program Enhancements (GDPE) significantly reduced delays due to compression—a process that is run periodically throughout the duration of a GDP. It reduces overall delays by identifying open arrival slots due to flight cancellations or delays and fills in the vacant slots by moving up operating flights that can use those slots. During the first two years of this program, almost 90,000 hours of scheduled delays were avoided due to compression, resulting in cost savings to the airline industry of more than $150 million. GDPE also has improved the flow of air traffic into airports; improved compliance to controlled times of departure; improved data quality and predictability; resulted in equity in delays across carriers; and often avoided the necessity to implement FAA ground delay programs, which can be disruptive to air carrier operations.

FLOW CONTROL

ATC provides IFR aircraft separation services for NAS users. Since the capabilities of IFR operators vary from airlines operating hundreds of complex jet aircraft to private pilots in single engine, piston-powered airplanes, the ATC system must accommodate the least sophisticated user. The lowest common denominator is the individual controller speaking to a single pilot on a VHF voice radio channel. While this commonality is desirable, it has led to a mindset where other opportunities to interact with NAS users have gone undeveloped. The greatest numbers of operations at the 20 busiest air carrier airports are commercial operators (airlines and commuters) operating IFR with some form of ground-based operational control. Since not all IFR operations have ground-based operational control, very little effort has been expended in developing ATC and Airline Operations Control Center (AOC) collaboration techniques, even though ground-based computer-to-computer links can provide great data transfer capacity. Until the relatively recent concept of Air Traffic Control-Traffic Flow Management (ATC-TFM), the primary purpose of ATC was aircraft separation, and the direct pilot-controller interaction was adequate to the task. Effective and efficient traffic flow management now requires a new level of control that includes the interaction of and information transfer among ATC, TFM, AOCs, and the cockpit. [Figure 1-17]

Figure 1-17. Flow Control Restrictions.

As the first step in modernizing the traffic flow management infrastructure, the FAA began reengineering traffic flow management software using commercial off-the-shelf products. In FY 1996, the FAA and NASA collaborated on new traffic flow management research and development efforts for the development of collaborative decision making tools that will enable FAA traffic flow managers to work cooperatively with airline personnel in responding to congested conditions. Additionally, the FAA provided a flight scheduling software system to nine airlines.

LAND AND HOLD SHORT OPERATIONS

Many older airports, including some of the most congested, have intersecting runways. Expanding the use of Land and Hold Short Operations (LAHSO) on intersecting runways is one of the ways to increase the number of arrivals and departures. Currently, LAHSO operations are permitted only on dry runways under acceptable weather conditions and limited to airports where a clearance depends on what is happening on the other runway, or where approved rejected landing procedures are in place. A dependent procedure example is when a landing airplane is a minimum distance from the threshold and an airplane is departing an intersecting runway, the LAHSO clearance can be issued because even in the event of a rejected landing, separation is assured. It is always the pilot's option to reject a LAHSO clearance.

Working with ICAO, pilot organizations, and industry groups, the FAA is developing new LAHSO procedures that will provide increased efficiency while maintaining safety. These procedures will address issues such as wet runway conditions, mixed commercial and general aviation operations, the frequency of missed approaches, and multi-stop runway locations. After evaluating the new procedures using independent case studies, the revised independent LAHSO procedures may be implemented in the near future.

SURFACE MOVEMENT GUIDANCE AND CONTROL SYSTEM

To enhance taxiing capabilities in low visibility conditions and reduce the potential for runway incursions, improvements have been made in signage, lighting, and markings. In addition to these improvements, airports have implemented the Surface Movement Guidance and Control System (SMGCS),[4] a strategy that requires a low visibility taxi plan for any airport with takeoff or landing operations with less than 1,200 feet RVR visibility conditions. This plan affects both aircrew and airport vehicle operators, as it specifically designates taxi routes to and from the SMGCS runways and displays them on a SMGCS Low Visibility Taxi Route chart.

---

[4] SMGCS, pronounced "SMIGS," is the Surface Movement Guidance and Control System. SMGCS provides for guidance and control or regulation for facilities, information, and advice necessary for pilots of aircraft and drivers of ground vehicles to find their way on the airport during low visibility operations and to keep the aircraft or vehicles on the surfaces or within the areas intended for their use. Low visibility operations for this system means reported conditions of RVR 1,200 or less.

SMGCS is an increasingly important element in a seamless, overall gate-to-gate management concept to ensure safe, efficient air traffic operations. It is the ground-complement for arrival and departure management and the en route components of free flight. The FAA has supported several major research and development efforts on SMGCS to develop solutions and prototype systems that support pilots and ATC in their control of aircraft ground operations.

EXPECT CHANGES IN THE ATC SYSTEM

To maintain air safety, ATC expects all aircraft to adhere to a set of rules based on established separation standards. Until recently, air traffic controllers followed established procedures based upon specific routes to maintain the desired separations needed for safety. This system has an excellent safety record for aircraft operations. Because of increases in the number of flights, the availability of more accurate and reliable technologies, and the inherent limitations of the existing system, there will be many changes in the near future. Use of the free flight concept where aircraft operators select paths, altitudes, and speeds in real time can maximize efficiency and minimize operating costs. New technologies and enhanced aircraft capabilities necessitate changes in procedures, an increase in the level of automation and control in the cockpit and in the ground system, and more human reliance on automated information processing, sophisticated displays, and faster data communication.

DISSEMINATING AERONAUTICAL INFORMATION

The system for disseminating aeronautical information is made up of two subsystems, the Airmen's Information System (AIS) and the Notice to Airman (NOTAM) System. The AIS consists of charts and publications. The NOTAM system is a telecommunication system and is discussed in later paragraphs. Aeronautical information disseminated through charts and publications includes aeronautical charts depicting permanent baseline data and flight information publications outlining baseline data.

IFR aeronautical charts include en route high altitude conterminous U.S., and en route low altitude conterminous U.S., plus Alaska charts and Pacific Charts. Additional charts include U.S. terminal procedures, consisting of departure procedures (DPs), standard terminal arrivals (STARs), and standard instrument approach procedures (SIAPs).

Flight information publications outlining baseline data in addition to the Notices to Airmen Publication (NTAP) include the Airport/Facility Directory (A/FD), a Pacific Chart Supplement, an Alaska Supplement, an Alaska Terminal publication, and the Aeronautical Information Manual (AIM).

PUBLICATION CRITERIA

The following conditions or categories of information are forwarded to the National Flight Data Center (NFDC) for inclusion in flight information publications and aeronautical charts:

- NAVAID commissioning, decommissioning, outages, restrictions, frequency changes, changes in monitoring status and monitoring facility used in the NAS.

- Commissioning, decommissioning, and changes in hours of operation of FAA air traffic control facilities.

- Changes in hours of operations of surface areas and airspace.

- RCO and RCAG commissioning, decommissioning, and changes in voice control or monitoring facility.

- Weather reporting station commissioning, decommissioning, failure, and nonavailability or unreliable operations.

- Public airport commissioning, decommissioning, openings, closings, abandonments, and some airport operating area (AOA) changes.

- Aircraft Rescue & Fire Fighting (ARFF) capability, including restrictions to air carrier operations.

- Changes to runway identifiers, dimensions, threshold placements, and surface compositions.

- NAS lighting system commissioning, decommissioning, outages, and change in classification or operation.

- IFR Area Charts.

A wide variety of additional flight information publications are available online at the FAA website http://www.faa.gov and can be found with the "Library" link, and the tabs for both "Education and Research" and "Regulation and Policies." Electronic flight publications include electronic bulletin boards, advisory circulars, the AC checklist, Federal Aviation Regulations, the Federal Register, and notices of proposed rulemaking (NPRM).

When planning a flight, you can obtain information on the real-time status of the national airspace system by accessing the Air Traffic Control System Command Center's Operational Information System (OIS) at http://www.fly.faa.gov/ois/. This data is updated every five minutes, and contains useful information on closures, delays, and other aspects of the system.

AERONAUTICAL CHARTS

Pilots can obtain most aeronautical charts and publications produced by the FAA National Aeronautical Charting Office (NACO). They are available by subscription or one-time sales through a network of FAA chart agents primarily located at or near major civil airports. Additionally, opportunities to purchase or download aeronautical publications online are expanding, which provides pilots quicker and more convenient access to the latest information. Civil aeronautical charts for the U.S. and its territories, and possessions are produced according to a 56-day IFR chart cycle by NACO, which is part of the FAA's Technical Ops Aviation Systems Standards (AJW-3). Comparable IFR charts and publications are available from commercial sources, including charted visual flight procedures, airport qualification charts, etc.

Most charts and publications described in this chapter can be obtained by subscription or one-time sales from NACO. Charts and publications are also available through a network of FAA chart agents primarily located at or near major civil airports. To order online, use the "Catalogs/Ordering Info" link at http://www.naco.faa.gov. Below is the contact information for NACO.

FAA, National Aeronautical Charting Office
Distribution Division AJW-3550
10201 Good Luck Road
Glenn Dale, MD 20769-9700
Telephone
(301) 436-8301
(800) 638-8972 toll free, U.S. only FAX
(301) 436-6829
Email: 9-AMC-chartsales@faa.gov

IFR charts are revised more frequently than VFR charts because chart currency is critical for safe operations. Selected NACO IFR charts and products available include IFR navigation charts, planning charts, supplementary charts and publications, and digital products. IFR navigation charts include the following:

- **IFR En route Low Altitude Charts**
  (Conterminous U.S. and Alaska): En route low altitude charts provide aeronautical information for navigation under IFR conditions below 18,000 feet MSL. This four-color chart series includes airways; limits of controlled airspace; VHF NAVAIDs with frequency, identification, channel, geographic coordinates; airports with terminal air/ground communications; minimum en route and obstruction clearance altitudes; airway distances; reporting points; special use airspace; and military training routes. Scales vary from 1 inch = 5 NM to 1 inch = 20 NM. The size is 50 x 20 inches folded to 5 x 10 inches. The charts are revised every 56 days. Area charts show congested terminal areas at a large scale. They are included with subscriptions to any conterminous U.S. Set Low (Full set, East or West sets). [Figure 1-18]

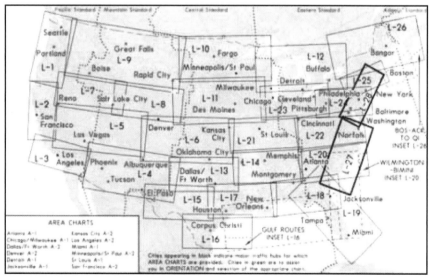

Figure 1-18. En route Low Altitude Charts.

- **IFR En route High Altitude Charts**
  (Conterminous U.S. and Alaska): En route high altitude charts are designed for navigation at or above 18,000 feet MSL. This four-color chart series includes the jet route structure; VHF NAVAIDs with frequency, identification, channel, geographic coordinates; selected airports; and reporting points. The chart scales vary from 1 inch = 45 NM to 1 inch = 18 NM. The size is 55 x 20 inches folded to 5 x 10 inches. Revised every 56 days. [Figure 1-19]

- **U.S. Terminal Procedures Publication (TPP)**
  TPPs are published in 20 loose-leaf or perfect bound volumes covering the conterminous U.S., Puerto Rico, and the Virgin Islands. A Change Notice is published at the midpoint between revisions in bound volume format. [Figure 1-20]

  - Instrument Approach Procedure (IAP) Charts: IAP charts portray the aeronautical data that is required to execute instrument approaches to airports. Each chart depicts the IAP, all related navigation data, communications information, and an airport sketch. Each procedure is designated for use with a specific electronic navigational aid, such as an ILS, VOR, NDB, RNAV, etc.

  - Instrument Departure Procedure (DP) Charts: There are two types of departure procedures; Standard Instrument Departures (SIDs) and Obstacle Departure Procedures (ODPs). SIDs will always be in a graphic format and are designed to assist ATC by expediting clearance delivery and to facilitate transition between takeoff and en route operations. ODPs are established to ensure proper obstacle clearance and are either textual or graphic, depending on complexity.

  - Standard Terminal Arrival (STAR) Charts: STAR charts are designed to expedite ATC arrival procedures and to facilitate transition between en route and instrument approach operations. They depict preplanned IFR ATC arrival procedures in graphic and textual form. Each STAR procedure is presented as a separate chart and may serve either a single airport or more than one airport in a given geographic area.

  - Airport Diagrams: Full page airport diagrams are designed to assist in the movement of ground traffic at locations with complex runway and taxiway configurations and provide information for updating geodetic position navigational systems aboard aircraft.

- **Alaska Terminal Procedures Publication**: This publication contains all terminal flight procedures for civil and military aviation in Alaska. Included are IAP charts, DP charts, STAR charts, airport diagrams, radar minimums, and supplementary support data such as IFR alternate minimums, take-off minimums, rate of descent tables, rate of climb tables, and inoperative components tables. The volume is 5-3/8 x 8-1/4 inches top bound, and is revised every 56 days with provisions for a Terminal Change Notice, as required.

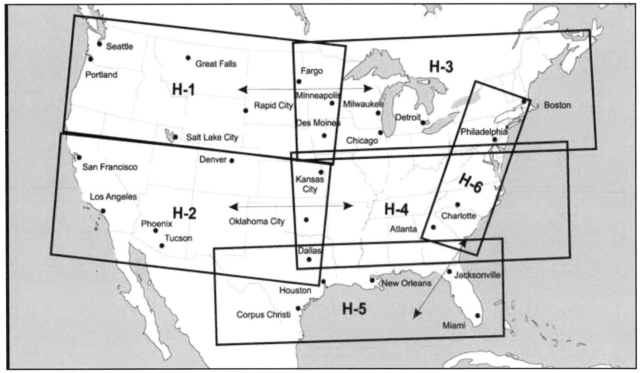

Figure 1-19. En route High Altitude Charts.

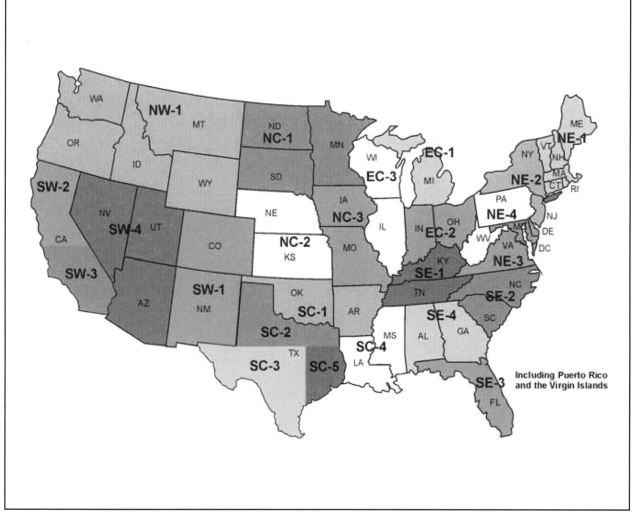

Figure 1-20. Terminal Procedures Publication.

• **U.S. IFR/VFR Low Altitude Planning Chart**: This chart is designed for preflight and en route flight planning for IFR/VFR flights. Depiction includes low altitude airways and mileage, NAVAIDs, airports, special use airspace, cities, time zones, major drainage, a directory of airports with their airspace classification, and a mileage table showing great circle distances between major airports. The chart scale is 1 inch = 47 NM/1:3,400,000, and is revised annually, available either folded or unfolded for wall mounting.

Supplementary charts and publications include:

• *Airport/Facility Directory* (**A/FD**): This seven volume booklet series contains data on airports, seaplane bases, heliports, NAVAIDs, communications data, weather data sources, airspace, special notices, and operational procedures. The coverage includes the conterminous U.S., Puerto Rico, and the Virgin Islands. The A/FD shows data that cannot be readily depicted in graphic form; e.g., airport hours of operations, types of fuel available, runway widths, lighting codes, etc. The A/FD also provides a means for pilots to update visual charts between edition dates, and is published every 56 days. The volumes are side-bound 5-3/8 x 8-1/4 inches.

• **Supplement Alaska**: This is a civil/military flight information publication issued by the FAA every 56 days. This booklet is designed for use with appropriate IFR or VFR charts. The Supplement Alaska contains an airport/facility directory, airport sketches, communications data, weather data sources, airspace, listing of navigational facilities, and special notices and procedures. The volume is side-bound 5-3/8 x 8-1/4 inches.

- **Chart Supplement Pacific**: This supplement is designed for use with appropriate VFR or IFR en route charts. Included in this booklet are the airport/facility directory, communications data, weather data sources, airspace, navigational facilities, special notices, and Pacific area procedures. IAP charts, DP charts, STAR charts, airport diagrams, radar minimums, and supporting data for the Hawaiian and Pacific Islands are included. The manual is published every 56 days. The volume is side-bound 5-3/8 x 8-1/4 inches.

- **North Pacific Route Charts**: These charts are designed for FAA controllers to monitor transoceanic flights. They show established intercontinental air routes, including reporting points with geographic positions. The Composite Chart scale is 1 inch = 164 NM/1:12,000,000. 48 x 41-1/2 inches. Area Chart scales are 1 inch = 95.9 NM/1:7,000,000. The size is 52 x 40-1/2 inches. All charts shipped unfolded. The charts are revised every 56 days.

- **North Atlantic Route Chart**: Designed for FAA controllers to monitor transatlantic flights, this five-color chart shows oceanic control areas, coastal navigation aids, oceanic reporting points, and NAVAID geographic coordinates. The full size chart scale is 1 inch = 113.1 NM/1:8,250,000, shipped flat only. The half size chart scale is 1 inch = 150.8 NM/1:11,000,000. The size is 29-3/4 x 20-1/2 inches, shipped folded to 5 x 10 inches only, and is revised every 56 weeks.

- **FAA Aeronautical Chart User's Guide**: This publication is designed to be used as a teaching aid and reference document. It describes the substantial amount of information provided on the FAA's aeronautical charts and publications. It includes explanations and illustrations of chart terms and symbols organized by chart type. It is available online at:

    http://www.naco.faa.gov/index.asp?xml=naco/online/aero_guide

- **Airport/Facility Directory (A/FD)**

    Digital products include:

- **The NAVAID Digital Data File**: This file contains a current listing of NAVAIDs that are compatible with the NAS. Updated every 56 days, the file contains all NAVAIDs including ILS and its components, in the U.S., Puerto Rico, and the Virgin Islands plus bordering facilities in Canada, Mexico, and the Atlantic and Pacific areas. The file is available by subscription only, on a 3.5-inch, 1.4 megabyte diskette.

- **The Digital Obstacle File**: This file describes all obstacles of interest to aviation users in the U.S., with limited coverage of the Pacific, Caribbean, Canada, and Mexico. The obstacles are assigned unique numerical identifiers, accuracy codes, and listed in order of ascending latitude within each state or area. The file is updated every 56 days, and is available on 3.5-inch, 1.4 megabyte diskettes.

- **The Digital Aeronautical Chart Supplement (DACS)**: The DACS is a subset of the data provided to FAA controllers every 56 days. It reflects digitally what is shown on the en route high and low charts. The DACS is designed to be used with aeronautical charts for flight planning purposes only. It should not be used as a substitute for a chart. The DACS is available on two 3.5-inch diskettes, compressed format. The supplement is divided into the following nine individual sections:

Section 1: High Altitude Airways, Conterminous U.S.

Section 2: Low Altitude Airways, Conterminous U.S.

Section 3: Selected Instrument Approach Procedure NAVAID and Fix Data

Section 4: Military Training Routes

Section 5: Alaska, Hawaii, Puerto Rico, Bahamas, and Selected Oceanic Routes

Section 6: STARs, Standard Terminal Arrivals

Section 7: DPs, Instrument Departure Procedures

Section 8: Preferred IFR Routes (low and high altitude)

Section 9: Air Route and Airport Surveillance Radar Facilities

NOTICE TO AIRMEN

Since the NAS is continually evolving, Notices to Airmen (NOTAM) provide the most current essential flight operation information available, not known sufficiently in advance to publicize in the most recent aeronautical charts or A/FD. NOTAMs provide information on airports and changes that affect the NAS that are time critical and in particular are of concern to IFR operations. Published FAA domestic/international NOTAMs are available by subscription and on the Internet. Each NOTAM is classified as a NOTAM (D), a NOTAM (L), or an FDC NOTAM. [Figure 1-21]

```
                            NOTAM(D)

        DEN 09/080 DEN 17L IS LLZ OTS WEF 0209141200-0210012359

                            NOTAM(L)

   TWY C (BTN TWYS L/N); TWY N (BTN TWY C AND RWY10L/28R); TWY P (BTN
   TWY C AND RWY10L/28R) - CLSD DLY
   1615-2200.

                            FDC NOTAM

   FDC 2/9651 DFW FI/P DALLAS/FORT WORTH INTL, DALLAS/FORT WORTH, TX
   CORRECT U.S. TERMINAL PROCEDURES SOUTH CENTRAL (SC) VOL 2 OF 5.
   EFFECTIVE 8 AUGUST 2002, PAGE 192.
   CHANGE RADIAL FROM RANGER (FUZ) VORTAC TO EPOVE INT TO READ
   352 VICE 351.
```

Figure 1-21. NOTAM Examples.

A NOTAM (D) or distant NOTAM is given dissemination beyond the area of responsibility of a Flight Service Station (AFSS/FSS). Information is attached to hourly weather reports and is available at AFSSs/FSSs. AFSSs/FSSs accept NOTAMs from the following personnel in their area of responsibility: Airport Manager, Airways Facility SMO, Flight Inspection, and Air Traffic. They are disseminated for all navigational facilities that are part of the U.S. NAS, all public use airports, seaplane bases, and heliports listed in the A/FD. The complete NOTAM (D) file is maintained in a computer database at the National Weather Message Switching Center (WMSC) in Atlanta, Georgia. Most air traffic facilities, primarily AFSSs/FSSs, have access to the entire database of NOTAM (D)s, which remain available for the duration of their validity, or until published.

A NOTAM (L) or local NOTAM requires dissemination locally, but does not qualify as NOTAM (D) information. These NOTAMs usually originate with the Airport Manager and are issued by the FSS/AFSS. A NOTAM (L) contains information such as taxiway closures, personnel and equipment near or crossing runways, and airport rotating beacon and lighting aid outages. A separate file of local NOTAMs is maintained at each FSS/AFSS for facilities in the area. NOTAM (L) information for other FSS/AFSS areas must be specifically requested directly from the FSS/AFSS that has responsibility for the airport concerned. Airport/Facility Directory listings include the associated FSS/AFSS and NOTAM file identifiers. [Figure 1-22]

FDC NOTAMs are issued by the National Flight Data Center (NFDC) and contain regulatory information such as temporary flight restrictions or amendments to instrument approach procedures and other current aeronautical charts. FDC NOTAMs are available through all air traffic facilities with telecommunications access. Information for instrument charts is supplied by Aviation System Standards (AVN) and much of the other FDC information is extracted from the NOTAM (D) System.

The *Notices to Airmen Publication* (NTAP) is published by Air Traffic Publications every 28 days and contains all current NOTAM (D)s and FDC NOTAMs (except FDC NOTAMs for temporary flight restrictions) available for publication. Federal airway changes, which are identified as Center Area NOTAMs, are included with the NOTAM (D) listing. Published NOTAM (D) information is not provided during pilot briefings unless requested. Data of a permanent nature are sometimes printed in the NOTAM publication as an interim step prior to publication on the appropriate aeronautical chart or in the A/FD. The NTAP is divided into four parts:

- Notices in part one are provided by the National Flight Data Center, and contain selected NOTAMs that are expected to be in effect on the

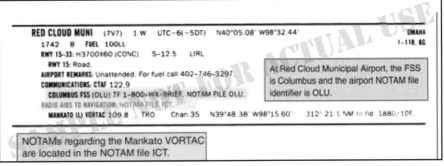

Figure 1-22. NOTAM File Reference in A/FD.

effective date of the publication. This part is divided into three sections:

a. Airway NOTAMs reflecting airway changes that fall within an ARTCC's airspace;

b. Airports/facilities, and procedural NOTAMs;

c. FDC general NOTAMs containing NOTAMs that are general in nature and not tied to a specific airport/facility, i.e. flight advisories and restrictions.

- Part two contains revisions to minimum en route IFR altitudes and changeover points.

- Part three, International, contains flight prohibitions, potential hostile situations, foreign notices, and oceanic airspace notices.

- Part four contains special notices and graphics pertaining to almost every aspect of aviation; such as, military training areas, large scale sporting events, air show information, and airport-specific information. Special traffic management programs (STMPs) are published in part four.

If you plan to fly internationally, you can benefit by accessing Class I international ICAO System NOTAMs, that include additional information. These help you differentiate IFR versus VFR NOTAMs, assist pilots who are not multilingual with a standardized format, and may include a "Q" line, or qualifier line that allows computers to read, recognize, and process NOTAM content information.

NAVIGATION DATABASES

The FAA updates and distributes the National Flight Database (NFD), a navigation database that is published by NACO every 28 days. This helps pilots and aircraft owners maintain current information in onboard navigation databases, such as those used in GPS and RNAV equipment. Current data elements include airports and heliports, VHF and NDB navigation aids, fixes/waypoints, airways, DPs, STARs, and GPS and RNAV (GPS) standard instrument approach procedures

(SIAPs) with their associated minimum safe altitude (MSA) data, runways for airports that have a SIAP coded in the NFD, and special use airspace (SUA) including military operation areas (MOA) and national security areas (NSA).

Future data elements to be added are:

- Air Traffic Service (ATS) routes

- Class B, C, and D Airspace

- Terminal Navigation Aids

- ILS and LOC SIAPs with Localizer and Glideslope records

- FIR/UIR Airspace

- Communication

Details about the NFD can be found at:
http://www.naco.faa.gov/index.asp?xml=naco/catalog/charts/digital/nfd

The FAA has developed an implementation and development plan that will provide users with data in an acceptable, open-industry standard for use in GPS/RNAV systems. The established aviation industry standard database model, Aeronautical Radio, Incorporated (ARINC 424) format, includes the essential information necessary for IFR flight in addition to those items necessary for basic VFR navigation. Essentially the new FAA database will fulfill requirements for operations within the NAS while still providing the opportunity for private entities to build upon the basic navigation database and provide users with additional services when desired. Refer to Appendix A, Airborne Navigation Databases for more detailed information.

As FAA and other government websites are continuously being changed and updated, be ready to use the search feature to find the information or publications you need.

CHAPTER 2
TAKEOFFS AND DEPARTURES

SAFETY IN THE DEPARTURE ENVIRONMENT

Thousands of IFR takeoffs and departures occur daily in the National Airspace System (NAS). In order to accommodate this volume of Instrument Flight Rule (IFR) traffic, Air Traffic Control (ATC) must rely on pilots to use charted airport sketches and diagrams as well as standard instrument departures (SIDs) and obstacle departure procedures (ODPs). While many charted (and uncharted) departures are based on radar vectors, the bulk of IFR departures in the NAS require pilots to navigate out of the terminal environment to the en route phase.

IFR takeoffs and departures are fast-paced phases of flight, and pilots often are overloaded with critical flight information. During takeoff, pilots are busy requesting and receiving clearances, preparing their aircraft for departure, and taxiing to the active runway. During IFR conditions, they are doing this with minimal visibility, and they may be without constant radio communication if flying out of a non-towered airport. Historically, takeoff minimums for commercial operations have been successively reduced through a combination of improved signage, runway markings and lighting aids, and concentrated pilot training and qualifications. Today at major terminals, some commercial operators with appropriate equipment, pilot qualifications, and approved Operations Specifications (OpsSpecs) may takeoff with visibility as low as runway visual range (RVR) 3, or 300 feet runway visual range. One of the consequences of takeoffs with reduced visibility is that pilots are challenged in maintaining situational awareness during taxi operations.

SURFACE MOVEMENT SAFETY

One of the biggest safety concerns in aviation is the surface movement accident. As a direct result, the FAA has rapidly expanded the information available to pilots including the addition of taxiway and runway information in FAA publications, particularly the *IFR U.S. Terminal Procedures Publication* (TPP) booklets and *Airport/Facility Directory* (A/FD) volumes. The FAA has also implemented new procedures and created edu-

cational and awareness programs for pilots, air traffic controllers, and ground operators. By focusing resources to attack this problem head on, the FAA hopes to reduce and eventually eliminate surface movement accidents.

AIRPORT SKETCHES AND DIAGRAMS

Airport sketches and **airport diagrams** provide pilots of all levels with graphical depictions of the airport layout. The National Aeronautical Charting Office (NACO) provides an airport sketch on the lower left or right portion of every instrument approach chart. [Figure 2-1] This sketch depicts the runways, their length, width, and slope, the touchdown zone elevation, the lighting system installed on the end of the runway, and taxiways.

For select airports, typically those with heavy traffic or complex runway layouts, NACO also prints an airport diagram. The diagram is located in the IFR TPP booklet following the instrument approach chart for a particular airport. It is a full-page depiction of the airport that includes the same features of the airport sketch plus additional details such as taxiway identifiers, airport latitude and longitude, and building identification. The airport diagrams are also available in the A/FD and on the NACO website,

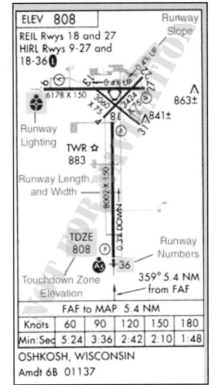

Figure 2-1. Airport Sketch Included on the KOSH ILS RWY 36 Approach Chart.

http://naco.faa.gov. by selecting "Online *digital* - TPP." [Figure 2-2]

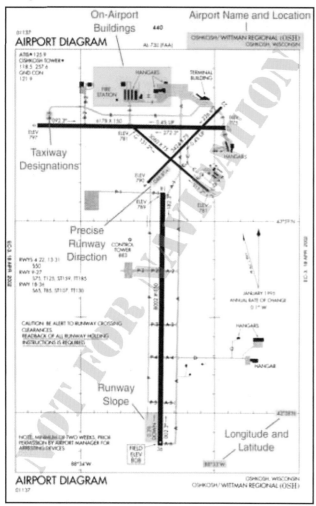

Figure 2-2. Airport Diagram for KOSH.

AIRPORT/FACILITY DIRECTORY

The *Airport/Facility Directory* (**A/FD**), published in regional booklets by NACO, provides textual information about all airports, both VFR and IFR. The A/FD includes runway length and width, runway surface, load bearing capacity, runway slope, airport services, and hazards such as birds and reduced visibility. [Figure 2-3] Sketches of airports also are being added to aid VFR pilots in surface movement activities. In support of the FAA Runway Incursion Program, full-page airport diagrams are included in the A/FD. These charts are the same as those published in the IFR TPP and are printed for airports with complex runway or taxiway layouts.

SURFACE MOVEMENT GUIDANCE CONTROL SYSTEM

The **Surface Movement Guidance Control System** (**SMGCS**) was developed in 1992 to facilitate the safe movement of aircraft and vehicles at airports where scheduled air carriers were conducting authorized operations. This program was designed to provide guidelines for the creation of low visibility taxi plans for all airports with takeoff or landing operations using visibility minimums less than 1,200 feet RVR. For landing operations, this would be pertinent only to those operators whose OpsSpecs permit them to land with lower than standard minimums. For departures, however, since there are no regulatory takeoff minimums for Title 14 Code of Federal Regulations (14 CFR) Part 91 operators, the SMGCS information is pertinent to all departing traffic operating in Instrument Meteorological Conditions (IMC). *Advisory Circular (AC) 120-57A, Surface Movement Guidance and Control System*, outlines the SMGCS program in its entirety including standards and guidelines for establishment of a low visibility taxi plan.

The SMGCS low visibility taxi plan includes the improvement of taxiway and runway signs, markings, and lighting, as well as the creation of SMGCS low visibility taxi route charts. [Figure 2-4 on page 2-4] The plan also clearly identifies taxi routes and their supporting facilities and equipment. Airport enhancements that are part of the SMGCS program include (but are not limited to):

- **Stop bars** consist of a row of red unidirectional, in-pavement lights installed along the holding position marking. When extinguished by the controller, they confirm clearance for the pilot or vehicle operator to enter the runway. They are required at intersections of an illuminated taxiway and active runway for operations less than 600 feet RVR.

- **Taxiway centerline lights**, which work in conjunction with stop bars, are green in-pavement lights that guide ground traffic under low visibility conditions and during darkness.

- **Runway guard lights**, either elevated or in-pavement, will be installed at all taxiways that provide access to an active runway. They consist of alternately flashing yellow lights, used to denote both the presence of an active runway and identify the location of a runway holding position marking.

- **Geographic position markings**, used as hold points or for position reporting, enable ATC to verify the position of aircraft and vehicles. These checkpoints or "pink spots" are outlined with a black and white circle and designated with a number, a letter, or both.

- **Clearance bars** consist of three yellow in-pavement lights used to denote holding positions for aircraft and vehicles. When used for hold points, they are co-located with geographic position markings.

Additional information concerning airport lighting, markings, and signs can be found in the *Aeronautical Information Manual* (AIM), as well as on the FAA's website at:

http://www.faa.gov/library/manuals/aviation.

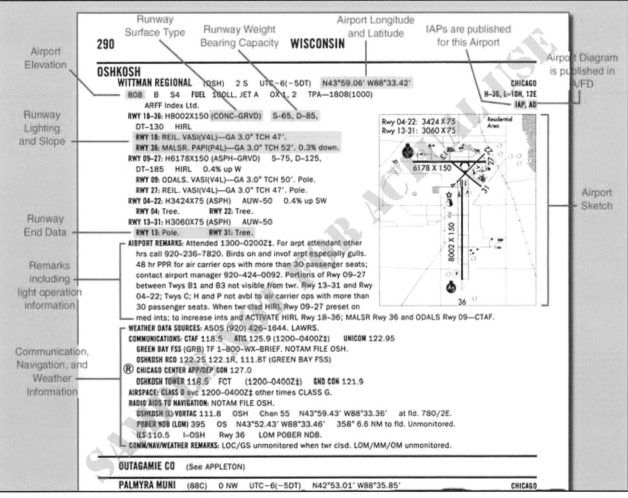

Figure 2-3. Excerpt from Airport/Facility Directory for Oshkosh/Wittman Field.

Both flight and ground crews are required to comply with SMGCS plans when implemented at their specific airport. All airport tenants are responsible for disseminating information to their employees and conducting training in low visibility operating procedures. Anyone operating in conjunction with the SMGCS plan must have a copy of the low visibility taxi route chart for their given airport as these charts outline the taxi routes and other detailed information concerning low visibility operations. These charts are available from private sources outside of the FAA. Part 91 operators are expected to comply with the guidelines listed in the AC to the best of their ability and should expect "Follow Me" service when low visibility operations are in use. Any SMGCS outage that would adversely affect operations at the airport is issued as a Notice to Airmen (NOTAM).

AIRPORT SIGNS, LIGHTING, and MARKING

Flight crews use airport lighting, markings, and signs to help maintain situational awareness when operating on the ground and in the air. These visual aids provide information concerning the aircraft's location on the airport, the taxiway in use, and the runway entrance being used. Overlooking this information can lead to

ground accidents that are entirely preventable. If you encounter unfamiliar markings or lighting, contact ATC for clarification and, if necessary, request progressive taxi instructions. Pilots are encouraged to notify the appropriate authorities of erroneous, misleading, or decaying signs or lighting that would contribute to the failure of safe ground operations.

RUNWAY INCURSIONS

A runway incursion is any occurrence at an airport involving aircraft, ground vehicles, people, or objects on the ground that creates a collision hazard or results in the loss of separation with an aircraft taking off, intending to take off, landing, or intending to land. Primarily, runway incursions are caused by errors resulting from a misunderstanding of the given clearance, failure to communicate effectively, failure to navigate the airport correctly, or failure to maintain positional awareness. Figure 2-5 on page 2-5 highlights several steps that reduce the chances of being involved in a runway incursion.

In addition to the SMGCS program, the FAA has implemented additional programs to reduce runway incursions and other surface movement issues. They

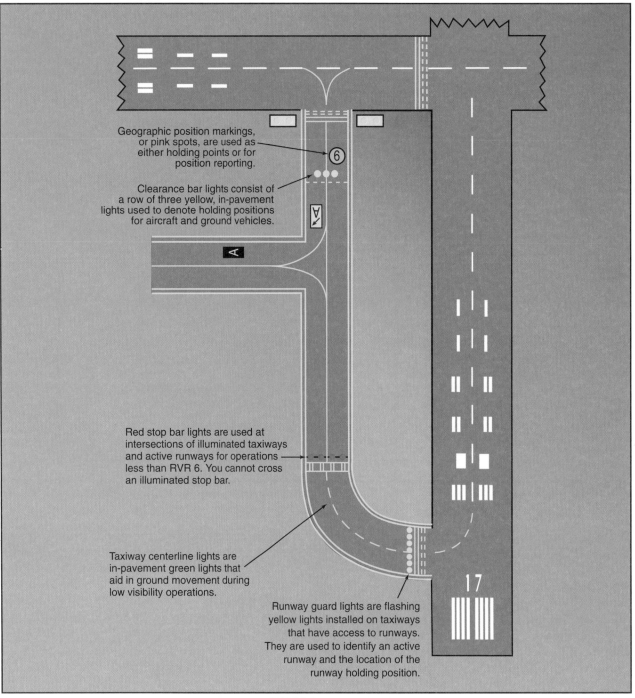

Geographic position markings, or pink spots, are used as either holding points or for position reporting.

Clearance bar lights consist of a row of three yellow, in-pavement lights used to denote holding positions for aircraft and ground vehicles.

Red stop bar lights are used at intersections of illuminated taxiways and active runways for operations less than RVR 6. You cannot cross an illuminated stop bar.

Taxiway centerline lights are in-pavement green lights that aid in ground movement during low visibility operations.

Runway guard lights are flashing yellow lights installed on taxiways that have access to runways. They are used to identify an active runway and the location of the runway holding position.

Figure 2-4. SMGCS Signage and Lighting.

identified runway hotspots, designed standardized taxi routes, and instituted the Runway Safety Program.

RUNWAY HOTSPOTS

Runway hotspots (some FAA Regions refer to them as high alert areas) are locations on particular airports that historically have hazardous intersections. These hotspots are depicted on some airport charts as circled areas. FAA Regions, such as the Western Pacific, notify pilots of these areas by Letter to Airmen. The FAA Office of Runway Safety website (www.faa.gov/runwaysafety) has links to the FAA regions that maintain a complete list of airports with runway hotspots. Also,

charts provided by private sources show these locations. Hotspots alert pilots to the fact that there may be a lack of visibility at certain points or the tower may be unable to see that particular intersection. Whatever the reason, pilots need to be aware that these hazardous intersections exist and they should be increasingly vigilant when approaching and taxiing through these intersections.

STANDARDIZED TAXI ROUTES

Standard taxi routes improve ground management at high-density airports, namely those that have airline service. At these airports, typical taxiway traffic patterns used to move aircraft between gate and runway

The FAA recommends that you:

- Receive and understand all NOTAMs, particularly those concerning airport construction and lighting.
- Read back, in full, all clearances involving holding short, taxi into position and hold, and crossing active runways to insure proper understanding.
- Abide by the sterile cockpit rule.
- Develop operational procedures that minimize distractions during taxiing.
- Ask ATC for directions if you are lost or unsure of your position.
- Adhere to takeoff and runway crossing clearances in a timely manner.
- Position your aircraft so landing traffic can see you.
- Monitor radio communications to maintain a situational awareness of other aircraft.
- Remain on frequency until instructed to change.
- Make sure you know the reduced runway distances and whether or not you can comply before accepting a land and hold short clearance.
- Report confusing airport diagrams to the proper authorities.
- Use exterior taxi and landing lights when practical.

Note: The sterile cockpit rule refers to a concept outlined in Parts 121.542 and 135.100 that requires flight crews to refrain from engaging in activities that could distract them from the performance of their duties during critical phases of flight. This concept is explained further in Chapter 4.

Figure 2-5. FAA Recommendations for Reducing Runway Incursions.

are laid out and coded. The ATC specialist (ATCS) can reduce radio communication time and eliminate taxi instruction misinterpretation by simply clearing the pilot to taxi via a specific, named route. An example of this would be Chicago O'Hare, where the Silver Alpha taxi route is used to transition to Runway 4L. [Figure 2-6] The "Silver A" route requires you to taxi via taxiway Alpha to Alpha Six, then taxiway Juliet, then taxiway Whiskey to Runway 4L. These routes are issued by ground control, and if unable to comply, pilots must advise ground control on initial contact. If for any reason the pilot becomes uncertain as to the correct taxi route, a request should be made for progressive taxi instructions. These step-by-step routing directions are also issued if the controller deems it necessary due to traffic, closed taxiways, airport construction, etc. It is the pilot's responsibility to

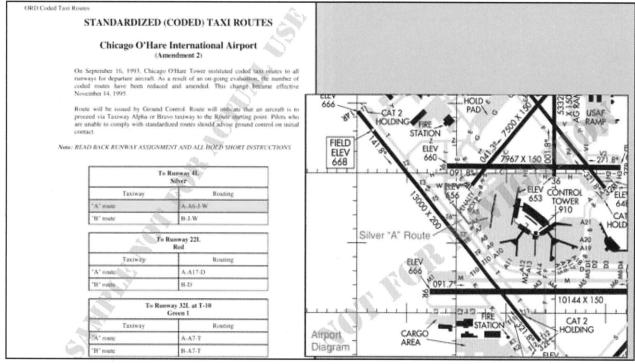

Figure 2-6. Chicago O'Hare Silver Standardized Taxi Route and NACO Airport Diagram.

know if a particular airport has preplanned taxi routes, to be familiar with them, and to have the taxi descriptions in their possession. Specific information about airports that use coded taxiway routes is included in the Notice to Airmen Publication (NTAP).

RUNWAY SAFETY PROGRAM

On any given day, the NAS may handle almost 200,000 takeoffs and landings. Due to the complex nature of the airport environment and the intricacies of the network of people that make it operate efficiently, the FAA is constantly looking to maintain the high standard of safety that exists at airports today. Runway safety is one of its top priorities. The **Runway Safety Program (RSP)** is designed to create and execute a plan of action that reduces the number of runway incursions at the nation's airports.

The RSP office has created a National Blueprint for Runway Safety. [Figure 2-7] In that document, the FAA has identified four types of runway surface events:

- Surface Incident – an event during which authorized or unauthorized/unapproved movement occurs in the movement area or an occurrence in the movement area associated with the operation of an aircraft that affects or could affect the safety of flight.

- **Runway Incursion** – an occurrence at an airport involving an aircraft, vehicle, person, or object on the ground that creates a collision hazard or results in a loss of separation with an aircraft that is taking off, intending to take off, landing, or intending to land.

- Collision Hazard – a condition, event, or circumstance that could induce an occurrence of a collision or surface accident or incident.

- Loss of Separation – an occurrence or operation that results in less than prescribed separation between aircraft, or between an aircraft and a vehicle, pedestrian, or object.

Runway incursions are further identified by four categories: ATC operational error, pilot deviation, vehicle/pedestrian deviation, and miscellaneous errors that cannot be attributed to the previous categories.

Since runway incursions cannot be attributed to one single group of people, everyone involved in airport operations must be equally aware of the necessity to improve runway safety. As a result, the RSP created goals to develop refresher courses for ATC, promote educational awareness for air carriers, and require flight training that covers more in depth material concerning ground operations. Beyond the human aspect of runway safety, the FAA is also reviewing technology, communications, operational procedures, airport signs, markings, lighting, and analyzing causal factors to find areas for improvement.

Runway safety generates much concern especially with the continued growth of the aviation industry. The takeoff and departure phases of flight are critical portions of the flight since the majority of this time is spent on the ground with multiple actions occurring. It is the desire of the FAA and the aviation industry to reduce runway surface events of all types, but it cannot be done simply through policy changes and educational programs. Pilots must take responsibility for ensuring safety during surface operations and continue to educate themselves through government (www.faa.gov/runwaysafety) and industry runway safety programs.

TAKEOFF MINIMUMS

While mechanical failure is potentially hazardous during any phase of flight, a failure during takeoff under instrument conditions is extremely critical. In the event of an emergency, a decision must be made to either return to the departure airport or fly directly to a takeoff alternate. If the departure weather were below the landing minimums for the departure airport, the flight would be unable to return for landing, leaving few options and little time to reach a takeoff alternate.

In the early years of air transportation, landing minimums for commercial operators were usually lower than takeoff minimums. Therefore, it was possible that minimums allowed pilots to land at an airport but not depart from that airport. Additionally, all takeoff minimums once included ceiling as well as visibility

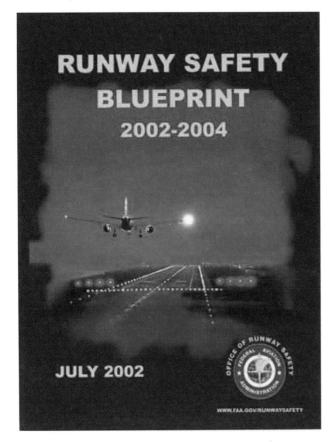

Figure 2-7. National Blueprint for Runway Safety.

requirements. Today, takeoff minimums are typically lower than published landing minimums and ceiling requirements are only included if it is necessary to see and avoid obstacles in the departure area.

The FAA establishes takeoff minimums for every airport that has published Standard Instrument Approaches. These minimums are used by commercially operated aircraft, namely Part 121 and 135 operators. At airports where minimums are not established, these same carriers are required to use FAA designated standard minimums (1 statute mile [SM] visibility for single- and twin-engine aircraft, and 1/2 SM for helicopters and aircraft with more than two engines).

Aircraft operating under Part 91 are not required to comply with established takeoff minimums. Legally, a zero/zero departure may be made, but it is never advisable. If commercial pilots who fly passengers on a daily basis must comply with takeoff minimums, then good judgment and common sense would tell all instrument pilots to follow the established minimums as well.

NACO charts list takeoff minimums only for the runways at airports that have other than standard minimums. These takeoff minimums are listed by airport in alphabetical order in the front of the TPP booklet. If an airport has non-standard takeoff minimums, a ▼ (referred to by some as either the "triangle T" or "trouble T") will be placed in the notes sections of the instrument procedure chart. In the front of the TPP booklet, takeoff minimums are listed before the obstacle departure procedure. Some departure procedures allow a departure with standard minimums provided specific aircraft performance requirements are met. [Figure 2-8]

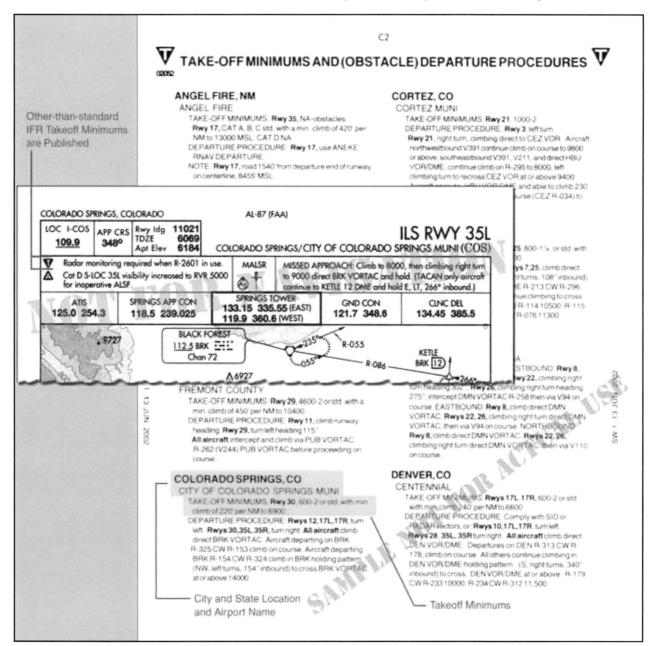

Figure 2-8. Takeoff minimums are listed in the front of each NACO U.S. Terminal Procedures booklet.

TAKEOFF MINIMUMS FOR COMMERCIAL OPERATORS

While Part 121 and 135 operators are the primary users of takeoff minimums, they may be able to use alternative takeoff minimums based on their individual OpsSpecs. Through these OpsSpecs, operators are authorized to depart with lower-than-standard minimums provided they have the necessary equipment and crew training.

OPERATIONS SPECIFICATIONS

Operations specifications (OpsSpecs) are required by Part 119.5 to be issued to commercial operators to define the appropriate authorizations, limitations, and procedures based on their type of operation, equipment, and qualifications. The OpsSpecs can be adjusted to accommodate the many variables in the air transportation industry, including aircraft and aircraft equipment, operator capabilities, and changes in aviation technology. The OpsSpecs are an extension of the CFR; therefore, they are legal, binding contracts between a properly certificated air transportation organization and the FAA for compliance with the CFR's applicable to their operation. OpsSpecs are designed to provide specific operational limitations and procedures tailored to a specific operator's class and size of aircraft and types of operation, thereby meeting individual operator needs.

Part 121 and 135 operators have the ability, through the use of approved OpsSpecs, to use lower-than-standard takeoff minimums. Depending on the equipment installed in a specific type of aircraft, the crew training, and the type of equipment installed at a particular airport, these operators can depart from appropriately equipped runways with as little as 300 feet RVR. Additionally, OpsSpecs outline provisions for approach minimums, alternate airports, and weather services in Part 119 and FAA Order 8400.10, *Air Transportation Operations Inspector's Handbook.*

HEAD-UP GUIDANCE SYSTEM

As technology improves over time, the FAA is able to work in cooperation with specific groups desiring to use these new technologies. **Head-up guidance system (HGS)** is an example of an advanced system currently being used by some airlines. Air carriers have requested the FAA to approve takeoff minimums at 300 feet RVR. This is the lowest takeoff minimum approved by OpsSpecs. As stated earlier, only specific air carriers with approved, installed equipment, and trained pilots are allowed to use HGS for decreased takeoff minimums. [Figure 2-9]

CEILING AND VISIBILITY REQUIREMENTS

All takeoffs and departures have visibility minimums (some may have minimum ceiling requirements) incorporated into the procedure. There are a number of methods to report visibility, and a variety of ways to distribute these reports, including automated weather observations. Flight crews should always check the weather, including ceiling and visibility information,

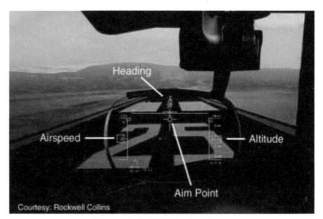

Figure 2-9. HGS Technology.

prior to departure. Never launch an IFR flight without obtaining current visibility information immediately prior to departure. Further, when ceiling and visibility minimums are specified for IFR departure, both are applicable.

Weather reporting stations for specific airports across the country can be located by reviewing the A/FD. Weather sources along with their respective phone numbers and frequencies are listed by airport. Frequencies for weather sources such as automatic terminal information service (ATIS), digital automatic terminal information service (D-ATIS), Automated Weather Observing System (AWOS), Automated Surface Observing System (ASOS), and FAA Automated Flight Service Station (AFSS) are published on approach charts as well. [Figure 2-10]

RUNWAY VISUAL RANGE

Runway visual range (RVR) is an instrumentally derived value, based on standard calibrations, that represents the horizontal distance a pilot will see down the runway from the approach end. It is based on the sighting of either high intensity runway lights or on the

Figure 2-10. Frequencies for Weather Information are listed on Approach and Airport Charts.

visual contrast of other targets whichever yields the greater visual range. RVR, in contrast to prevailing or runway visibility, is based on what a pilot in a moving aircraft should see looking down the runway. RVR is reported in hundreds of feet, so the values must be converted to statute miles if the visibility in statute miles is not reported. [Figure 2-11] This visibility measurement is updated every minute; therefore, the most accurate visibility report will come from the local controller instead of a routine weather report. Transmissometers near the runway measure visibility for the RVR report. If multiple transmissometers are installed, they provide reports for multiple locations, including **touchdown RVR, mid-RVR,** and **rollout RVR.** RVR visibility may be reported as RVR 5-5-5. This directly relates to the multiple locations from which RVR is reported and indicates 500 feet visibility at touchdown RVR, 500 feet at mid-RVR, and 500 feet at the rollout RVR stations.

Conversion

RVR (FT)	Visibility (SM)
1,600	1/4
2,400	1/2
3,200	5/8
4,000	3/4
4,500	7/8
5,000	1
6,000	1 1/4

Figure 2-11. RVR Conversion Table.

RVR is the primary visibility measurement used by Part 121 and 135 operators, with specific visibility reports and controlling values outlined in their respective OpsSpecs. Under their OpsSpecs agreements, the operator must have specific, current RVR reports, if available, to proceed with an instrument departure. OpsSpecs also outline which visibility report is controlling in various departure scenarios.

RUNWAY VISIBILITY VALUE

Runway visibility value (RVV) is the distance down the runway that a pilot can see unlighted objects. It is reported in statute miles for individual runways. RVV, like RVR, is derived from a transmissometer for a particular runway. RVV is used in lieu of prevailing visibility in determining specific runway minimums.

PREVAILING VISIBILITY

Prevailing visibility is the horizontal distance over which objects or bright lights can be seen and identified over at least half of the horizon circle. If the prevailing visibility varies from area to area, the visibility of the majority of the sky is reported. When critical differences exist in various sectors of the sky and the prevailing visibility is less than three miles, these differences will be reported at manned stations. Typically, this is referred to as sector visibility in the remarks section of a METAR report. Prevailing visibility is reported in statute miles or fractions of miles.

TOWER VISIBILITY

Tower visibility is the prevailing visibility as determined from the air traffic control tower (ATCT). If visibility is determined from only one point on the airport and it is the tower, then it is considered the usual point of observation. Otherwise, when the visibility is measured from multiple points, the control tower observation is referred to as the tower visibility. It too is measured in statute miles or fractions of miles.

ADEQUATE VISUAL REFERENCE

Another set of lower-than-standard takeoff minimums is available to Part 121 and 135 operations as outlined in their respective OpsSpecs document. When certain types of visibility reports are unavailable or specific equipment is out of service, the flight can still depart the airport if the pilot can maintain adequate visual reference. An appropriate visual aid must be available to ensure the takeoff surface can be continuously identified and directional control can be maintained throughout the takeoff run. Appropriate visual aids include high intensity runway lights, runway centerline lights, runway centerline markings, or other runway lighting and markings. A visibility of 1600 feet RVR or 1/4 SM is below standard and may be considered adequate for specific commercial operators if contained in an OpsSpecs approval.

AUTOMATED WEATHER SYSTEM

An **automated weather system** consists of any of the automated weather sensor platforms that collect weather data at airports and disseminate the weather information via radio and/or landline. The systems consist of the **Automated Surface Observing System (ASOS)/Automated Weather Sensor System (AWSS),** and the Automated Weather Observation System (AWOS). These systems are installed and maintained at airports across the United States (U.S.) by both government (FAA and NWS) and private entities. They are relatively inexpensive to operate because they require no outside observer, and they provide invaluable weather information for airports without operating control towers. [Figure 2-12 on page 2-10]

AWOS and ASOS/AWSS offer a wide variety of capabilities and progressively broader weather reports. Automated systems typically transmit weather every one to two minutes

Figure 2-12. ASOS Station Installation.

so the most up-to-date weather information is constantly broadcast. Basic AWOS includes only altimeter setting, wind speed, wind direction, temperature, and dew point information. More advanced systems such as the ASOS/AWSS and AWOS-3 are able to provide additional information such as cloud and ceiling data and precipitation type. ASOS/AWSS stations providing service levels A or B also report RVR. The specific type of equipment found at a given facility is listed in the A/FD. [Figure 2-13]

Automated weather information is available both over a radio frequency specific to each site and via telephone. When an automated system is brought online, it first goes through a period of testing. Although you can listen to the reports on the radio and over the phone during the test phase, they are not legal for use until they are fully operational, and the test message is removed.

The use of the aforementioned visibility reports and weather services are not limited for Part 91 operators. Part 121 and 135 operators are bound by their individual OpsSpecs documents and are required to use weather reports that come from the National Weather Service or other approved sources. While every operator's specifications are individually tailored, most operators are required to use ATIS information, RVR reports, and selected reports from automated weather stations. All reports coming from an AWOS-3 station are usable for Part 121 and 135 operators. Each type of automated station has different levels of approval as outlined in FAA Order 8400.10

and individual OpsSpecs. Ceiling and visibility reports given by the tower with the departure information are always considered official weather, and RVR reports are typically the controlling visibility reference.

AUTOMATIC TERMINAL INFORMATION SERVICE AND DIGITAL ATIS

The **automatic terminal information service (ATIS)** is another valuable tool for gaining weather information. ATIS is available at most airports that have an operating control tower, which means the reports on the ATIS frequency are only available during the regular hours of tower operation. At some airports that operate part-time towers, ASOS/AWSS information is broadcast over the ATIS frequency when the tower is closed. This service is available only at those airports that have both an ASOS/AWSS on the field and an ATIS-ASOS/AWSS interface switch installed in the tower.

Each ATIS report includes crucial information about runways and instrument approaches in use, specific outages, and current weather conditions including visibility. Visibility is reported in statute miles and may be omitted if the visibility is greater than five miles. ATIS weather information comes from a variety of sources depending on the particular airport and the equipment installed there. The reported weather may come from a manual weather observer, weather instruments located in the tower, or from automated weather stations. This information, no matter the origin, must be from National Weather Service approved weather sources for it to be used in the ATIS report.

The **digital ATIS (D-ATIS)** is an alternative method of receiving ATIS reports. The service provides text messages to aircraft, airlines, and other users outside the standard reception range of conventional ATIS via landline and data link communications to the cockpit. Aircraft equipped with data link services are capable of receiving ATIS information over their Aircraft Communications Addressing and Reporting System (ACARS) unit. This allows the pilots to read and print out the ATIS report inside the aircraft, thereby increasing report accuracy and decreasing pilot workload.

Also, the service provides a computer-synthesized voice message that can be transmitted to all aircraft within range of existing transmitters. The Terminal Data Link System (TDLS) D-ATIS application uses weather inputs from local automated weather sources or manually entered meteorological data together with preprogrammed menus to provide standard information to users. Airports with D-ATIS capability are listed in the A/FD.

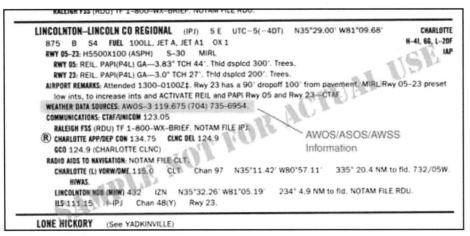

Figure 2-13. A/FD Entry for an AWOS Station.

It is important to remember that ATIS information is updated hourly and anytime a significant change in the weather occurs. As a result, the information is not the most current report available. Prior to departing the airport, you need to get the latest weather information from the tower. ASOS/AWSS and AWOS also provide a source of current weather, but their information should not be substituted for weather reports from the tower.

IFR ALTERNATE REQUIREMENTS

The requirement for an alternate depends on the aircraft category, equipment installed, approach NAVAID and forecast weather. For example, airports with only a global positioning system (GPS) approach procedure cannot be used as an alternate by TSO-C129/129A users even though the ▲N/A has been removed from the approach chart. For select RNAV (GPS) and GPS approach procedures the ▲ N/A is being removed so they may be used as an alternate by aircraft equipped with an approach approved WAAS receiver. Because GPS is not authorized as a substitute means of navigation guidance when conducting a conventional approach at an alternate airport, if the approach procedure requires either DME or ADF, the aircraft must be equipped with the appropriate DME or ADF avionics in order to use the approach as an alternate.

For airplane Part 91 requirements, an alternate airport must be listed on IFR flight plans if the forecast weather at the destination airport, from a time period of plus or minus one hour from the estimated time of arrival (ETA), includes ceilings lower than 2,000 feet and/or visibility less than 3 SM. A simple way to remember the rules for determining the necessity of filing an alternate for airplanes is the "1, 2, 3 Rule." For helicopter Part 91, similar alternate filing requirements apply. An alternate must be listed on an IFR flight plan if the forecast weather at the destination airport or heliport, from the ETA and for one hour after the ETA, includes ceilings lower than 1,000 feet, or less than 400 feet above the lowest applicable approach minima, whichever is higher, and the visibility less than 2 SM.

Not all airports can be used as alternate airports. An airport may not be qualified for alternate use if the airport NAVAID is unmonitored, or if it does not have weather reporting capabilities. For an airport to be used as an alternate, the forecast weather at that airport must meet certain qualifications at the estimated time of arrival. Standard alternate minimums for a precision approach are a 600-foot ceiling and a 2 SM visibility. For a non-precision approach, the minimums are an 800-foot ceiling and a 2 SM visibility. Standard alternate minimums apply unless higher alternate minimums are listed for an airport.

On NACO charts, standard alternate minimums are not published. If the airport has other than standard alternate minimums, they are listed in the front of the approach chart booklet. The presence of a triangle with an ▲ on the approach chart indicates the listing of alternate minimums should be consulted. Airports that do not qualify for use as an alternate airport are designated with an ▲ N/A. [Figure 2-14]

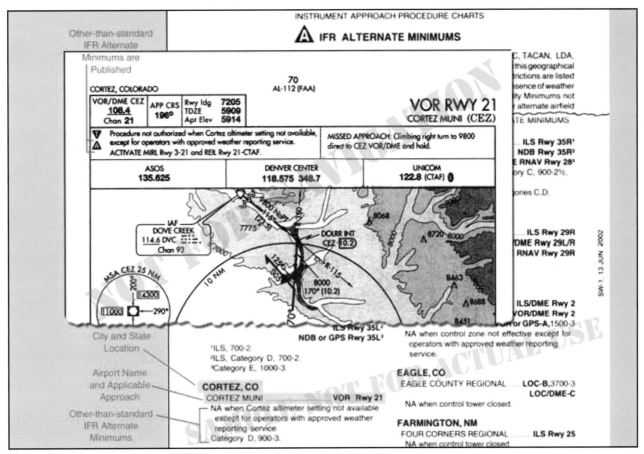

Figure 2-14. IFR Alternate Minimums.

ALTERNATE MINIMUMS FOR COMMERCIAL OPERATORS

IFR alternate minimums for Part 121 and 135 operators are very specific and have more stringent requirements than Part 91 operators.

Part 121 operators are required by their OpsSpecs and Parts 121.617 and 121.625 to have a takeoff alternate airport for their departure airport in addition to their airport of intended landing if the weather at the departure airport is below the landing minimums in the certificate holder's OpsSpecs for that airport. The alternate must be within two hours flying time for an aircraft with three or more engines with an engine out in normal cruise in still air. For two engine aircraft, the alternate must be within one hour. The airport of intended landing may be used in lieu of an alternate providing it meets all the requirements. Part 121 operators must also file for alternate airports when the weather at their destination airport, from one hour before to one hour after their ETA, is forecast to be below a 2,000-foot ceiling and/or less than 3 miles visibility.

For airports with at least one operational navigational facility that provides a straight-in non-precision approach, a straight-in precision approach, or a circling maneuver from an instrument approach procedure determine the ceiling and visibility by:

- Adding 400 feet to the authorized CAT I HAA/HAT for ceiling.

- Adding one mile to the authorized CAT I visibility for visibility minimums.

This is but one example of the criteria required for Part 121 operators when calculating minimums. Part 135 operators are also subject to their own specific rules regarding the selection and use of alternate minimums as outlined in their OpsSpecs and Part 135.219 through Part 135.225, and they differ widely from those used by Part 121 operators.

Typically, dispatchers who plan flights for these operators are responsible for planning alternate airports. The dispatcher considers aircraft performance, aircraft equipment and its condition, and route of flight when choosing alternates. In the event changes need to be made to the flight plan en route due to deteriorating weather, the dispatcher will maintain contact with the flight crew and will reroute their flight as necessary. Therefore, it is the pilot's responsibility to execute the flight as planned by the dispatcher; this is especially true for Part 121 pilots. To aid in the planning of alternates, dispatchers have a list of airports that are approved as alternates so they can quickly determine which airports should be used for a particular flight. Dispatchers also use flight-planning software that plans routes including alternates for the flight. This type of software is tailored for individual operators and includes their normal flight paths and approved airports. Flight planning software and services are provided through private sources.

Though the pilot is the final authority for the flight and ultimately has full responsibility, the dispatcher is responsible for creating flight plans that are accurate and comply with the CFRs. Alternate minimum criteria are only used as planning tools to ensure the pilot-in-command and dispatcher are thinking ahead to the approach phase of flight. In the event the flight would actually need to divert to an alternate, the published approach minimums or lower-than-standard minimums must be used as addressed in OpsSpecs documents.

DEPARTURE PROCEDURES

Departure procedures are preplanned routes that provide transitions from the departure airport to the en route structure. Primarily, these procedures are designed to provide obstacle protection for departing aircraft. They also allow for efficient routing of traffic and reductions in pilot/controller workloads. These procedures come in many forms, but they are all based on the design criteria outlined in TERPS and other FAA orders. The A/FD includes information on high altitude redesign RNAV routing pitch points, preferred IFR routings, or other established routing programs where a flight can begin a segment of nonrestrictive routing.

DESIGN CRITERIA

The design of a departure procedure is based on TERPS, a living document that is updated frequently. Departure design criterion assumes an initial climb of 200 feet per nautical mile (NM) after crossing the departure end of the runway (DER) at a height of at least 35 feet. [Figure 2-15] The aircraft climb path assumption provides a minimum of 35 feet of additional obstacle clearance above the required obstacle clearance (ROC), from the DER outward, to absorb variations ranging from the distance of the static source to the landing gear, to differences in establishing the minimum 200 feet per NM climb gradient, etc. The ROC is the planned separation between the obstacle clearance surface (OCS) and the required climb gradient of 200 feet per NM. The ROC value is zero at the DER elevation and increases along the departure route until the appropriate ROC value is attained to allow en route flight to commence. It is typically about 25 NM for 1,000 feet of ROC in non-mountainous areas, and 46 NM for 2,000 feet of ROC in mountainous areas.

Recent changes in TERPS criteria make the OCS lower and more restrictive. [Figure 2-16 on page 2-14] However, there are many departures today that were evaluated under the old criteria [Figure 2-15] that allowed some obstacle surfaces to be as high as 35 feet at the DER. Since there is no way for the pilot to determine whether the departure was evaluated using the

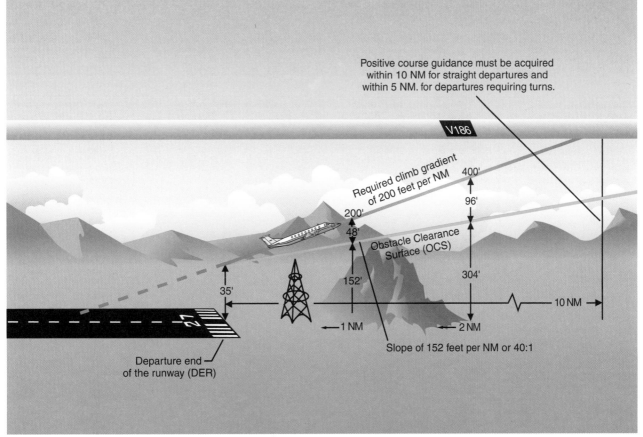

Positive course guidance must be acquired within 10 NM for straight departures and within 5 NM. for departures requiring turns.

V186

Required climb gradient of 200 feet per NM

400'

96'

200'

48'

Obstacle Clearance Surface (OCS)

304'

152'

35'

10 NM

1 NM

2 NM

Slope of 152 feet per NM or 40:1

Departure end of the runway (DER)

Figure 2-15. Previous TERPS Design Criteria for Departure Procedures.

previous or current criteria and until all departures have been evaluated using the current criteria, pilots need to be very familiar with the departure environment and associated obstacles especially if crossing the DER at less than 35 feet.

Assuming a 200-foot per NM climb, the departure is structured to provide at least 48 feet per NM of clearance above objects that do not penetrate the obstacle slope. The slope, known as the OCS, is based on a 40 to 1 ratio, which is the equivalent of a 2.5 percent or a 152-foot per NM slope. As a result, a departure is designed using the OCS as the minimum obstacle clearance, and then by requiring a minimum climb gradient of 200 feet per NM, additional clearance is provided. The departure design must also include the acquisition of positive course guidance (PCG) typically within 5 to 10 NM of the DER for straight departures and within 5 NM after turn completion on departures requiring a turn. Even when aircraft performance greatly exceeds the minimum climb gradient, the published departure routing must always be flown.

Airports declaring that the sections of a runway at one or both ends are not available for landing or takeoff publish the declared distances in the A/FD. These include takeoff runway available (TORA), takeoff distance available (TODA), accelerate-stop distance available (ASDA), and landing distance available (LDA). These distances are calculated by adding to the full length of paved runway, any applicable clearway or stopway, and subtracting from that sum the sections of the runway unsuitable for satisfying the required takeoff run, takeoff, accelerate/stop, or landing distance, as shown in Figure 2-16 on page 2-14.

In a perfect world, the 40 to 1 slope would work for every departure design; however, due to terrain and man-made obstacles, it is often necessary to use alternative requirements to accomplish a safe, obstacle-free departure design. In such cases, the design of the departure may incorporate a climb gradient greater than 200 feet per NM, an increase in the standard takeoff minimums to allow the aircraft to "see and avoid" the obstacles, standard minimums combined with a climb gradient of 200 feet per NM or greater with a specified reduced runway length, or a combination of these options and a specific departure route. If a departure route is specified, it must be flown in conjunction with the other options. A published climb gradient in this case is based on the ROC 24 percent rule. To keep the same ROC ratio as standard, when the required climb gradient is greater than 200 feet per NM, 24 percent of the total height

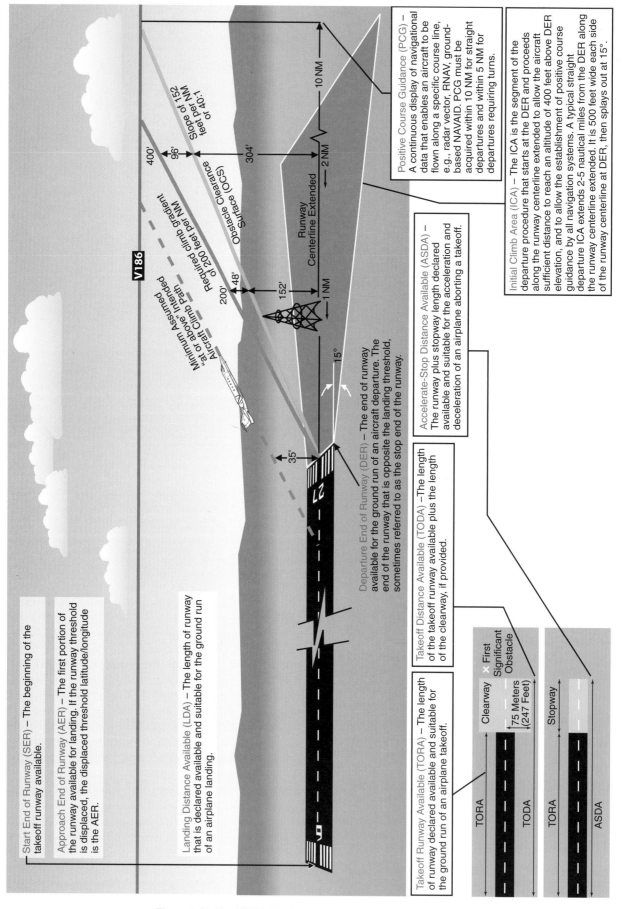

Figure 2-16. New TERPS Design Criteria for Departure Procedures.

Positive Course Guidance (PCG) – A continuous display of navigational data that enables an aircraft to be flown along a specific course line, e.g., radar vector, RNAV, ground-based NAVAID. PCG must be acquired within 10 NM for straight departures and within 5 NM for departures requiring turns.

Initial Climb Area (ICA) – The ICA is the segment of the departure procedure that starts at the DER and proceeds along the runway centerline extended to allow the aircraft sufficient distance to reach an altitude of 400 feet above DER elevation, and to allow the establishment of positive course guidance by all navigation systems. A typical straight departure ICA extends 2-5 nautical miles from the DER along the runway centerline extended. It is 500 feet wide each side of the runway centerline at DER, then splays out at 15°.

Accelerate-Stop Distance Available (ASDA) – The runway plus stopway length declared available and suitable for the acceleration and deceleration of an airplane aborting a takeoff.

Departure End of Runway (DER) – The end of runway available for the ground run of an aircraft departure. The end of the runway that is opposite the landing threshold, sometimes referred to as the stop end of the runway.

Takeoff Distance Available (TODA) –The length of the takeoff runway available plus the length of the clearway, if provided.

Takeoff Runway Available (TORA) – The length of runway declared available and suitable for the ground run of an airplane takeoff.

Landing Distance Available (LDA) – The length of runway that is declared available and suitable for the ground run of an airplane landing.

Approach End of Runway (AER) – The first portion of the runway available for landing. If the runway threshold is displaced, the displaced threshold latitude/longitude is the AER.

Start End of Runway (SER) – The beginning of the takeoff runway available.

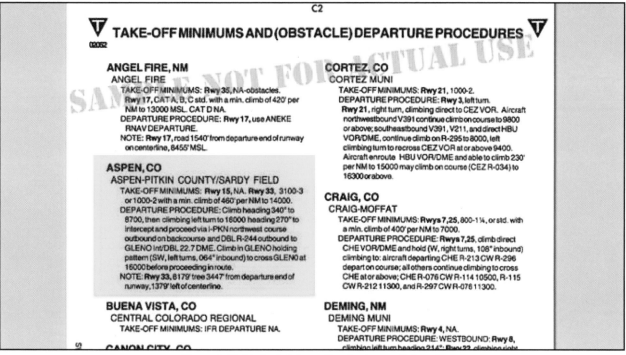

Figure 2-17. Obstacle Information for Aspen, Colorado.

above the starting elevation gained by an aircraft departing to a minimum altitude to clear an obstacle that penetrates the OCS is the ROC. The required climb gradient depicted in ODPs is obtained by using the formulas:

Standard Formula

$$CG = \frac{O - E}{0.76\,D}$$

DoD Option*

$$CG = \frac{(48D + O) - E}{D}$$

where O = obstacle MSL elevation

E = climb gradient starting MSL elevation

D = distance (NM) from DER to the obstacle

Examples:

$$\frac{2049 - 1221}{0.76 \times 3.1} = 351.44$$

Round to 352 ft/NM

$$\frac{(48 \times 3.1 + 2049) - 1221}{3.1} = 315.10$$

Round to 316 ft/NM

*Military only

These formulas are published in TERPS Volume 4 for calculating the required climb gradient to clear obstacles.

The following formula is used for calculating climb gradients for other than obstacles, i.e., ATC requirements:

$$CG = \frac{A - E}{D}$$

where A = "climb to" altitude

E = climb gradient starting MSL elevation

D = distance (NM) from the beginning of the climb

Example:

$$\frac{3000 - 1221}{5} = 355.8 \text{ round to } 356 \text{ ft/NM}$$

NOTE: The climb gradient must be equal to or greater than the gradient required for obstacles along the route of flight.

Obstacles that are located within 1 NM of the DER and penetrate the 40:1 OCS are referred to as "low, close-in obstacles." The standard ROC of 48 feet per NM to clear these obstacles would require a climb gradient greater than 200 feet per NM for a very short distance, only until the aircraft was 200 feet above the DER. To eliminate publishing an excessive climb gradient, the obstacle AGL/MSL height and location relative to the DER is noted in the Take-off Minimums and (OBSTACLE) Departure Procedures section of a given TPP booklet. The purpose of this note is to identify the obstacle and alert the pilot to the height and location of the obstacle so they can be avoided. [Figure 2-17]

Departure design, including climb gradients, does not take into consideration the performance of the aircraft; it only considers obstacle protection for all aircraft. TERPS criteria assumes the aircraft is operating with all available engines and systems fully functioning. When a climb gradient is required for a specific departure, it is vital that pilots fully understand the performance of their aircraft and determine if it can comply with the required climb. The standard climb of 200 feet per NM is not an issue for most aircraft. When an increased climb gradient is specified due to obstacle issues, it is important to calculate aircraft performance, particularly when flying out of airports at higher altitudes on warm days. To aid in the calculations, the front matter of every TPP booklet contains a rate of climb table that relates specific climb gradients and typical airspeeds. [Figure 2-18 on page 2-16]

A **visual climb over airport (VCOA)** is an alternate departure method for aircraft unable to meet required climb gradients and for airports at which a conventional instrument departure procedure is impossible to design due to terrain or other obstacle hazard. The development

D1

01081
CLIMB TABLE

RATE OF CLIMB TABLE

A rate of climb table is provided for use in planning and executing
takeoff procedures under known or approximate ground speed conditions.

(ft. per min.)

SE-3, 13 JUN 2002

REQUIRED GRADIENT RATE (ft. per NM)	GROUND SPEED (KNOTS)						
	30	60	80	90	100	120	140
200	100	200	267	300	333	400	467
250	125	250	333	375	417	500	583
300	150	300	400	450	500	600	700
350	175	350	467	525	583	700	816
400	200	400	533	600	667	800	933
450	225	450	600	675	750	900	1050
500	250	500	667	750	833	1000	1167
550	275	550	733	825	917	1100	1283
600	300	600	800	900	1000	1200	1400
650	325	650	867	975	1083	1300	1516
700	350	700	933	1050	1167	1400	1633

SE-3, 13 JUN 2002

Ground Speed is 180 knots

REQUIRED GRADIENT RATE (ft. per NM)	GROUND SPEED (KNOTS)					
	150	180	210	240	270	300
200	500	600	700	800	900	1000
250	625	750	875	1000	1125	1250
300	750	900	1050	1200	1350	1500
350	875	1050	1225	1400	1575	1750
400	1000	1200	1400	1600	1700	2000
450	1125	1350	1575	1800	2025	2250
500	1250	1500	1750	2000	2250	2500
550	1375	1650	1925	2200	2475	2750
600	1500	1800	2100	2400	2700	3000
650	1625	1950	2275	2600	2925	3250
700	1750	2100	2450	2800	3150	3500

Required climb gradient of 300 feet per NM

Given the parameters, you would need to climb at a rate of 900 feet per minute to maintain the required climb gradient.

Figure 2-18. Rate of Climb Table.

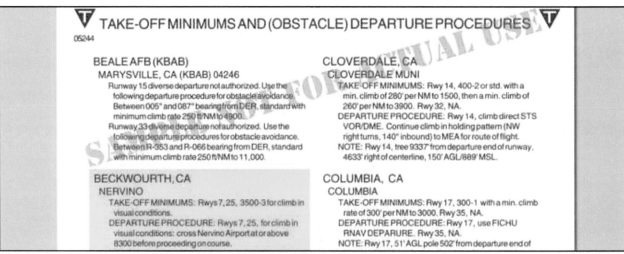

Figure 2-19. Beckwourth, CA.

of this type of procedure is required when obstacles more than 3 SM from the DER require a greater than 200 feet per NM climb gradient. An example of this procedure is visible at Nervino Airport in Beckwourth, California. [Figure 2-19]

The procedure for climb in visual conditions requires crossing Nervino Airport at or above 8,300 feet before proceeding on course. Additional instructions often complete the departure procedure and transition the flight to the en route structure. VCOA procedures are available on specific departure procedures, but are not established in conjunction with SIDs or RNAV obstacle departure procedures. Pilots must know if their specific flight operations allow VCOA procedures on IFR departures.

AIRPORT RUNWAY ANALYSIS

It may be necessary for pilots and aircraft operators to consult an aircraft performance engineer and airport/runway analysis service for information regarding the clearance of specific obstacles during IFR departure procedures to help maximize aircraft payload while complying with engine-out performance regulatory requirements. Airport/runway analysis involves the complex application of extensive airport databases and terrain information to generate computerized computations for aircraft performance in a specific configuration. This yields maximum allowable takeoff and landing weights for particular aircraft/engine configurations for a specific airport, runway, and range of temperatures. The computations also consider flap settings, various aircraft characteristics, runway conditions, obstacle clearance, and weather conditions. Data also is available for operators who desire to perform their own analysis.

When a straight-out departure is not practical or recommended, a turn procedure can be developed for the engine-out flight path for each applicable runway designed to maximize the allowable takeoff weights and ultimately, aircraft payload. Engine-out graphics are available, giving the pilot a pictorial representation of each procedure. Airport/runway analysis also is helpful for airline dispatchers, flight operations officers, engineering staff, and others to ensure that a flight does not exceed takeoff and landing limit weights.

CAUTION: Pilots and aircraft operators have the responsibility to consider obstacles and to make the necessary adjustments to their departure procedures to ensure safe clearance for aircraft over those obstacles.

Information on obstacle assessment, controlling obstacles, and other obstacles that may affect a pilot's IFR departure may not be depicted or noted on a chart and may be outside the scope of IFR departure procedure obstacle assessment criteria. Departure criteria is predicated on normal aircraft operations for considering obstacle clearance requirements. Normal aircraft opera-

tion means all aircraft systems are functioning normally, all required navigational aids (NAVAIDS) are performing within flight inspection parameters, and the pilot is conducting instrument operations utilizing instrument procedures based on the TERPS standard to provide ROC.

SID VERSUS DP

In 2000, the FAA combined into a single product both textual IFR departure procedures that were developed by the National Flight Procedures Office (NFPO) under the guidance of the Flight Standards Service (AFS) and graphic standard instrument departures (SIDs) that were designed and produced under the direction of the Air Traffic Organization (ATO). This combined product introduced the new term departure procedures (DPs) to the pilot and ATC community, and the aforementioned terms IFR departure procedure and SID were eliminated. The FAA also provided for the graphic publication of IFR departure procedures, as well as all area navigation (RNAV) DPs, to facilitate pilot understanding of the procedure. This includes both those developed solely for obstruction clearance and those developed for system enhancement. Elimination of the term SID created undue confusion in both the domestic and international aviation communities. Therefore, in the interest of international harmonization, the FAA reintroduced the term SID while also using the term obstacle departure procedure (ODP) to describe certain procedures.

There are two types of DPs: those developed to assist pilots in obstruction avoidance, ODP, and those developed to communicate air traffic control clearances, SID. DPs and/or takeoff minimums must be established for those airports with approved instrument approach procedures. ODPs are developed by the NFPO at locations with instrument procedure development responsibility. ODPs may also be required at private airports where the FAA does not have instrument procedure development responsibility. It is the responsibility of non-FAA proponents to ensure a TERPS diverse departure obstacle assessment has been accomplished and an ODP developed, where applicable. DPs are also categorized by equipment requirements as follows:

- **Non-RNAV DP.** Established for aircraft equipped with conventional avionics using ground-based NAVAIDs. These DPs may also be designed using dead reckoning navigation. A flight management system (FMS) may be used to fly a non-RNAV DP if the FMS unit accepts inputs from conventional avionics sources such as DME, VOR, and LOC. These inputs include radio tuning and may be applied to a navigation solution one at a time or in combination. Some FMSs provide for the detection and isolation of faulty navigation information.

- **RNAV DP**. Established for aircraft equipped with RNAV avionics; e.g., GPS, VOR/DME, DME/DME, etc. Automated vertical navigation is not required, and all RNAV procedures not requiring GPS must be annotated with the note: "RADAR REQUIRED." Prior to using GPS for RNAV departures, approach RAIM availability should be checked for that location with the navigation receiver or a Flight Service Station.

- **Radar DP**. Radar may be used for navigation guidance for SID design. Radar SIDs are established when ATC has a need to vector aircraft on departure to a particular ATS Route, NAVAID, or Fix. A fix may be a ground-based NAVAID, a waypoint, or defined by reference to one or more radio NAVAIDS. Not all fixes are waypoints since a fix could be a VOR or VOR/DME, but all waypoints are fixes. Radar vectors may also be used to join conventional or RNAV navigation SIDs. SIDs requiring radar vectors must be annotated "RADAR REQUIRED."

OBSTACLE DEPARTURE PROCEDURES

The term **Obstacle Departure Procedure (ODP)** is used to define procedures that simply provide obstacle clearance. ODPs are only used for obstruction clearance and do not include ATC related climb requirements. In fact, the primary emphasis of ODP design is to use the least onerous route of flight to the en route structure or at an altitude that allows random (diverse) IFR flight, while attempting to accommodate typical departure routes.

An ODP must be developed when obstructions penetrate the 40:1 departure OCS, using a complex set of ODP development combinations to determine each situation and required action. Textual ODPs are only issued by ATC controllers when required for traffic. If they are not issued by ATC, textual ODPs are at the pilot's option to fly or not fly the textual ODP, even in less than VFR weather conditions, for FAR Part 91 operators, military, and public service. As a technique, the pilot may enter "will depart (airport) (runway) via textual ODP" in the remarks section of the flight plan, this information to the controller clarifies the intentions of the pilot and helps prevent a potential pilot/controller misunderstanding.

ODPs are textual in nature, however, due to the complex nature of some procedures, a visual presentation may be necessary for clarification and understanding. Additionally, all newly developed area navigation (RNAV) ODPs are issued in graphic form. If necessary, an ODP is charted graphically just as if it were a SID and the chart itself includes "Obstacle" in parentheses in the title. A graphic ODP may also be filed in an instrument flight plan by using the computer code included in the procedure title.

Only one ODP is established for a runway. It is considered to be the default IFR departure procedure and is intended for use in the absence of ATC radar vectors or a SID assignment. ODPs use ground based NAVAIDS, RNAV, or dead reckoning guidance wherever possible, without the use of radar vectors for navigation.

Military departure procedures are not handled or published in the same manner as civil DPs. Approval authority for DPs at military airports rests with the military. The FAA develops U.S. Army and U.S. Air Force DPs for domestic civil airports. The National Geospatial-Intelligence Agency (NGA) publishes all military DPs. The FAA requires that all military DPs be coordinated with FAA ATC facilities or regions when those DPs affect the NAS.

All ODP procedures are listed in the front of the NACO approach chart booklets under the heading Takeoff Minimums and Obstacle Departure Procedures. Each procedure is listed in alphabetical order by city and state. The ODP listing in the front of the booklet will include a reference to the graphic chart located in the main body of the booklet if one exists. Pilots do not need ATC clearance to use an ODP and they are responsible for determining if the departure airport has this type of published procedure. [Figure 2-20]

FLIGHT PLANNING CONSIDERATIONS

During planning, pilots need to determine whether or not the departure airport has an ODP. Remember, an ODP can only be established at an airport that has instrument approach procedures (IAPs). An ODP may drastically affect the initial part of the flight plan. Pilots may have to depart at a higher than normal climb rate, or depart in a direction opposite the intended heading and maintain that for a period of time, any of which would require an alteration in the flight plan and initial headings. Considering the forecast weather, departure runway, and existing ODP, plan the flight route, climb performance, and fuel burn accordingly to compensate for the departure procedure.

Additionally, when close-in obstacles are noted in the Takeoff Minimums and (Obstacle) Departure Procedures section, it may require the pilot to take action to avoid these obstacles. Consideration must be given to decreased climb performance from an inoperative engine or to the amount of runway used for takeoff. Aircraft requiring a short takeoff roll on a long runway may have little concern. On the other hand, airplanes that use most of the available runway for takeoff may not have the standard ROC when climbing at the normal 200 feet per NM.

Another factor to consider is the possibility of an engine failure during takeoff and departure. During the preflight planning, use the aircraft performance charts to determine if the aircraft can still maintain the required climb performance. For high performance aircraft, an engine failure may not impact the ability to maintain the prescribed climb gradients. Aircraft that are performance limited may have diminished capability and may be

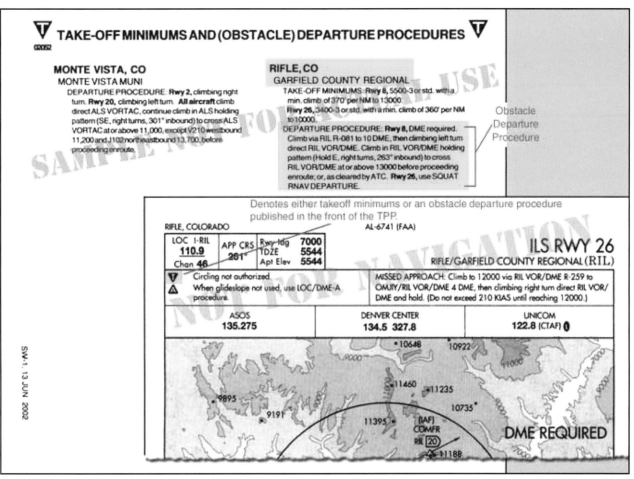

Figure 2-20. Graphic ODP/Booklet Front Matter.

unable to maintain altitude, let alone complete a climb to altitude. Based on the performance expectations for the aircraft, construct an emergency plan of action that includes emergency checklists and the actions to take to ensure safety in this situation.

STANDARD INSTRUMENT DEPARTURES

A **Standard Instrument Departure (SID)** is an ATC requested and developed departure route, typically used in busy terminal areas. It is designed at the request of ATC in order to increase capacity of terminal airspace, effectively control the flow of traffic with minimal communication, and reduce environmental impact through noise abatement procedures.

While obstacle protection is always considered in SID routing, the primary goal is to reduce ATC/pilot workload while providing seamless transitions to the en route structure. SIDs also provide additional benefits to both the airspace capacity and the airspace users by reducing radio congestion, allowing more efficient airspace use, and simplifying departure clearances. All of the benefits combine to provide effective, efficient terminal operations, thereby increasing the overall capacity of the NAS.

If you cannot comply with a SID, if you do not possess SID charts or textual descriptions, or if you simply do not wish to use standard instrument departures, include

the statement "NO SIDs" in the remarks section of your flight plan. Doing so notifies ATC that they cannot issue you a clearance containing a SID, but instead will clear you via your filed route to the extent possible, or via a **Preferential Departure Route (PDR)**. It should be noted that SID usage not only decreases clearance delivery time, but also greatly simplifies your departure, easing you into the IFR structure at a desirable location and decreasing your flight management load. While you are not required to depart using a SID, it may be more difficult to receive an "as filed" clearance when departing busy airports that frequently use SID routing.

SIDs are always charted graphically and are located in the TPP after the last approach chart for an airport. The SID may be one or two pages in length, depending on the size of the graphic and the amount of space required for the departure description. Each chart depicts the departure route, navigational fixes, transition routes, and required altitudes. The departure description outlines the particular procedure for each runway. [Figure 2-21 on page 2-20]

Charted transition routes allow pilots to transition from the end of the basic SID to a location in the en route structure. Typically, transition routes fan out in various directions from the end of the basic SID to allow pilots to choose the transition route that takes them in the

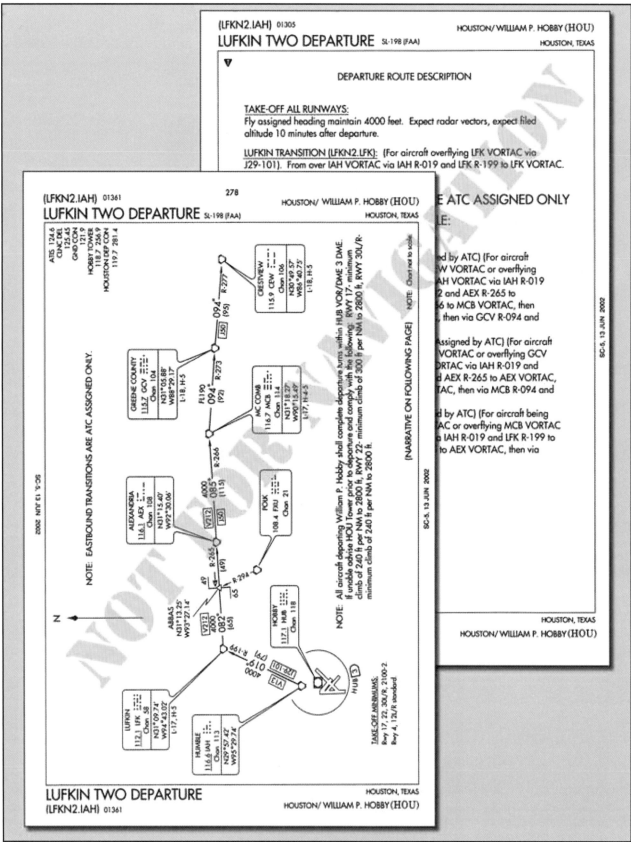

Figure 2-21. SID Chart

direction of intended departure. A transition route includes a course, a minimum altitude, and distances between fixes on the route. When filing a SID for a specific transition route, include the transition in the flight plan, using the correct departure and transition code. ATC also assigns transition routes as a means of putting the flight on course to the destination. In any case, the pilot must receive an ATC clearance for the departure and the associated transition, and the clearance from ATC will include both the departure name and transition e.g., Joe Pool Nine Departure, College Station Transition. [Figure 2-22]

PILOT NAV AND VECTOR SIDS

SIDs are categorized by the type of navigation used to fly the departure, so they are considered either pilot navigation or vector SIDs. Pilot navigation SIDs are

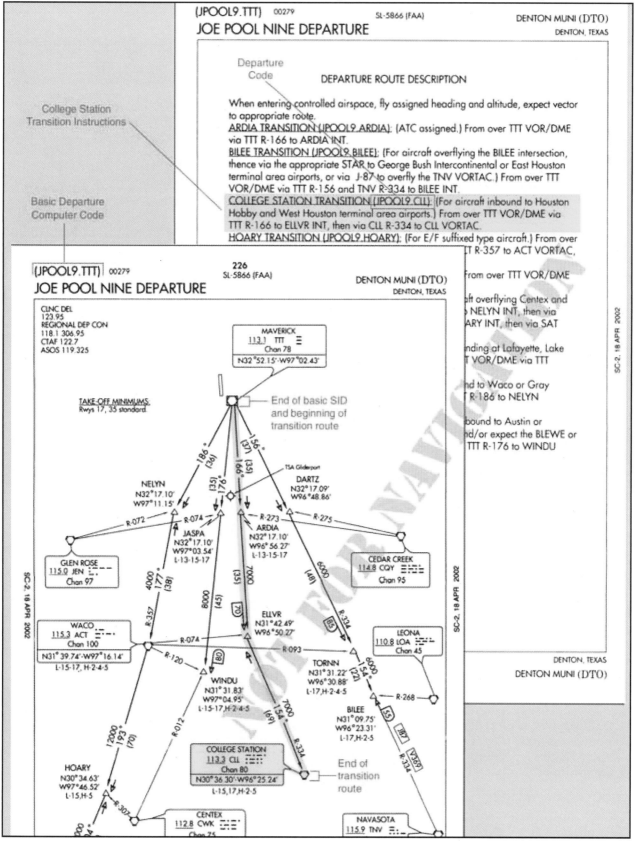

Figure 2-22. Transition Routes as Depicted on SIDs.

designed to allow you to provide your own navigation with minimal radio communication. This type of procedure usually contains an initial set of departure instructions followed by one or more transition routes. A pilot navigation SID may include an initial segment requiring radar vectors to help the flight join the procedure, but the majority of the navigation will remain the pilot's responsibility. These are the most common type of SIDs because they reduce the workload for ATC by requiring minimal communication and navigation support. [Figure 2-23].

A Vector SID usually requires ATC to provide radar vectors from just after takeoff (ROC is based on a climb to 400 feet above the DER elevation before making the initial turn) until reaching the assigned route or a fix depicted on the SID chart. However, some textual ODPs originate in uncontrolled airspace, while the SID begins in controlled airspace. Vector SIDs do not include departure routes or transition routes because independent pilot navigation is not involved. The procedure sets forth an initial set of departure instructions that typically include an initial heading and altitude. ATC must have radar contact with the aircraft to be able to provide vectors. ATC expects you to immediately comply with radar vectors and they expect you to notify them if you are unable to fulfill their request. ATC also expects you to make contact immediately if an instruction will cause you to compromise safety due to obstructions or traffic.

It is prudent to review vector SID charts prior to use because this type of procedure often includes nonstandard lost communication procedures. If you were to lose radio contact while being vectored by ATC, you would be expected to comply with the lost communication procedure as outlined on the chart, not necessarily those procedures outlined in the AIM. [Figure 2-24 on page 2-24]

FLIGHT PLANNING CONSIDERATIONS

Take into consideration the departure paths included in the SIDs and determine if you can use a standardized departure procedure. You have the opportunity to choose the SID that best suits your flight plan. During the flight planning phase, you can investigate each departure and determine which procedure allows you to depart the airport in the direction of your intended flight. Also consider how a climb gradient to a specific altitude will affect the climb time and fuel burn portions of the flight plan. If ATC assigns you a SID, you may need to quickly recalculate your performance numbers.

PROCEDURAL NOTES

Another important consideration to make during your flight planning is whether or not you are able to fly your chosen departure procedure as charted. Notes giving procedural requirements are listed on the graphic

portion of a departure procedure, and they are mandatory in nature. [Figure 2-25 on page 2-25] Mandatory procedural notes may include:

- Aircraft equipment requirements (DME, ADF, etc.).

- ATC equipment in operation (RADAR).

- Minimum climb requirements.

- Restrictions for specific types of aircraft (TURBOJET ONLY).

- Limited use to certain destinations.

There are numerous procedural notes requiring specific compliance on your part. Carefully review the charts for the SID you have selected to ensure you can use the procedures. If you are unable to comply with a specific requirement, you must not file the procedure as part of your flight plan, and furthermore, you must not accept the procedure if ATC assigns it. Cautionary statements may also be included on the procedure to notify you of specific activity, but these are strictly advisory. [Figure 2-26 on page 2-26]

DP RESPONSIBILITY

Responsibility for the safe execution of departure procedures rests on the shoulders of both ATC and the pilot. Without the interest and attention of both parties, the IFR system cannot work in harmony, and achievement of safety is impossible.

ATC, in all forms, is responsible for issuing clearances appropriate to the operations being conducted, assigning altitudes for IFR flight above the minimum IFR altitudes for a specific area of controlled airspace, ensuring the pilot has acknowledged the clearance or instructions, and ensuring the correct read back of instructions. Specifically related to departures, ATC is responsible for specifying the direction of takeoff or initial heading when necessary, obtaining pilot concurrence that the procedure complies with local traffic patterns, terrain, and obstruction clearance, and including departure procedures as part of the ATC clearance when pilot compliance for separation is necessary.

The pilot has a number of responsibilities when simply operating in conjunction with ATC or when using departure procedures under an IFR clearance:

- Acknowledge receipt and understanding of an ATC clearance.

- Read back any part of a clearance that contains "hold short" instructions.

- Request clarification of clearances.

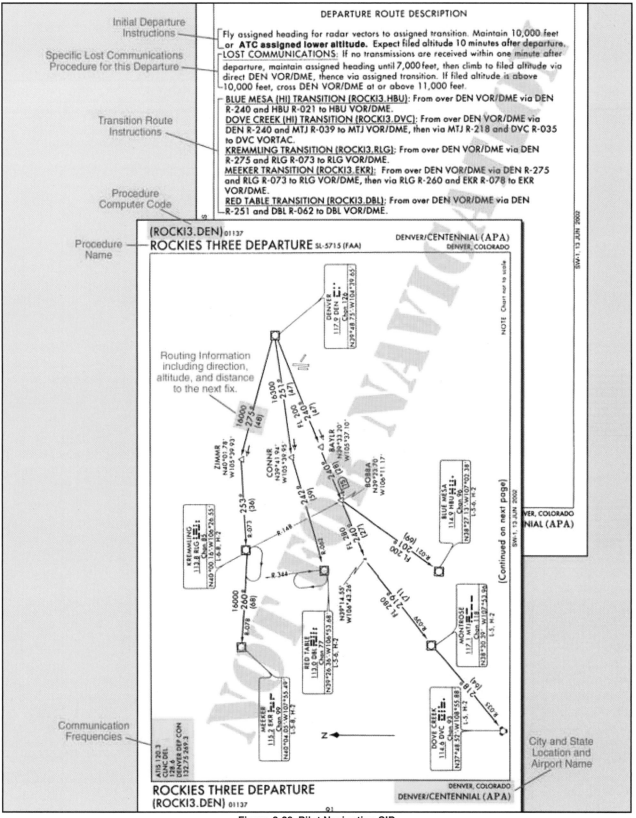

DEPARTURE ROUTE DESCRIPTION

Initial Departure Instructions — Fly assigned heading for radar vectors to assigned transition. Maintain 10,000 feet or **ATC assigned lower altitude.** Expect filed altitude 10 minutes after departure.

Specific Lost Communications Procedure for this Departure — LOST COMMUNICATIONS: If no transmissions are received within one minute after departure, maintain assigned heading until 7,000 feet, then climb to filed altitude via direct DEN VOR/DME, thence via assigned transition. If filed altitude is above 10,000 feet, cross DEN VOR/DME at or above 11,000 feet.

Transition Route Instructions — BLUE MESA (HI) TRANSITION (ROCKI3.HBU): From over DEN VOR/DME via DEN R-240 and HBU R-021 to HBU VOR/DME.
DOVE CREEK (HI) TRANSITION (ROCKI3.DVC): From over DEN VOR/DME via DEN R-240 and MTJ R-039 to MTJ VOR/DME, then via MTJ R-218 and DVC R-035 to DVC VORTAC.
KREMMLING TRANSITION (ROCKI3.RLG): From over DEN VOR/DME via DEN R-275 and RLG R-073 to RLG VOR/DME.
MEEKER TRANSITION (ROCKI3.EKR): From over DEN VOR/DME via DEN R-275 and RLG R-073 to RLG VOR/DME, then via RLG R-260 and EKR R-078 to EKR VOR/DME.

Procedure Computer Code

RED TABLE TRANSITION (ROCKI3.DBL): From over DEN VOR/DME via DEN R-251 and DBL R-062 to DBL VOR/DME.

Procedure Name — (ROCKI3.DEN) 01137
ROCKIES THREE DEPARTURE SL-5715 (FAA)

DENVER/CENTENNIAL (APA)
DENVER, COLORADO

Routing Information including direction, altitude, and distance to the next fix.

(Continued on next page)

Communication Frequencies

City and State Location and Airport Name

ROCKIES THREE DEPARTURE
(ROCKI3.DEN) 01137

DENVER, COLORADO
DENVER/CENTENNIAL (APA)

Figure 2-23. Pilot Navigation SID.

- Request an amendment to a clearance if it is unacceptable from a safety perspective.

- Promptly comply with ATC requests. Advise ATC immediately if unable to comply with a clearance.

When planning for a departure, pilots should:

- Consider the type of terrain and other obstructions in the vicinity of the airport.

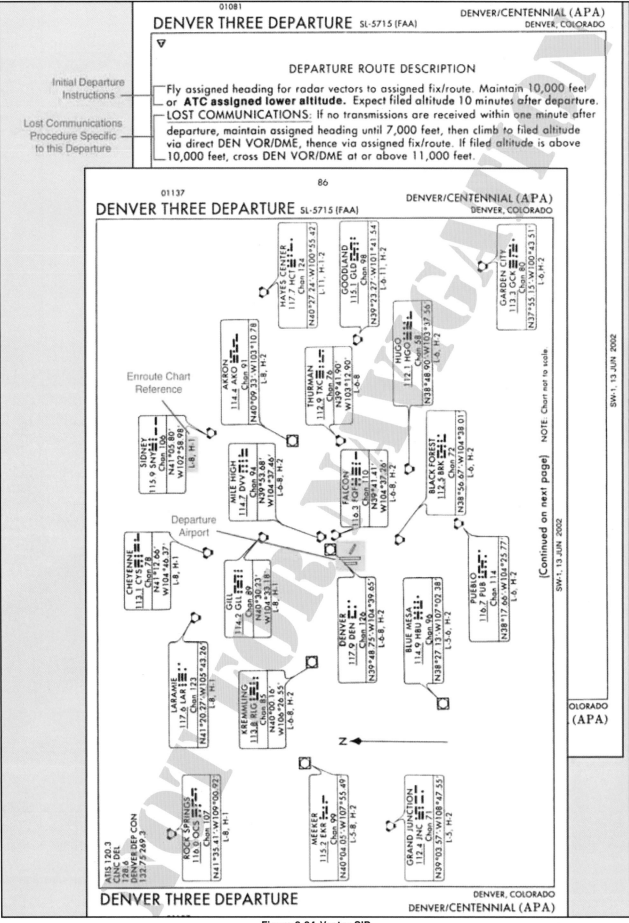

Figure 2-24. Vector SID.

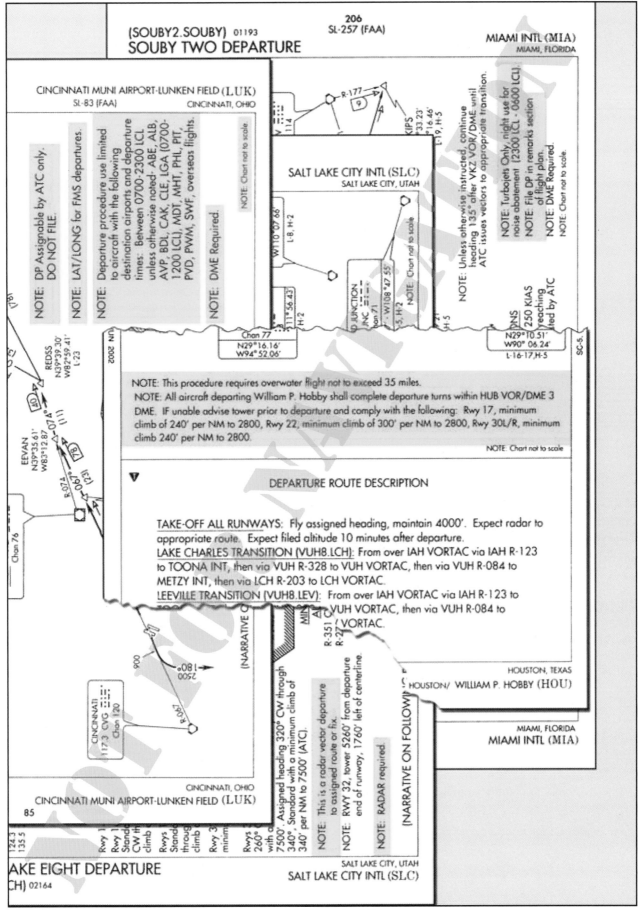

Figure 2-25. Procedural Notes.

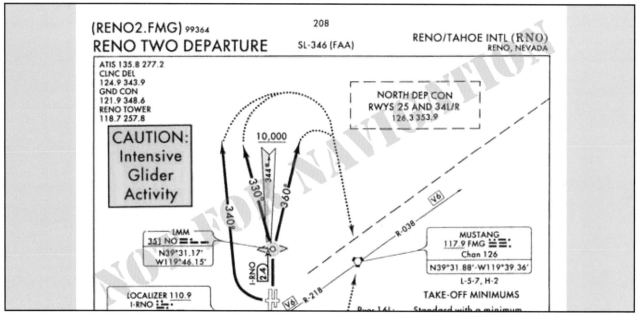

Figure 2-26. Cautionary Statements.

- Determine if obstacle clearance can be maintained visually, or if they need to make use of a departure procedure.

- Determine if an ODP or SID is available for the departure airport.

- Determine what actions allow for a safe departure out of an airport that does not have any type of affiliated departure procedures.

By simply complying with departure procedures in their entirety as published, obstacle clearance is guaranteed. Depending on the type of departure used, responsibility for terrain clearance and traffic separation may be shared between pilots and controllers.

PROCEDURES ASSIGNED BY ATC

ATC can assign SIDs or radar vectors as necessary for traffic management and convenience. You can also request a SID in your initial flight plan, or from ATC. To fly a SID, you must receive approval to do so in a clearance. In order to accept a clearance that includes a SID, you must have at least a textual description of the SID in your possession at the time of departure. It is your responsibility as pilot in command to accept or reject the issuance of a SID by ATC. You must accept or reject the clearance based on:

- The ability to comply with the required performance.

- Possession of at least the textual description of the SID.

- Personal understanding of the SID in its entirety.

When you accept a clearance to depart using a SID or radar vectors, ATC is responsible for traffic separation.

ATC is also responsible for obstacle clearance. When departing with a SID, ATC expects you to fly the procedure as charted because the procedure design considers obstacle clearance. It is also expected that you will remain vigilant in scanning for traffic when departing in visual conditions. Furthermore, it is your responsibility to notify ATC if your clearance would endanger your safety or the safety of others.

PROCEDURES NOT ASSIGNED BY ATC

Obstacle departure procedures are not assigned by ATC unless absolutely necessary to achieve aircraft separation. It is the pilot's responsibility to determine if there is an ODP published for that airport. If a Part 91 pilot is not given a clearance containing an ODP, SID, or radar vectors and an ODP exists, compliance with such a procedure is the pilot's choice. If he/she chooses not to use the ODP, the pilot must be operating under visual meteorological conditions (VMC), which permits the avoidance of obstacles during the departure.

DEPARTURES FROM TOWER-CONTROLLED AIRPORTS

Departing from a tower-controlled airport is relatively simple in comparison to departing from an airport that isn't tower controlled. Normally you request your IFR clearance through ground control or clearance delivery. Communication frequencies for the various controllers are listed on departure, approach, and airport charts as well as the A/FD. At some airports, you may have the option of receiving a pre-taxi clearance. This program allows you to call ground control or clearance delivery no more than ten minutes prior to beginning taxi operations and receive your IFR clearance. A pre-departure clearance (PDC) program that allows pilots to receive a clearance via data link from a dispatcher is available for Part 121 and 135 operators. A clearance is given to the dispatcher who in turn relays it to the crew, enabling the crew to bypass communication with clearance delivery, thus reducing frequency congestion. Once you have received your clearance, it is your responsibility to comply with the instructions as given and notify ATC if you are unable to comply with the clearance. If you do not understand the clearance, or if you think that you have missed a portion of the clearance, contact ATC immediately for clarification.

DEPARTURES FROM AIRPORTS WITHOUT AN OPERATING CONTROL TOWER

There are hundreds of airports across the U.S. that operate successfully everyday without the benefit of a control tower. While a tower is certainly beneficial when departing IFR, most other departures can be made with few challenges. As usual, you must file your flight plan at least 30 minutes in advance. During your planning phase, investigate the departure airport's method for receiving an instrument clearance. You can contact the Automated Flight Service Station (AFSS) on the ground by telephone and they will request your clearance from ATC. Typically, when a clearance is given in this manner, the clearance includes a void time. You must depart the airport before the clearance void time; if you fail to depart, you must contact ATC by a specified notification time, which is within 30 minutes of the original void time. After the clearance void time, your reserved space within the IFR system is released for other traffic.

There are several other ways to receive a clearance at a non-towered airport. If you can contact the AFSS or ATC on the radio, you can request your departure clearance. However, these frequencies are typically congested and they may not be able to provide you with a clearance via the radio. You also can use a **Remote Communications Outlet (RCO)** to contact an AFSS if one is located nearby. Some airports have licensed UNICOM operators that can also contact ATC on your behalf and in turn relay your clearance from ATC. You are also allowed to depart the airport VFR if conditions permit and contact the controlling authority and request your clearance in the air. As technology improves, new methods for delivery of clearances at non-towered airports are being created.

GROUND COMMUNICATIONS OUTLETS

A new system, called a **Ground Communication Outlet (GCO)**, has been developed in conjunction with the FAA to provide pilots flying in and out of non-towered airports with the capability to contact ATC and AFSS via Very High Frequency (VHF) radio to a telephone connection. This lets pilots obtain an instrument clearance or close a VFR/IFR flight plan. You can use

four key clicks on your VHF radio to contact the nearest ATC facility and six key clicks to contact the local AFSS, but it is intended to be used only as a ground operational tool. A GCO is an unstaffed, remote controlled ground-to-ground communication facility that is relatively inexpensive to install and operate. Installations of these types of outlets are scheduled at instrument airports around the country.

GCOs are manufactured by different companies including ARINC and AVTECH, each with different operating characteristics but with the ability to accomplish the same goal. This latest technology has proven to be an incredibly useful tool for communicating with the appropriate authorities when departing IFR from a non-towered airport. The GCO should help relieve the need to use the telephone to call ATC and the need to depart into marginal conditions just to achieve radio contact. GCO information is listed on airport charts and instrument approach charts with other communications frequencies. Signs may also be located on an airport to notify you of the frequency and proper usage.

OBSTACLE AVOIDANCE

Safety is always the foremost thought when planning and executing an IFR flight. As a result, the goal of all departure procedures is to provide a means for departing an airport in the safest manner possible. It is for this reason that airports and their surroundings are reviewed and documented and that procedures are put in place to prevent flight into terrain or other man-made obstacles. To aid in the avoidance of obstacles, takeoff minimums and departure procedures use minimum climb gradients and "see and avoid" techniques.

CLIMB GRADIENTS AND CLIMB RATES

You are required to contact ATC if you are unable to comply with climb gradients and climb rates. It is also expected that you are capable of maintaining the climb gradient outlined in either a standard or non-standard SID or ODP. If you cannot comply with the climb gradient in the SID, you should not accept a clearance for that SID. If you cannot maintain a standard climb gradient or the climb gradient specified in an ODP, you must wait until you can depart under VMC.

Climb gradients are developed as a part of a departure procedure to ensure obstacle protection as outlined in TERPS. Once again, the rate of climb table depicted in Figure 2-18, used in conjunction with the performance specifications in your airplane flight manual (AFM), can help you determine your ability to comply with climb gradients.

SEE AND AVOID TECHNIQUES

Meteorological conditions permitting, you are required to use "see and avoid" techniques to avoid traffic, terrain, and other obstacles. To avoid obstacles during a departure, the takeoff minimums may include a non-standard ceiling and visibility minimum. These are given to pilots so they can depart an airport without being able to meet the established climb gradient. Instead, they must see and avoid obstacles in the departure path. In these situations, ATC provides radar traffic information for radar-identified aircraft outside controlled airspace, workload permitting, and safety alerts to pilots believed to be within an unsafe proximity to obstacles or aircraft.

AREA NAVIGATION DEPARTURES

In the past, area navigation (RNAV) was most commonly associated with the station-mover/phantom waypoint technology developed around ground-based Very High Frequency Omni-directional Range (VOR) stations. RNAV today, however, refers to a variety of navigation systems that provide navigation beyond VOR and NDB. RNAV is a method of navigation which permits aircraft operation on any desired flight path within the coverage of station-referenced navigation aids or within the limits of the capability of self-contained aids, or a combination of these. The term also has become synonymous with the concept of "free flight," the goal of which is to provide easy, direct, efficient, cost-saving traffic management as a result of the inherent flexibility of RNAV.

In the past, departure procedures were built around existing ground-based technology and were typically designed to accommodate lower traffic volumes. Often, departure and arrival routes use the same navigation aids creating interdependent, capacity diminishing routes. As a part of the evolving RNAV structure, the FAA has developed departure procedures for pilots flying aircraft equipped with some type of RNAV technology. RNAV allows for the creation of new departure routes that are independent of present fixes and navigation aids. RNAV routing is part of the National Airspace Redesign and is expected to reduce complexity and increase efficiency of terminal airspace.

When new RNAV departure procedures are designed with all interests in mind, they require minimal vectoring and communications between pilots and ATC. Usually, each departure procedure includes position, time, and altitude, which increase the ability to predict what the pilot will actually do. All told, RNAV departure procedures have the ability to increase the capacity of terminal airspace by increasing on-time departures, airspace utilization, and improved predictability.

If unable to comply with the requirements of an RNAV or required navigation performance (RNP) procedure, pilots need to advise ATC as soon as possible. For example, ". . .N1234, failure of GPS system, unable RNAV, request amended clearance." Pilots are not authorized to fly a published RNAV or RNP procedure unless it is retrievable by the procedure name from the navigation database and conforms to the charted procedure. Pilots shall not change any database waypoint type from a fly-by to fly-over, or vice versa. No other modification of database waypoints or creation of user-defined waypoints on published RNAV or RNP procedures is permitted, except to change altitude and/or airspeed waypoint constraints to comply with an ATC clearance/instruction, or to insert a waypoint along the published route to assist in complying with an ATC instruction, for example, "Climb via the WILIT departure except cross 30 north of CHUCK at/or above FL 210." This is limited only to systems that allow along track waypoint construction.

Pilots of aircraft utilizing DME/DME for primary navigation updating shall ensure any required DME stations are in service as determined by NOTAM, ATIS, or ATC advisory. No pilot monitoring of an FMS navigation source is required. While operating on RNAV segments, pilots are encouraged to use the flight director in lateral navigation mode. RNAV terminal procedures may be amended by ATC issuing radar vectors and/or clearances direct to a waypoint. Pilots should avoid premature manual deletion of waypoints from their active "legs" page to allow for rejoining procedures. While operating on RNAV segments, pilots operating /R aircraft shall adhere to any flight manual limitation or operating procedure required to maintain the RNP value specified for the procedure.

RNAV DEPARTURE PROCEDURES

There are two types of public RNAV SIDs and graphic ODPs. Type A procedures generally start with a heading or vector from the DER, and have an initial RNAV fix around 15 NM from the departure airport. In addition, these procedures require system performance currently met by GPS, DME/DME, or DME/DME/Inertial Reference Unit (IRU) RNAV systems that satisfy the criteria discussed in AC 90-100, U.S. Terminal and En Route Area Navigation (RNAV) Operations. Type A terminal procedures require that the aircraft's track keeping accuracy remain bounded by ±2 NM for 95 percent of the total flight time. For type A procedure RNAV engagement altitudes, the pilot must be able to engage RNAV equipment no later than 2,000 feet above airport elevation. For Type A RNAV DPs, it is recommended that pilots use a CDI/flight director and/or autopilot in lateral navigation mode.

Type B procedures generally start with an initial RNAV leg near the DER. In addition, these procedures require system performance currently met by GPS or DME/DME/IRU RNAV systems that satisfy the criteria discussed in AC 90-100. Type B procedures require the aircraft's track keeping accuracy remain bounded by ±1 NM for 95 percent of the total flight time. For type B procedures, the pilot must be able to engage RNAV equipment no later than 500 feet above airport elevation. For Type B RNAV DPs, pilots must use a CDI/flight director and/or autopilot in lateral navigation mode. For Type A RNAV DPs and STARs, these procedures are recommended. Other methods providing an equivalent level of performance may also be acceptable. For Type B RNAV

DPs, pilots of aircraft without GPS using DME/DME/IRU must ensure that the aircraft navigation system position is confirmed, within 1,000 feet, at the start point of take-off roll. The use of an automatic or manual runway update is an acceptable means of compliance with this requirement. Other methods providing an equivalent level of performance may also be acceptable.

For procedures requiring GPS and/or aircraft approvals requiring GPS, if the navigation system does not automatically alert the flight crew of a loss of GPS, aircraft operators must develop procedures to verify correct GPS operation. If not equipped with GPS, or for multi-sensor systems with GPS that do not alert upon loss of GPS, aircraft must be capable of navigation system updating using DME/DME or DME/DME/IRU for type A and B procedures. AC 90-100 may be used as operational guidance for RNAV ODPs. Pilots of FMS-equipped aircraft, who are assigned an RNAV DP procedure and subsequently receive a change of runway, transition, or procedure, must verify that the appropriate changes are loaded and available for navigation.

RNAV departure procedures are developed as SIDs and ODPs—both are charted graphically. An RNAV departure is identifiable by the inclusion of the term RNAV in the title of the departure. From an RNP standpoint, RNAV departure routes are designed with a 1 or 2 NM performance standard. This means you as the pilot and your aircraft equipment must be able to maintain the aircraft within 1 NM or 2 NM either side of route centerline. [Figure 2-27]

Additionally, new waypoint symbols are used in conjunction with RNAV charts. There are two types of waypoints currently in use: fly-by (FB) and fly-over (FO). A fly-by waypoint typically is used in a position at which a change in the course of procedure occurs. Charts represent them with four-pointed stars. This type of waypoint is designed to allow you to anticipate and begin your turn prior to reaching the waypoint, thus providing smoother transitions. Conversely, RNAV charts show a fly-over waypoint as a four-pointed star enclosed in a circle. This type of waypoint is used to denote a missed approach point, a missed approach holding point, or other specific points in space that must be flown over. [Figure 2-28 on page 2-30]

RNAV departure procedures are being developed at a rapid pace to provide RNAV capabilities at all airports. With every chart revision cycle, new RNAV departures are being added for small and large airports. These departures are flown in the same manner as traditional navigation-based departures; you are provided headings, altitudes, navigation waypoint, and departure descriptions. RNAV SIDs are found in the TPP with traditional departure procedures. On the plan view of this procedure, in the lower left corner of the chart, the previous aircraft equipment suffix code and equipment notes have been replaced with note 3, the new type code, Type B RNAV departure procedure. Additionally, ATC has the aircraft equipment suffix code on file from the flight plan. [Figure 2-29 on page 2-31]

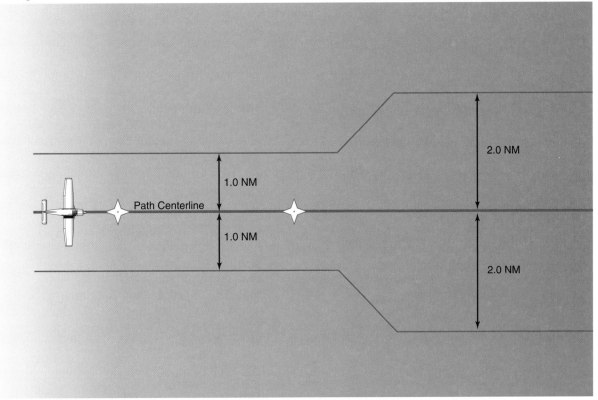

Figure 2-27. RNP Departure Levels.

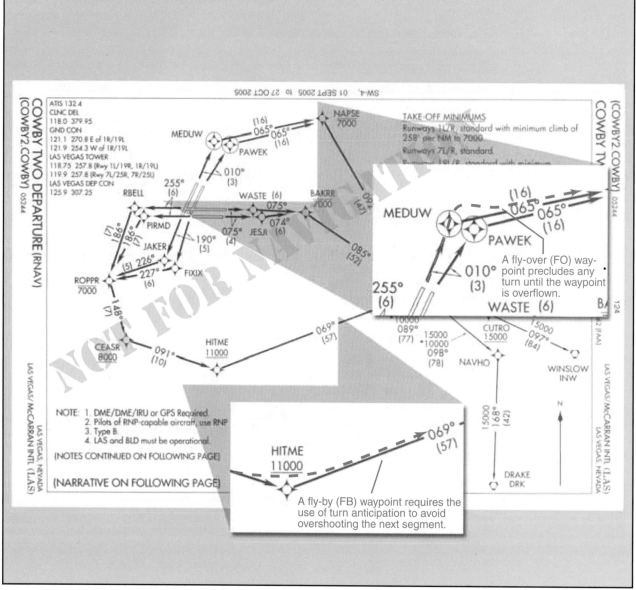

Figure 2-28. Fly-Over and Fly-By Waypoints.

RNAV ODPs are always charted graphically, and like other ODPs, a note in the Takeoff Minimums and IFR Obstacle Departure Procedures section refers you to the graphic ODP chart contained in the main body of the TPP. [Figure 2-30 on page 2-32]

There are specific requirements, however, that must be met before using RNAV procedures. Every RNAV departure chart lists general notes and may include specific equipment and performance requirements, as well as the type of RNAV departure procedure in the chart plan view. New aircraft equipment suffix codes are used to denote capabilities for advanced RNAV navigation, for flight plan filing purposes. [Figure 2-31 on page 2-33]

The chart notes may also include operational information for certain types of equipment, systems, and performance requirements, in addition to the type of RNAV departure procedure. DME/DME navigation system updating may require specific DME facilities to meet performance stan-

dards. Based on DME availability evaluations at the time of publication, current DME coverage is not sufficient to support DME/DME RNAV operations everywhere without IRU augmentation or use of GPS. [Figure 2-32 on page 2-33]

PILOT RESPONSIBILITY FOR USE OF RNAV DEPARTURES

RNAV usage brings with it multitudes of complications as it is being implemented. It takes time to transition, to disseminate information, and to educate current and potential users. As a current pilot using the NAS, you need to have a clear understanding of the aircraft equipment requirements for operating in a given RNP environment. You must understand the type of navigation system installed in your aircraft, and furthermore, you must know how your system operates to ensure that you can comply with all RNAV requirements. Operational information should be included in your AFM or its supplements. Additional information concerning how to use your

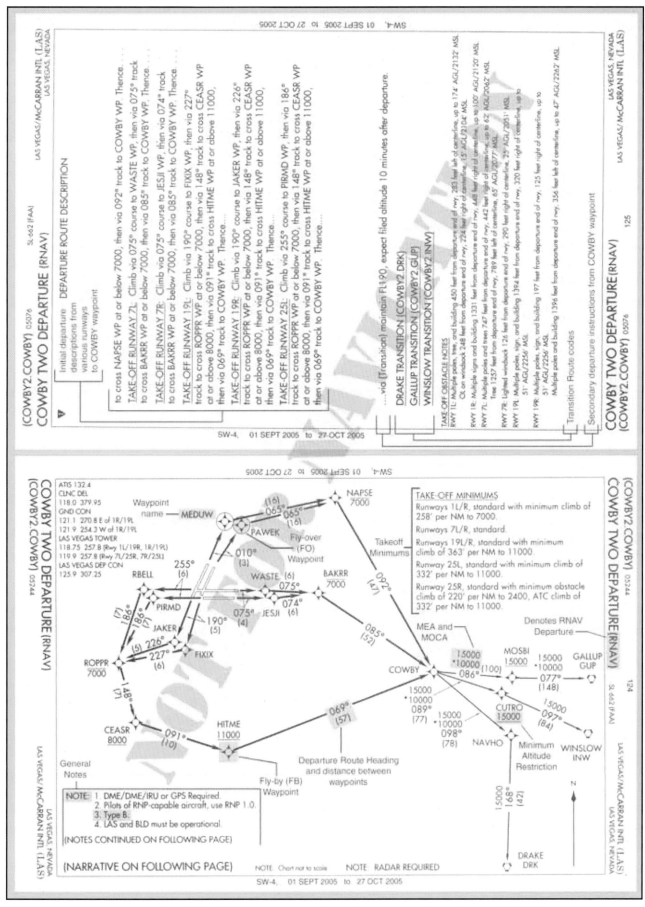

Figure 2-29. The COWBY TWO Departure, Las Vegas, Nevada, is an Example of an RNAV SID.

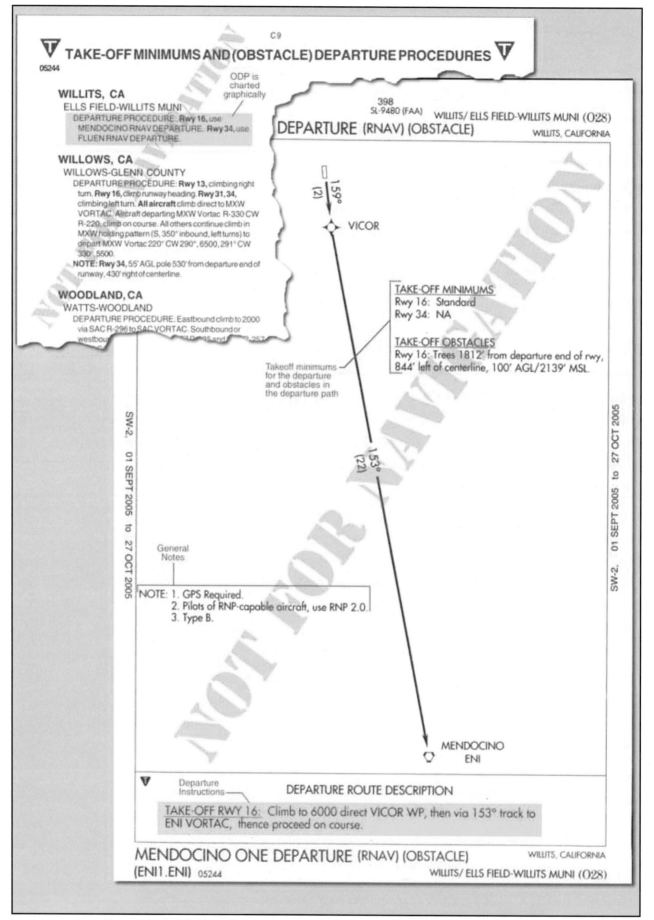

TAKE-OFF MINIMUMS AND (OBSTACLE) DEPARTURE PROCEDURES

05244

ODP is charted graphically

WILLITS, CA
ELLS FIELD-WILLITS MUNI
DEPARTURE PROCEDURE: Rwy 16, use MENDOCINO RNAV DEPARTURE. Rwy 34, use FLUEN RNAV DEPARTURE.

WILLOWS, CA
WILLOWS-GLENN COUNTY
DEPARTURE PROCEDURE: Rwy 13, climbing right turn. Rwy 16, climb runway heading. Rwy 31, 34, climbing left turn. All aircraft climb direct to MXW VORTAC. Aircraft departing MXW Vortac R-330 CW R-220, climb on course. All others continue climb in MXW holding pattern (S, 350° inbound, left turns) to depart MXW Vortac 220° CW 290°, 6500, 291° CW 330°, 5500.
NOTE: Rwy 34, 55' AGL pole 530' from departure end of runway, 430' right of centerline.

WOODLAND, CA
WATTS-WOODLAND
DEPARTURE PROCEDURE: Eastbound climb to 2000 via SAC R-296 to SAC VORTAC. Southbound or westbou...

398
SL-9480 (FAA)   WILLITS/ ELLS FIELD-WILLITS MUNI (O28)

DEPARTURE (RNAV) (OBSTACLE)
WILLITS, CALIFORNIA

159°
(2)

VICOR

Takeoff minimums for the departure and obstacles in the departure path

TAKE-OFF MINIMUMS
Rwy 16: Standard
Rwy 34: NA

TAKE-OFF OBSTACLES
Rwy 16: Trees 1812' from departure end of rwy, 844' left of centerline, 100' AGL/2139' MSL.

153°
(22)

General Notes

NOTE: 1. GPS Required.
2. Pilots of RNP-capable aircraft, use RNP 2.0.
3. Type B.

MENDOCINO
ENI

Departure Instructions

DEPARTURE ROUTE DESCRIPTION

TAKE-OFF RWY 16: Climb to 6000 direct VICOR WP, then via 153° track to ENI VORTAC, thence proceed on course.

MENDOCINO ONE DEPARTURE (RNAV) (OBSTACLE)
WILLITS, CALIFORNIA

(ENI1.ENI) 05244
WILLITS/ ELLS FIELD-WILLITS MUNI (O28)

SW-2, 01 SEPT 2005 to 27 OCT 2005

SW-2, 01 SEPT 2005 to 27 OCT 2005

Figure 2-30. MENDOCINO ONE Departure, Willits, California, is an Example of an RNAV ODP.

RNAV Equipment Codes

ADVANCED RNAV WITH TRANSPONDER AND MODE C (If an aircraft is unable to operate with a transponder and/or Mode C, it will revert to the appropriate code listed above under Area Navigation.)

/E   FMS with DME/DME and IRU position updating

/F   FMS with DME/DME position updating

/G   Global Navigation Satellite System (GNSS), including GPS or WAAS, with en route and terminal capability.

/R   RNP. The aircraft meets the RNP type prescribed for the route segment(s), route(s) and/or area concerned.

Reduced Vertical Separation Minimum (RVSM). Prior to conducting RVSM operations within the U.S., the operator must obtain authorization from the FAA or from the responsible authority, as appropriate.

/J   /E with RVSM

/K   /F with RVSM

/L   /G with RVSM

/Q   /R with RVSM

/W   RVSM

Figure 2-31. RNAV Equipment Codes.

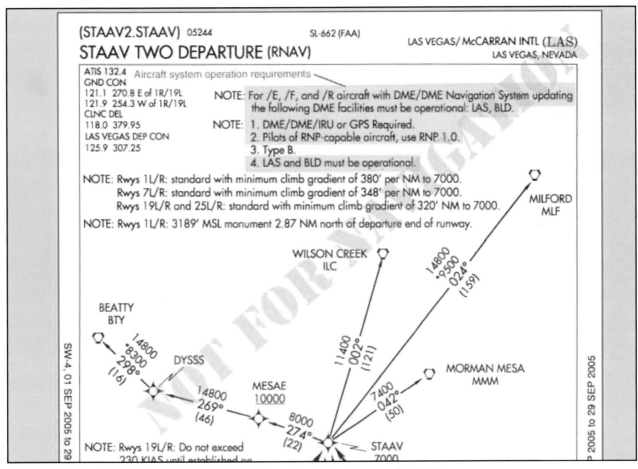

Figure 2-32. Operational Requirements for RNAV.

equipment to its fullest capacity, including "how to" training may be gathered from your avionics manufacturer. If you are in doubt about the operation of your avionics system and its ability to comply with RNAV requirements, contact the FAA directly through your local Flight Standards District Office (FSDO). In-depth information regarding navigation databases is included in Appendix A—Airborne Navigation Databases.

RADAR DEPARTURE

A radar departure is another option for departing an airport on an IFR flight. You might receive a radar departure if the airport does not have an established departure procedure, if you are unable to comply with a departure procedure, or if you request "No SIDs" as a part of your flight plan. Expect ATC to issue an initial departure heading if you are being radar vectored after takeoff, however, do not expect to be given a purpose for the specific vector heading. Rest assured that the controller knows your flight route and will vector you into position. By nature of the departure type, once you are issued your clearance, the responsibility for coordination of your flight rests with ATC, including the tower controller and, after handoff, the departure controller who will remain with you until you are released on course and allowed to "resume own navigation."

For all practical purposes, a radar departure is the easiest type of departure to use. It is also a good alternative to a published departure procedure, particularly when none of the available departure procedures are conducive to your flight route. However, it is advisable to always maintain a detailed awareness of your location as you are being radar vectored by ATC. If for some reason radar contact is lost, you will be asked to provide position reports in order for ATC to monitor your flight progress. Also, ATC may release you to "resume own navigation" after vectoring you off course momentarily for a variety of reasons including weather or traffic.

Upon initial contact, state your aircraft or flight number, the altitude you are climbing through, and the altitude to which you are climbing. The controller will verify that your reported altitude matches that emitted by your transponder. If your altitude does not match, or if you do not have Mode C capabilities, you will be continually required to report your position and altitude for ATC.

The controller is not required to provide terrain and obstacle clearance just because ATC has radar contact with your aircraft. It remains your responsibility until the controller begins to provide navigational guidance in the form of radar vectors. Once radar vectors are given, you are expected to promptly comply with headings and altitudes as assigned. Question any assigned heading if you believe it to be incorrect or if it would cause a violation of a regulation, then advise ATC immediately and obtain a revised clearance.

DIVERSE VECTOR AREA

ATC may establish a minimum vectoring altitude (MVA) around certain airports. This altitude is based on terrain and obstruction clearance and provides controllers with minimum altitudes to vector aircraft in and around a particular location. However, it may be necessary to vector aircraft below this altitude to assist in the efficient flow of departing traffic. For this reason, an airport may have established a **Diverse Vector Area (DVA)**. DVA design requirements are outlined in TERPS and allow for the vectoring of aircraft off the departure end of the runway below the MVA. The presence of a DVA is not published for pilots in any form, so the use of a textual ODP in a DVA environment could result in a misunderstanding between pilots and controllers. ATC instructions take precedence over an ODP. Most DVAs exist only at the busiest airports. [Figure 2-33]

VFR DEPARTURE

There may be times when you need to fly an IFR flight plan due to the weather you will encounter at a later time (or if you simply wish to fly IFR to remain proficient), but the weather outside is clearly VFR. It may be that you can depart VFR, but you need to get an IFR clearance shortly after departing the airport. A VFR departure can be used as a tool that allows you to get off the ground without having to wait for a time slot in the IFR system, however, departing VFR with the intent of receiving an IFR clearance in the air can also present serious hazards worth considering.

A VFR departure dramatically changes the takeoff responsibilities for you and for ATC. Upon receiving clearance for a VFR departure, you are cleared to depart; however, you must maintain separation between yourself and other traffic. You are also responsible for maintaining terrain and obstruction clearance as well as remaining in VFR weather conditions. You cannot fly in IMC without first receiving your IFR clearance. Likewise, a VFR departure relieves ATC of these duties, and basically requires them only to provide you with safety alerts as workload permits.

Maintain VFR until you have obtained your IFR clearance and have ATC approval to proceed on course in accordance with your clearance. If you accept this clearance and are below the minimum IFR altitude for operations in the area, you accept responsibility for terrain/obstruction clearance until you reach that altitude.

NOISE ABATEMENT PROCEDURES

As the aviation industry continues to grow and air traffic increases, so does the population of people and businesses around airports. As a result, noise abatement procedures have become commonplace at most of the nation's airports. Part 150 specifies the responsibilities of the FAA to investigate the recommendations of the airport operator in a noise compatibility program and approve or disapprove the noise abatement suggestions. This is a crucial step in ensuring that the airport is not unduly inhibited by noise requirements and that air traffic workload and efficiency are not significantly impacted, all while considering the noise problems addressed by the surrounding community.

While most departure procedures are designed for obstacle clearance and workload reduction, there are some SIDs that are developed solely to comply with noise abatement requirements. Portland International Jetport is an example of an airport where a SID was created strictly for noise abatement purposes as noted in the departure procedure. [Figure 2-34 on page 2-36] Typically, noise restrictions are incorporated into the main body of the SID. These types of restrictions require higher departure altitudes, larger climb gradients, reduced airspeeds, and turns to avoid specific areas.

Noise restrictions may also be evident during a radar departure. ATC may require you to turn away from your intended course or vector you around a particular area.

While these restrictions may seem burdensome, it is important to remember that it is your duty to comply with written and spoken requests from ATC.

Additionally, when required, departure instructions specify the actual heading to be flown after takeoff, as is the case in figure 2-34 under the departure route description, "Climb via heading 112 degrees..." Some existing procedures specify, "Climb runway heading." Over time, both of these departure instructions will be updated to read, "Climb heading 112 degrees...." **Runway Heading** is the magnetic direction that corresponds with the runway centerline extended (charted on the AIRPORT DIAGRAM), not the numbers painted on the runway. Pilots cleared to "fly or maintain runway heading" are expected to fly or maintain the published heading that corresponds with the extended centerline of the departure runway (until otherwise instructed by ATC), and are not to apply drift correction; e.g. RWY 11, actual magnetic heading of the runway centerline 112.2 degrees, "fly heading 112 degrees". In the event of parallel departures this prevents a loss of separation caused by only one aircraft applying a wind drift.

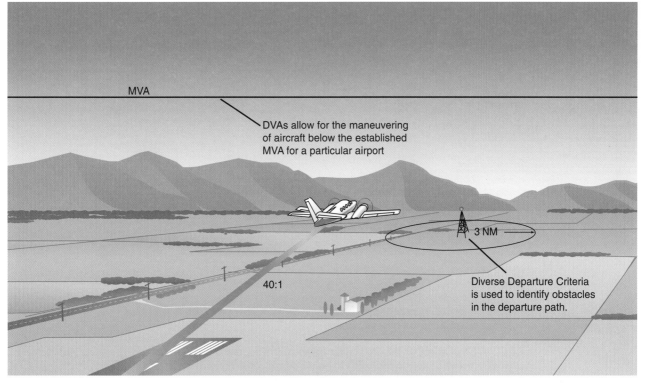

Figure 2-33. Diverse Vector Area Establishment Criteria.

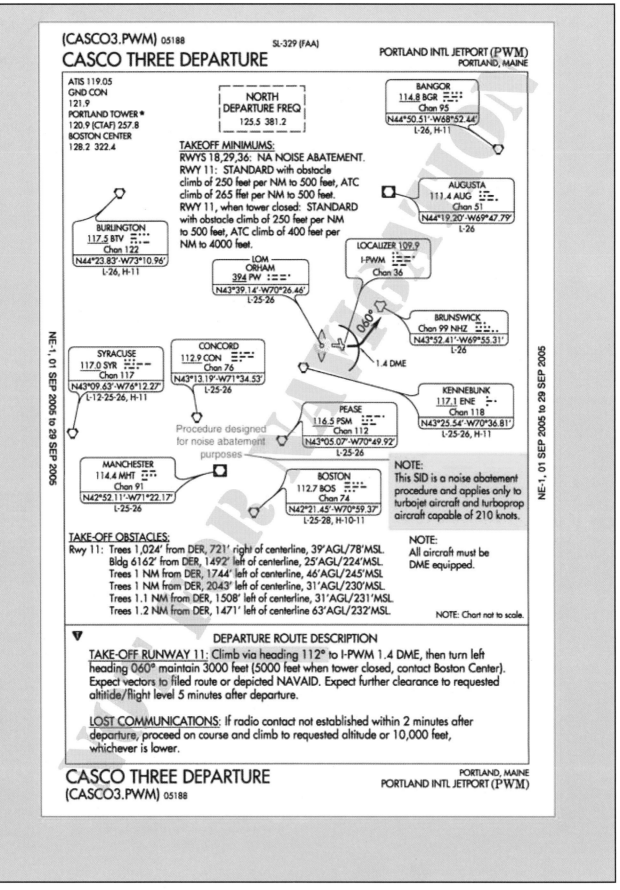

CASCO THREE DEPARTURE

ATIS 119.05
GND CON
121.9
PORTLAND TOWER *
120.9 (CTAF) 257.8
BOSTON CENTER
128.2 322.4

NORTH
DEPARTURE FREQ
125.5 381.2

BANGOR
114.8 BGR
Chan 95
N44°50.51'-W68°52.44'
L-26, H-11

TAKEOFF MINIMUMS:
RWYS 18,29,36: NA NOISE ABATEMENT.
RWY 11: STANDARD with obstacle climb of 250 feet per NM to 500 feet, ATC climb of 265 ffet per NM to 500 feet.
RWY 11, when tower closed: STANDARD with obstacle climb of 250 feet per NM to 500 feet, ATC climb of 400 feet per NM to 4000 feet.

AUGUSTA
111.4 AUG
Chan 51
N44°19.20'-W69°47.79'
L-26

BURLINGTON
117.5 BTV
Chan 122
N44°23.83'-W73°10.96'
L-26, H-11

LOCALIZER 109.9
I-PWM
Chan 36

LOM
ORHAM
394 PW
N43°39.14'-W70°26.46'
L-25-26

060°

1.4 DME

BRUNSWICK
Chan 99 NHZ
N43°52.41'-W69°55.31'
L-26

SYRACUSE
117.0 SYR
Chan 117
N43°09.63'-W76°12.27'
L-12-25-26, H-11

CONCORD
112.9 CON
Chan 76
N43°13.19'-W71°34.53'
L-25-26

KENNEBUNK
117.1 ENE
Chan 118
N43°25.54'-W70°36.81'
L-25-26, H-11

Procedure designed for noise abatement purposes

PEASE
116.5 PSM
Chan 112
N43°05.07'-W70°49.92'
L-25-26

MANCHESTER
114.4 MHT
Chan 91
N42°52.11'-W71°22.17'
L-25-26

BOSTON
112.7 BOS
Chan 74
N42°21.45'-W70°59.37'
L-25-28, H-10-11

NOTE:
This SID is a noise abatement procedure and applies only to turbojet aircraft and turboprop aircraft capable of 210 knots.

NOTE:
All aircraft must be DME equipped.

TAKE-OFF OBSTACLES:
Rwy 11: Trees 1,024' from DER, 721' right of centerline, 39'AGL/78'MSL.
Bldg 6162' from DER, 1492' left of centerline, 25'AGL/224'MSL.
Trees 1 NM from DER, 1744' left of centerline, 46'AGL/245'MSL.
Trees 1 NM from DER, 2043' left of centerline, 31'AGL/230'MSL.
Trees 1.1 NM from DER, 1508' left of centerline, 31'AGL/231'MSL.
Trees 1.2 NM from DER, 1471' left of centerline 63'AGL/232'MSL.

NOTE: Chart not to scale.

DEPARTURE ROUTE DESCRIPTION

TAKE-OFF RUNWAY 11: Climb via heading 112° to I-PWM 1.4 DME, then turn left heading 060° maintain 3000 feet (5000 feet when tower closed, contact Boston Center). Expect vectors to filed route or depicted NAVAID. Expect further clearance to requested altitide/flight level 5 minutes after departure.

LOST COMMUNICATIONS: If radio contact not established within 2 minutes after departure, proceed on course and climb to requested altitude or 10,000 feet, whichever is lower.

CASCO THREE DEPARTURE
(CASCO3.PWM) 05188

PORTLAND, MAINE
PORTLAND INTL JETPORT (PWM)

NE-1, 01 SEP 2005 to 29 SEP 2005

NE-1, 01 SEP 2005 to 29 SEP 2005

Figure 2-34. Noise Abatement SIDs.

The en route phase of flight has seen some of the most dramatic improvements in the way pilots navigate from departure to destination. Developments in technology have played a significant role in most of these improvements. Computerized avionics and advanced navigation systems are commonplace in both general and commercial aviation.

The procedures employed in the en route phase of flight are governed by a set of specific flight standards established by Title 14 of the Code of Federal Regulations (14 CFR), Federal Aviation Administration (FAA) Order 8260.3, *United States Standard for Terminal Instrument Procedures* (TERPS), and related publications. These standards establish courses to be flown, obstacle clearance criteria, minimum altitudes, navigation performance, and communications requirements. For the purposes of this discussion, the en route phase of flight is defined as that segment of flight from the termination point of a departure procedure to the origination point of an arrival procedure.

EN ROUTE NAVIGATION

Part 91.181 is the basis for the course to be flown. To operate an aircraft within controlled airspace under instrument flight rules (IFR), pilots must either fly along the centerline when on a Federal airway or, on routes other than Federal airways, along the direct course between navigational aids or fixes defining the route. The regulation allows maneuvering to pass well clear of other air traffic or, if in visual flight rules (VFR) conditions, to clear the flight path both before and during climb or descent.

En route IFR navigation is evolving from the ground based navigational aid (NAVAID) airway system to a sophisticated satellite and computer-based system that can generate courses to suit the operational requirements of almost any flight. Although the promise of the new navigation systems is immense, the present system of navigation serves a valuable function and is expected to remain for a number of years.

The procedures pilots employ in the en route phase of flight take place in the structure of the National Airspace System (NAS) consisting of three strata. The first, or lower stratum is an airway structure that extends from the base of controlled airspace up to but

not including 18,000 feet mean sea level (MSL). The second stratum is an area containing identifiable jet routes as opposed to designated airways, and extends from 18,000 feet MSL to Flight Level (FL) 450. The third stratum, above FL 450 is intended for random, point-to-point navigation.

AIR ROUTE TRAFFIC CONTROL CENTERS

The **Air Route Traffic Control Center (ARTCC)** encompasses the en route air traffic control system air/ground radio communications, that provides safe and expeditious movement of aircraft operating on IFR within the controlled airspace of the Center. ARTCCs provide the central authority for issuing IFR clearances and nationwide monitoring of each IFR flight. This applies primarily to the en route phase of flight, and includes weather information and other inflight services. There are 20 ARTCCs in the conterminous United States (U.S.), and each Center contains between 20 to 80 sectors, with their size, shape, and altitudes determined by traffic flow, airway structure, and workload. Appropriate radar and communication sites are connected to the Centers by microwave links and telephone lines. [Figure 3-1 on page 3-2]

The CFRs require the pilot in command under IFR in controlled airspace to continuously monitor an appropriate Center or control frequency. When climbing after takeoff, an IFR flight is either in contact with a radar-equipped local departure control or, in some areas, an ARTCC facility. As a flight transitions to the en route phase, pilots typically expect a handoff from departure control to a Center frequency if not already in contact with the Center. The FAA National Aeronautical Charting Office (NACO) publishes en route charts depicting Centers and sector frequencies, as shown in Figure 3-2 on page 3-2. During handoff from one Center to another, the previous controller assigns a new frequency. In cases where flights may be still out of range, the Center frequencies on the face of the chart are very helpful. In Figure 3-2 on page 3-2, notice the boundary between Memphis and Atlanta Centers, and the remoted sites with discrete very high frequency (VHF) and ultra high frequency (UHF) for communicating with the appropriate ARTCC. These Center frequency boxes can be used for finding the nearest frequency within the aircraft range. They also can be used

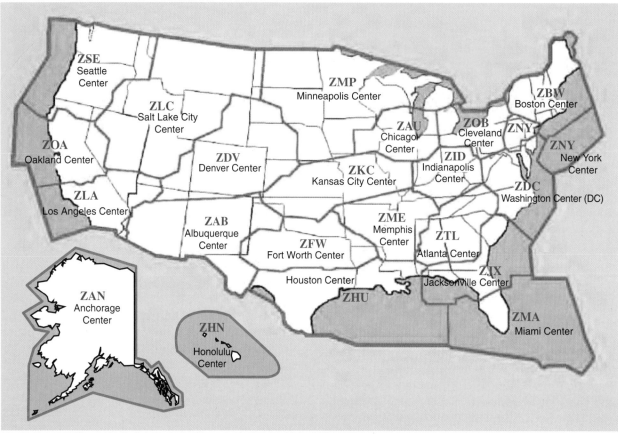

Figure 3-1. Air Route Traffic Control Centers.

for making initial contact with the Center for clearances. The exact location for the Center transmitter is not shown, although the frequency box is placed as close as possible to the known location.

During the en route phase, as a flight transitions from one Center facility to the next, a handoff or transfer of control is required as previously described. The handoff procedure is similar to the handoff between other

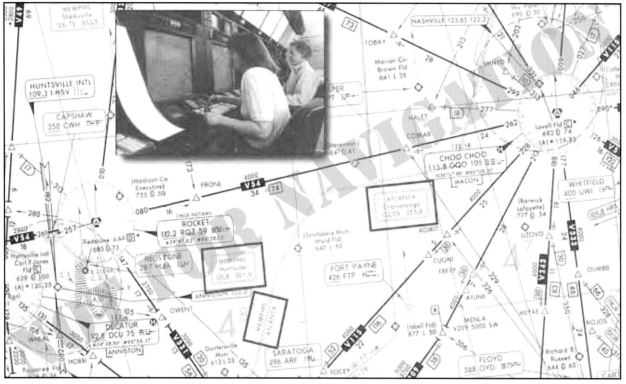

Figure 3-2. ARTCC Centers and Sector Frequencies.

radar facilities, such as departure or approach control. During the handoff, the controller whose airspace is being vacated issues instructions that include the name of the facility to contact, appropriate frequency, and other pertinent remarks.

Accepting radar vectors from controllers does not relieve pilots of their responsibility for safety of flight. Pilots must maintain a safe altitude and keep track of their position, and it is their obligation to question controllers, request an amended clearance, or, in an emergency, deviate from their instructions if they believe that the safety of flight is in doubt. Keeping track of altitude and position when climbing, and during all other phases of flight, are basic elements of situational awareness. Aircraft equipped with an enhanced ground proximity warning system (EGPWS) or terrain awareness and warning system (TAWS) and traffic alert and collision avoidance system (TCAS) help pilots detect and correct unsafe altitudes and traffic conflicts. Regardless of equipment, pilots must always maintain situational awareness regarding their location and the location of traffic in their vicinity.

PREFERRED IFR ROUTES

A system of **preferred IFR routes** helps pilots, flight crews, and dispatchers plan a route of flight to minimize route changes, and to aid in the efficient, orderly management of air traffic using Federal airways. Preferred IFR routes are designed to serve the needs of airspace users and to provide for a systematic flow of air traffic in the major terminal and en route flight environments. Cooperation by all pilots in filing preferred routes results in fewer air traffic delays and better efficiency for departure, en route, and arrival air traffic service. [Figure 3-3]

Preferred IFR routes are published in the *Airport/Facility Directory* for the low and high altitude stratum. If they begin or end with an airway number, it indicates that the airway essentially overlies the airport and flights normally are cleared directly on the airway. Preferred IFR routes beginning or ending with a fix indicate that pilots may be routed to or from these fixes via a standard instrument departure (SID) route, radar vectors, or a standard terminal arrival route (STAR). Routes for major terminals are listed alphabetically under the name of the departure airport. Where several airports are in proximity they are listed under the principal airport and categorized as a metropolitan area; e.g., New York Metro Area. One way preferred IFR routes are listed numerically showing the segment fixes and the direction and times effective. Where more than one route is listed, the routes have equal priority for use. Official location identifiers are used in the route description for very high frequency omnidirectional ranges (VORs) and very high frequency omnidirectional ranges/tactical air navigation (VORTACs), and intersection names are spelled out. The route is direct where two NAVAIDs, an intersection and a NAVAID, a NAVAID and a NAVAID radial and distance point, or any navigable combination of these route descriptions follow in succession.

SUBSTITUTE EN ROUTE FLIGHT PROCEDURES

Air route traffic control centers are responsible for specifying essential substitute airway and route segments and fixes for use during VOR/VORTAC shutdowns. Scheduled shutdowns of navigational facilities require planning and coordination to ensure an uninterrupted flow of air traffic. A schedule of proposed facility shutdowns within the region is maintained and forwarded as far in advance as possible to enable the substitute routes to be published. Substitute routes are normally based on VOR/VORTAC facilities established and published for use in the appropriate altitude strata. In the case of substitute routes in the upper airspace stratum, it may be necessary to establish routes by reference to VOR/VORTAC facilities used in the low altitude system. Nondirectional radio beacon (NDB) facilities may only be used where VOR/VORTAC coverage is inadequate and ATC requirements necessitate use of such NAVAIDs. Where operational necessity dictates, navigational aids may be used beyond their **standard service volume (SSV)** limits, provided that the routes can be given adequate frequency protection.

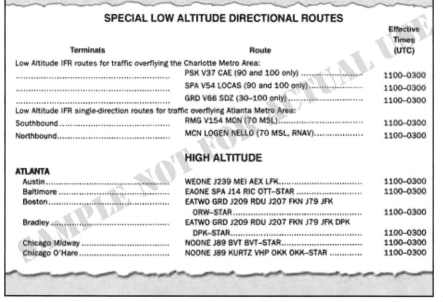

Figure 3-3. Preferred IFR Routes.

The centerline of substitute routes must be contained within controlled airspace, although substitute routes for off-airway routes may not be in controlled airspace. Substitute routes are flight inspected to verify clearance of controlling obstacles and to check for satisfactory facility performance. To provide pilots with necessary lead time, the substitute routes are submitted in advance of the en route chart effective date. If the lead time cannot be provided, the shutdown may be delayed or a special graphic NOTAM may be considered. Normally, shutdown of facilities scheduled for 28 days (half the life of the en route chart) or less will not be charted. The format for describing substitute routes is from navigational fix to navigational fix. A minimum en route altitude (MEA) and a maximum authorized altitude (MAA) is provided for each route segment. Temporary reporting points may be substituted for the out-of-service facility and only those other reporting points that are essential for air traffic control. Normally, temporary reporting points over intersections are not necessary where Center radar coverage exists. A minimum reception altitude (MRA) is established for each temporary reporting point.

TOWER EN ROUTE CONTROL

Within the NAS it is possible to fly an IFR flight without leaving approach control airspace, using **tower en route control (TEC)** service. This helps expedite air traffic and reduces air traffic control and pilot communication requirements. TEC is referred to as "tower en route," or "tower-to-tower," and allows flight beneath the en route structure. Tower en route control reallocates airspace both vertically and geographically to allow flight planning between city pairs while remaining with approach control airspace. All users are encouraged to use the TEC route descriptions in the *Airport/Facility Directory* when filing flight plans. All published TEC routes are designed to avoid en route airspace, and the majority are within radar coverage. [Figure 3-4]

The graphic depiction of TEC routes is not to be used for navigation or for detailed flight planning. Not all city pairs are depicted. It is intended to show geographic areas connected by tower en route control. Pilots should refer to route descriptions for specific flight planning. The word "DIRECT" appears as the route when radar vectors are used or no airway exists. Also, this indicates that a SID or STAR may be assigned by ATC. When a NAVAID or intersection identifier appears with no airway immediately preceding or following the identifier, the routing is understood to be direct to or from that point unless otherwise cleared by ATC. Routes beginning and ending with an airway indicate that the airway essentially overflies

the airport, or radar vectors will be issued. Where more than one route is listed to the same destination, ensure that the correct route for the type of aircraft classification has been filed. These are denoted after the route in the altitude column using J (jet powered), M (turbo props/special, cruise speed 190 knots or greater), P (non-jet, cruise speed 190 knots or greater), or Q (non-jet, cruise speed 189 knots or less). Although all airports are not listed under the destination column, IFR flights may be planned to satellite airports in the proximity of major airports via the same routing. When filing flight plans, the coded route identifier, i.e., BURL1, VTUL4, or POML3, may be used in lieu of the route of flight.

AIRWAY AND ROUTE SYSTEM

The present en route system is based on the VHF airway/route navigation system. Low frequency (LF) and integrated LF/VHF airways and routes have gradually been phased out in the conterminous U.S., with some remaining in Alaska.

MONITORING OF NAVIGATION FACILITIES

VOR, VORTAC, and instrument landing system (ILS) facilities, as well as most nondirectional radio beacons (NDBs) and marker beacons installed by the FAA, are provided with an internal monitoring feature. Internal monitoring is provided at the facility through the use of equipment that causes a facility shutdown if performance deteriorates below established tolerances. A remote status indicator also may be provided through the use of a signal-sampling receiver, microwave link, or telephone circuit. Older FAA NDBs and some non-Federal NDBs do not have the internal feature and monitoring is accomplished by manually checking the operation at least once each hour. FAA facilities such as automated flight service stations (AFSSs) and ARTCCs/sectors are usually the control point for NAVAID facility status. Pilots can query the appropriate FAA facility if they have questions in flight regarding NAVAID status, in addition to checking notices to airmen (NOTAMs) prior to flight, since NAVAIDs and associated monitoring equipment are continuously changing.

LF AIRWAYS/ROUTES

Numerous low frequency airways still exist in Alaska, as depicted in this NACO en route low altitude chart excerpt near Nome, Alaska. [Figure 3-5] Colored LF east and west airways G7, G212 (green), and R35 (red), are shown, along with north and south airways B2, B27 (blue), and A1 (amber), all based upon the Fort Davis NDB en route NAVAID. The nearby Nome VORTAC VHF en route NAVAID is used with victor airways V452, V333, V507, V506, and V440.

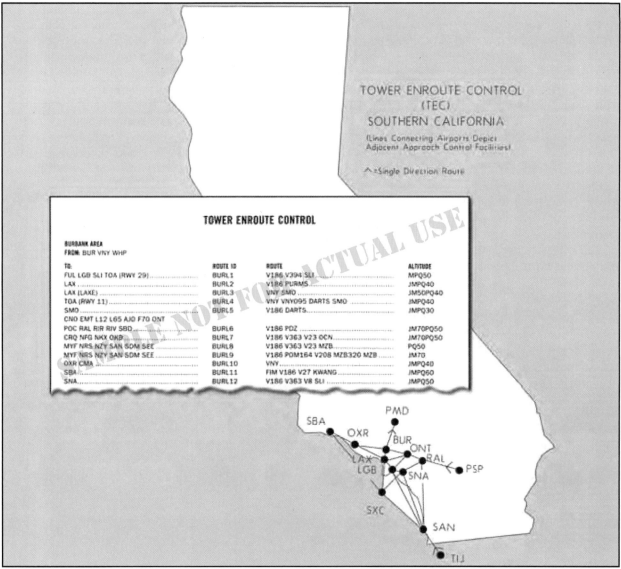

Figure 3-4. Tower En Route Control.

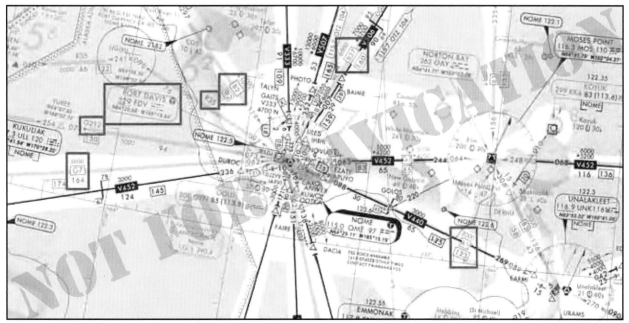

Figure 3-5. LF and VHF Airways — Alaska.

VHF AIRWAYS/ROUTES

Figure 3-6 depicts numerous arrowed, single direction jet routes on this excerpt from a NACO en route high altitude chart, effective at and above 18,000 feet MSL up to and including FL 450. Notice the MAAs of 41,000 and 29,000 associated with J24 and J193, respectively. Additionally, note the BAATT, NAGGI, FUMES, and MEYRA area navigation (RNAV) waypoints. Waypoints are discussed in detail later in this chapter.

VHF EN ROUTE OBSTACLE CLEARANCE AREAS

All published routes in the NAS are based on specific obstacle clearance criteria. An understanding of **en route obstacle clearance areas** helps with situational awareness and may help avoid controlled flight into terrain (CFIT). Obstacle clearance areas for the en route phase of flight are identified as primary, secondary, and turning areas.

The primary and secondary area obstacle clearance criteria, airway and route widths, and the ATC separation procedures for en route segments are a function of safety and practicality in flight procedures. These flight procedures are dependent upon the pilot, the aircraft, and the navigation system being used, resulting in a total VOR system accuracy factor, along with an associated probability factor. The pilot/aircraft information component of these criteria includes pilot ability to track the radial and the flight track resulting from turns at various speeds and altitudes under different wind conditions. The navigation system information includes navigation facility radial alignment displacement, transmitter monitor tolerance, and receiver accuracy. All of these factors were considered during development of en route criteria. From this analysis, the computations resulted in a total system accuracy of ±4.5° 95 percent of the time and ±6.7° 99 percent of the time. The 4.5° figure became the basis for primary area obstacle clearance criteria, airway and route widths, and the ATC separation procedures. The 6.7° value provides secondary obstacle clearance area dimensions. Figure 3-7 depicts the primary and secondary obstacle clearance areas.

PRIMARY AREA

The primary obstacle clearance area has a protected width of 8 nautical miles (NM) with 4 NM on each side of the centerline. The primary area has widths of route protection based upon system accuracy of a ±4.5° angle from the NAVAID. These 4.5° lines extend out from the NAVAID and intersect the boundaries of the primary area at a point approximately 51 NM from the NAVAID. Ideally, the 51 NM point is where pilots would change over from navigating away from the facility, to navigating toward the next facility, although this ideal is rarely achieved.

If the distance from the NAVAID to the changeover point (COP) is more than 51 NM, the outer boundary of the primary area extends beyond the 4 NM width along the 4.5° line when the COP is at midpoint. This

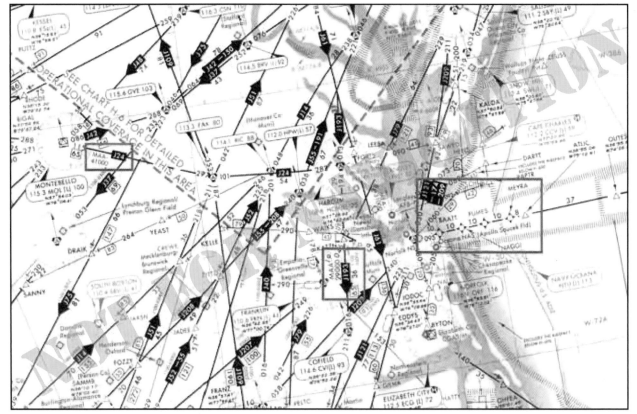

Figure 3-6. VHF Jet Routes.

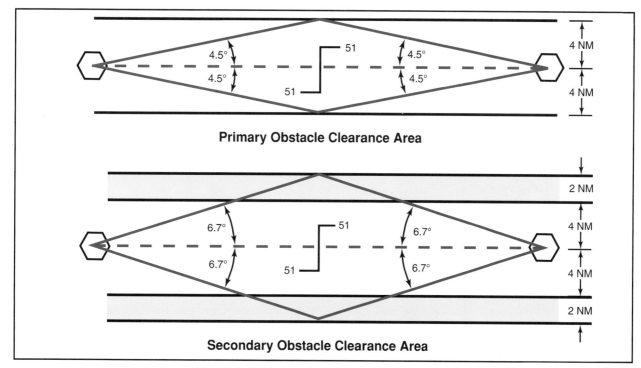

Primary Obstacle Clearance Area

Secondary Obstacle Clearance Area

Figure 3-7. VHF En Route Obstacle Clearance Areas.

means the primary area, along with its obstacle clearance criteria, is extended out into what would have been the secondary area. Additional differences in the obstacle clearance area result in the case of the effect of an offset COP or dogleg segment. For protected en route areas the minimum obstacle clearance in the primary area, not designated as mountainous under Part 95 — IFR altitude is 1,000 feet over the highest obstacle. [Figure 3-8]

Mountainous areas for the Eastern and Western U.S. are designated in Part 95, as shown in Figure 3-9 on page 3-8. Additional mountainous areas are designated for Alaska, Hawaii, and Puerto Rico. With some exceptions, the protected en route area minimum obstacle clearance over terrain and manmade obstacles in mountainous areas is 2,000 feet. Obstacle clearance is sometimes reduced to not less than 1,500 feet above terrain in the designated mountainous areas of the Eastern U.S., Puerto Rico, and Hawaii, and may be reduced to not less than 1,700 feet in mountainous areas of the Western U.S. and Alaska. Consideration is given to the following points before any altitudes providing less than 2,000 feet of terrain clearance are authorized:

- Areas characterized by precipitous terrain.

- Weather phenomena peculiar to the area.

- Phenomena conducive to marked pressure differentials.

- Type of and distance between navigational facilities.

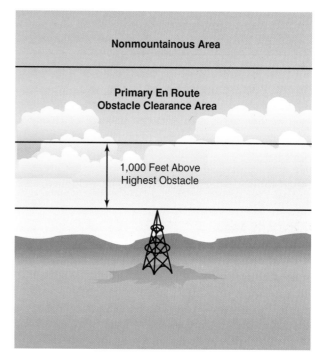

Figure 3-8. Obstacle Clearance - Primary Area.

- Availability of weather services throughout the area.

- Availability and reliability of altimeter resetting points along airways and routes in the area.

Altitudes providing at least 1,000 feet of obstacle clearance over towers and/or other manmade obstacles may be authorized within designated mountainous areas if the obstacles are not located on precipitous terrain where Bernoulli Effect is known or suspected to exist.

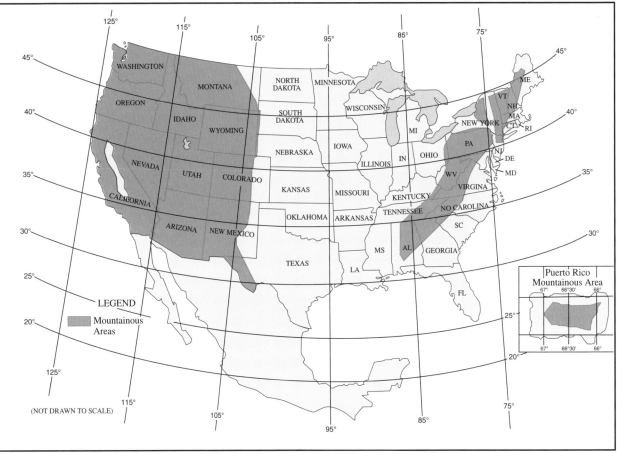

Figure 3-9. Designated Mountainous Areas.

Bernoulli Effect, atmospheric eddies, vortices, waves, and other phenomena that occur in conjunction with disturbed airflow associated with the passage of strong winds over mountains can result in pressure deficiencies manifested as very steep horizontal pressure gradients. Since downdrafts and turbulence are prevalent under these conditions, potential hazards may be multiplied.

SECONDARY AREA

The secondary obstacle clearance area extends along a line 2 NM on each side of the primary area. Navigation system accuracy in the secondary area has widths of route protection of a ±6.7° angle from the NAVAID. These 6.7° lines intersect the outer boundaries of the secondary areas at the same point as primary lines, 51 NM from the NAVAID. If the distance from the NAVAID to the COP is more than 51 NM, the secondary area extends along the 6.7° line when the COP is at midpoint. In all areas, mountainous and nonmountainous, obstacles that are located in secondary areas are considered as obstacles to air navigation if they extend above the secondary obstacle clearance plane. This plane begins at a point 500 feet above the obstacles upon which the primary obstacle clearance area is based, and slants upward at an angle that causes it to intersect the outer edge of the secondary area at a point 500 feet higher. [Figure 3-10]

The obstacle clearance areas for LF airways and routes are different than VHF, with the primary and secondary area route widths both being 4.34 NM. The accuracy lines are 5.0° in the primary obstacle clearance area and 7.5° in the secondary area. Obstacle clearance in the primary area of LF airways and routes is the same as that required for VHF, although the secondary area obstacle clearance requirements are based upon distance from the facility and location of the obstacle relative to the inside boundary of the secondary area.

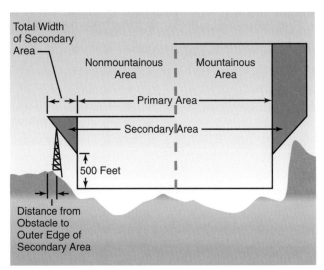

Figure 3-10. Obstacle Clearance - Secondary Area.

When a VHF airway or route terminates at a NAVAID or fix, the primary area extends beyond that termination point. Figure 3-11 and its inset show the construction of the primary and secondary areas at the termination point. When a change of course on VHF airways and routes is necessary, the en route obstacle clearance turning area extends the primary and secondary obstacle clearance areas to accommodate the turn radius of the aircraft. Since turns at or after fix passage may exceed airway and route boundaries, pilots are expected to adhere to airway and route protected airspace by leading turns early before a fix. The turn area provides obstacle clearance for both turn anticipation (turning prior to the fix) and flyover protection (turning after crossing the fix). This does not violate the requirement to fly the centerline of the airway. Many factors enter into the construction and application of the turning area to provide pilots with adequate obstacle clearance protection. These may include aircraft speed, the amount of turn versus NAVAID distance, flight track, curve radii, MEAs, and minimum turning altitude (MTA). A typical protected airspace is shown in Figure 3-11. Turning area system accuracy factors must be applied to the most adverse displacement of the NAVAID or fix and the airway or route boundaries at which the turn is made. If applying nonmountainous en route turning area criteria graphically, depicting the vertical obstruction clearance in a typical application, the template might appear as in Figure 3-12 on page 3-10.

Turns that begin at or after fix passage may exceed the protected en route turning area obstruction clearance.

Figure 3-13 on page 3-10 contains an example of a flight track depicting a turn at or after fix passage, together with an example of an early turn. Without leading a turn, an aircraft operating in excess of 290 knots true airspeed (TAS) can exceed the normal airway or route boundaries depending on the amount of course change required, wind direction and velocity, the character of the turn fix (DME, overhead navigation aid, or intersection), and pilot technique in making a course change. For example, a flight operating at 17,000 feet MSL with a TAS of 400 knots, a 25° bank, and a course change of more than 40° would exceed the width of the airway or route; i.e., 4 NM each side of centerline. Due to the high airspeeds used at 18,000 feet MSL and above, additional IFR separation protection for course changes is provided.

NAVAID SERVICE VOLUME

Each class of VHF NAVAID (VOR/DME/TACAN) has an established operational service volume to ensure adequate signal coverage and frequency protection from other NAVAIDs on the same frequency. The maximum distance at which NAVAIDs are usable varies with altitude and the class of the facility. When using VORs for direct route navigation, the following guidelines apply:

• For operations above FL 450, use aids not more than 200 NM apart. These are High Altitude (H) class facilities and are depicted on en route high altitude charts.

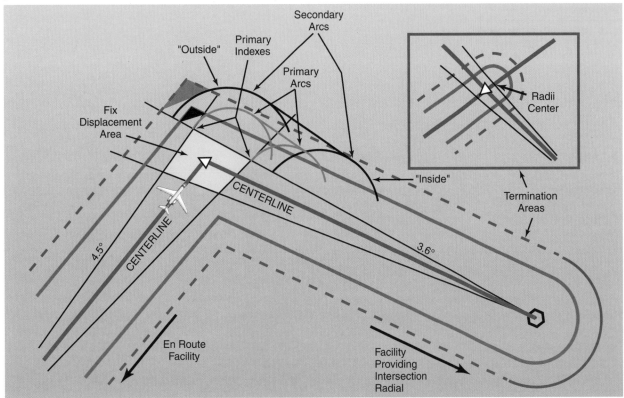

Figure 3-11. Turning Area, Intersection Fix, NAVAID Distance less than 51 NM.

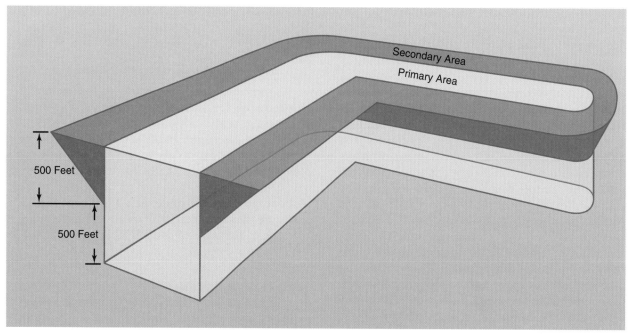

Figure 3-12. Turning Area Obstruction Clearance.

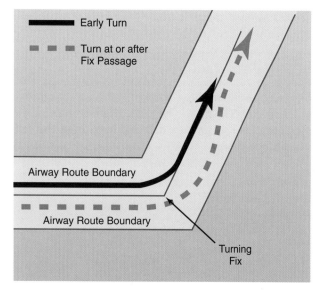

Figure 3-13. Adhering to Airway/Route Turning Area.

- For operations that are off established airways from 18,000 feet MSL to FL 450, use aids not more than 260 NM apart. These are High Altitude (H) class facilities and are depicted on en route high altitude charts.

- For operations that are off established airways below 18,000 feet MSL, use aids not more than 80 NM apart. These are Low Altitude (L) class facilities and are shown on en route low altitude charts.

- For operations that are off established airways between 14,500 feet MSL and 17,999 feet MSL in the conterminous United States, use H-class facilities not more than 200 NM apart.

The use of satellite based navigation systems has increased pilot requests for direct routes that take the aircraft outside ground based NAVAID service volume limits. These direct route requests are approved only in a radar environment, and approval is based on pilot responsibility for staying on the authorized direct route. ATC uses radar flight following for the purpose of aircraft separation. On the other hand, if ATC initiates a direct route that exceeds NAVAID service volume limits, ATC also provides radar navigational assistance as necessary. More information on direct route navigation is located in the En Route RNAV Procedures section later in this chapter.

NAVIGATIONAL GAPS
Where a navigational course guidance gap exists, referred to as an MEA gap, the airway or route segment may still be approved for navigation. The **navigational gap** may not exceed a specific distance that varies directly with altitude, from zero NM at sea level to 65 NM at 45,000 feet MSL and not more than one gap may exist in the airspace structure for the airway or route segment. Additionally, a gap usually does not occur at any airway or route turning point. To help ensure the maximum amount of continuous positive course guidance available when flying, there are established en route criteria for both straight and turning segments. Where large gaps exist that require altitude changes, MEA "steps" may be established at increments of not less than 2,000 feet below 18,000 feet MSL, or not less than 4,000 feet at 18,000 MSL and above, provided that a total gap does not exist for the entire segment within the airspace structure. MEA steps are limited to one step between any two facilities to eliminate continuous or repeated changes of altitude in problem areas. The allowable navigational gaps pilots can expect to see

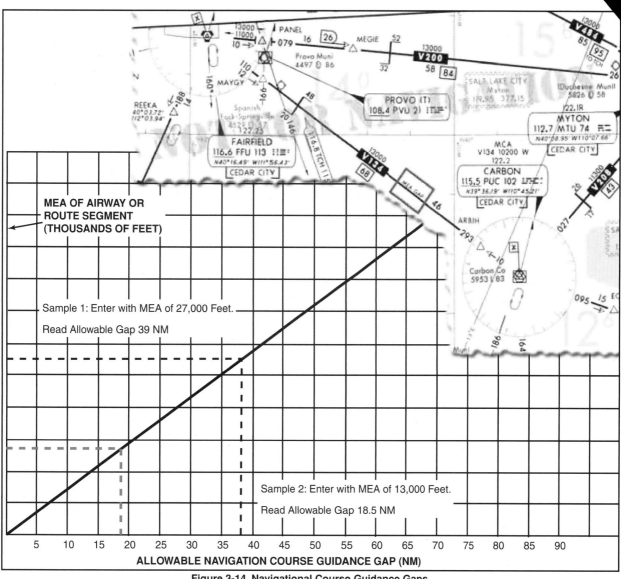

Figure 3-14. Navigational Course Guidance Gaps.

are determined, in part, by reference to the graph depicted in Figure 3-14. Notice the en route chart excerpt depicting that the MEA is established with a gap in navigation signal coverage northwest of the Carbon VOR/DME on V134. At the MEA of 13,000, the allowable navigation course guidance gap is approximately 18.5 NM, as depicted by Sample 2. The navigation gap area is not identified on the chart by distances from the navigation facilities.

CHANGEOVER POINTS

When flying airways, pilots normally change frequencies midway between navigation aids, although there are times when this is not practical. If the navigation signals cannot be received from the second VOR at the midpoint of the route, a **changeover point (COP)** is depicted and shows the distance in NM to each NAVAID, as depicted in Figure 3-15 on page 3-12. COPs indicate the point where a frequency change is necessary to receive course guidance from the facility ahead of the aircraft instead of the one behind. These changeover

points divide an airway or route segment and ensure continuous reception of navigation signals at the prescribed minimum en route IFR altitude. They also ensure that other aircraft operating within the same portion of an airway or route segment receive consistent azimuth signals from the same navigation facilities regardless of the direction of flight.

Where signal coverage from two VORs overlaps at the MEA, the changeover point normally is designated at the midpoint. Where radio frequency interference or other navigation signal problems exist, the COP is placed at the optimum location, taking into consideration the signal strength, alignment error, or any other known condition that affects reception. The changeover point has an effect on the primary and secondary obstacle clearance areas. On long airway or route segments, if the distance between two facilities is over 102 NM and the changeover point is placed at the midpoint, the system accuracy lines extend beyond the minimum widths of 8 and 12 NM, and a flare or spreading

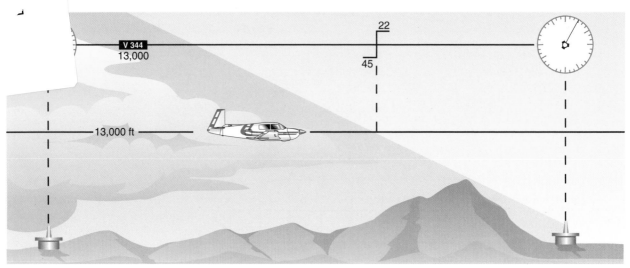

Figure 3-15. Changeover Points.

outward results at the COP, as shown in Figure 3-16. Offset changeover points and dogleg segments on airways or routes can also result in a flare at the COP.

IFR EN ROUTE ALTITUDES

Minimum en route altitudes, minimum reception altitudes, maximum authorized altitudes, minimum obstruction clearance altitudes, minimum crossing altitudes, and changeover points are established by the FAA for instrument flight along Federal airways, as well as some off-airway routes. The altitudes are established after it has been determined that the navigation aids to be used are adequate and so oriented on the airways or routes that signal coverage is acceptable, and that flight can be maintained within prescribed route widths.

For IFR operations, regulations require that pilots operate their aircraft at or above minimum altitudes. Except when necessary for takeoff or landing, pilots may not operate an aircraft under IFR below applicable minimum altitudes, or if no applicable minimum altitude is prescribed, in the case of operations over an area designated as mountainous, an altitude of 2,000 feet above

the highest obstacle within a horizontal distance of 4 NM from the course to be flown. In any other case, an altitude of 1,000 feet above the highest obstacle within a horizontal distance of 4 NM from the course to be flown must be maintained as a minimum altitude. If both a MEA and a minimum obstruction clearance altitude (MOCA) are prescribed for a particular route or route segment, pilots may operate an aircraft below the MEA down to, but not below, the MOCA, only when within 22 NM of the VOR. When climbing to a higher minimum IFR altitude (MIA), pilots must begin climbing immediately after passing the point beyond which that minimum altitude applies, except when ground obstructions intervene, the point beyond which that higher minimum altitude applies must be crossed at or above the applicable minimum crossing altitude (MCA) for the VOR.

If on an IFR flight plan, but cleared by ATC to maintain VFR conditions on top, pilots may not fly below minimum en route IFR altitudes. Minimum altitude rules are designed to ensure safe vertical separation between the aircraft and the terrain. These minimum altitude rules apply to all IFR flights, whether in IFR or VFR

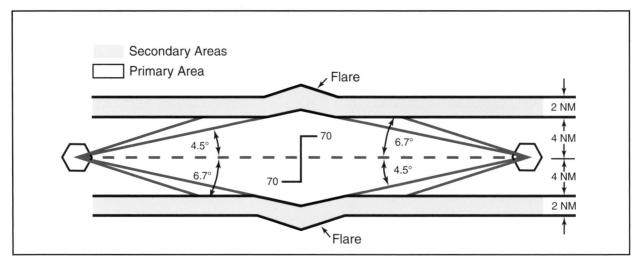

Figure 3-16. Changeover Point Effect on Long Airway or Route Segment.

weather conditions, and whether assigned a specific altitude or VFR conditions on top.

MINIMUM EN ROUTE ALTITUDE

The **minimum enroute altitude (MEA)** is the lowest published altitude between radio fixes that assures acceptable navigational signal coverage and meets obstacle clearance requirements between those fixes. The MEA prescribed for a Federal airway or segment, RNAV low or high route, or other direct route applies to the entire width of the airway, segment, or route between the radio fixes defining the airway, segment, or route. MEAs for routes wholly contained within controlled airspace normally provide a buffer above the floor of controlled airspace consisting of at least 300 feet within transition areas and 500 feet within control areas. MEAs are established based upon obstacle clearance over terrain and manmade objects, adequacy of navigation facility performance, and communications requirements, although adequate communication at the MEA is not guaranteed.

MINIMUM OBSTRUCTION CLEARANCE ALTITUDE

The **minimum obstruction clearance altitude (MOCA)** is the lowest published altitude in effect between fixes on VOR airways, off-airway routes, or route segments that meets obstacle clearance requirements for the entire route segment. This altitude also assures acceptable navigational signal coverage only within 22 NM of a VOR. The MOCA seen on the NACO en route chart, may have been computed by adding the required obstacle clearance (ROC) to the controlling obstacle in the primary area or computed by using a TERPS chart if the controlling obstacle is located in the secondary area. This figure is then rounded to the nearest 100 - foot increment, i.e., 2,049 feet becomes 2,000, and 2,050 feet becomes 2,100 feet. An extra 1,000 feet is added in mountainous areas, in most cases. The MOCA is based upon obstacle clearance over the terrain or over manmade objects, adequacy of navigation facility performance, and communications requirements.

ATC controllers have an important role in helping pilots remain clear of obstructions. Controllers are instructed to issue a safety alert if the aircraft is in a position that, in their judgment, places the pilot in unsafe proximity to terrain, obstructions, or other aircraft. Once pilots inform ATC of action being taken to resolve the situation, the controller may discontinue the issuance of further alerts. A typical terrain/obstruction alert may sound like this: *"Low altitude alert. Check your altitude immediately. The MOCA in your area is 12,000."*

MINIMUM VECTORING ALTITUDES

Minimum vectoring altitudes (MVAs) are established for use by ATC when radar ATC is exercised. The MVA provides 1,000 feet of clearance above the highest obstacle in nonmountainous areas and 2,000 feet above the highest obstacle in designated mountainous areas. Because of the ability to isolate specific obstacles, some MVAs may be lower than MEAs, MOCAs, or other minimum altitudes depicted on charts for a given location. While being radar vectored, IFR altitude assignments by ATC are normally at or above the MVA.

Controllers use MVAs only when they are assured an adequate radar return is being received from the aircraft. Charts depicting minimum vectoring altitudes are normally available to controllers but not available to pilots. Situational awareness is always important, especially when being radar vectored during a climb into an area with progressively higher MVA sectors, similar to the concept of minimum crossing altitude. Except where diverse vector areas have been established, when climbing, pilots should not be vectored into a sector with a higher MVA unless at or above the next sector's MVA. Where lower MVAs are required in designated mountainous areas to achieve compatibility with terminal routes or to permit vectoring to an instrument approach procedure, 1,000 feet of obstacle clearance may be authorized with the use of Airport Surveillance Radar (ASR). The MVA will provide at least 300 feet above the floor of controlled airspace. The MVA charts are developed to the maximum radar range. Sectors provide separation from terrain and obstructions. Each MVA chart has sectors large enough to accommodate vectoring of aircraft within the sector at the MVA. [Figure 3-17 on page 3-14]

MINIMUM RECEPTION ALTITUDE

Minimum reception altitudes (MRAs) are determined by FAA flight inspection traversing an entire route of flight to establish the minimum altitude the navigation signal can be received for the route and for off-course NAVAID facilities that determine a fix. When the MRA at the fix is higher than the MEA, an MRA is established for the fix, and is the lowest altitude at which an intersection can be determined.

MINIMUM CROSSING ALTITUDE

A **minimum crossing altitude (MCA)** is the lowest altitude at certain fixes at which the aircraft must cross when proceeding in the direction of a higher minimum en route IFR altitude, as depicted in Figure 3-18 on page 3-14. MCAs are established in all cases where obstacles intervene to prevent pilots from maintaining obstacle clearance during a normal climb to a higher MEA after passing a point beyond which the higher MEA applies. The same protected en route area vertical obstacle clearance requirements for the primary and secondary areas are considered in the determination of the MCA. The standard for determining the MCA is based upon the following climb gradients, and is computed from the flight altitude:

- Sea level through 5,000 feet MSL—150 feet per NM

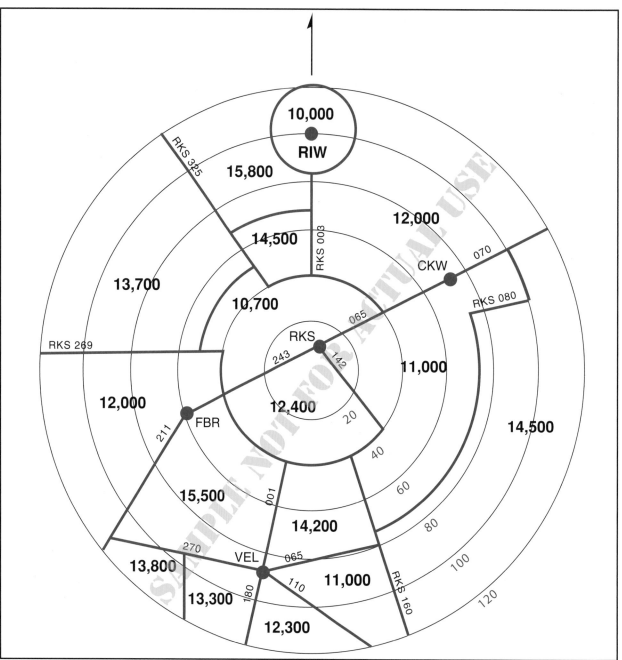

Figure 3-17. Example of an Air Route Traffic Control Center MVA Chart.

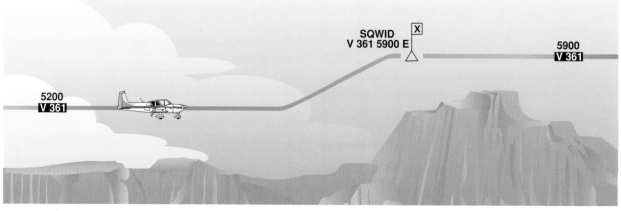

Figure 3-18. Minimum Crossing Altitude.

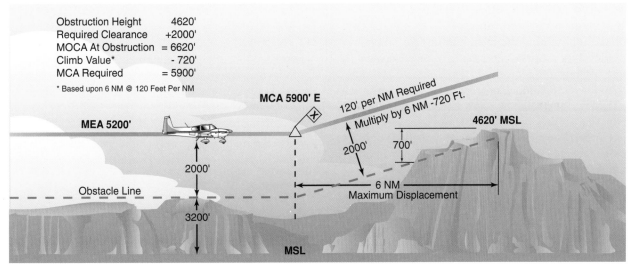

Obstruction Height	4620'
Required Clearance	+2000'
MOCA At Obstruction	= 6620'
Climb Value*	- 720'
MCA Required	= 5900'

* Based upon 6 NM @ 120 Feet Per NM

MCA 5900' E

MEA 5200'

120' per NM Required

Multiply by 6 NM -720 Ft.

4620' MSL

2000'

700'

2000'

Obstacle Line

6 NM
Maximum Displacement

3200'

MSL

Figure 3-19. MCA Determination Point.

- 5000 feet through 10,000 feet MSL — 120 feet per NM

- 10,000 feet MSL and over — 100 feet per NM

To determine the MCA seen on a NACO en route chart, the distance from the obstacle to the fix is computed from the point where the centerline of the en route course in the direction of flight intersects the farthest displacement from the fix, as shown in Figure 3-19. When a change of altitude is involved with a course change, course guidance must be provided if the change of altitude is more than 1,500 feet and/or if the course change is more than 45 degrees, although there is an exception to this rule. In some cases, course changes of up to 90 degrees may be approved without course guidance provided that no obstacles penetrate the established MEA requirement of the previous airway or route segment. Outside U. S. airspace, pilots may encounter different flight procedures regarding MCA and transitioning from one MEA to a higher MEA. In this case, pilots are expected to be at the higher MEA crossing the fix, similar to an MCA. Pilots must thoroughly review flight procedure differences when flying outside U.S. airspace. On NACO en route charts, routes and associated data outside the conterminous U.S. are shown for transitional purposes only and are not part of the high altitude jet route and RNAV route systems. [Figure 3-20]

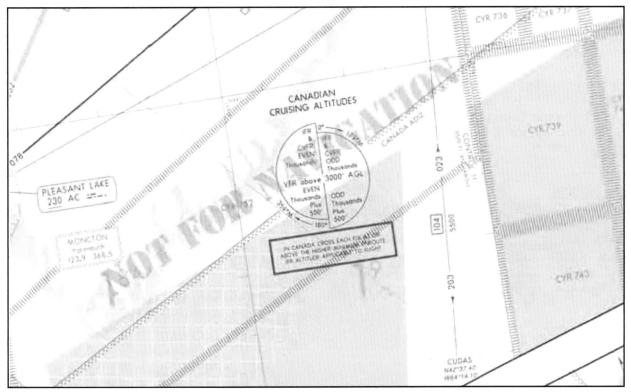

Figure 3-20. Crossing a Fix to a Higher MEA in Canada.

MAXIMUM AUTHORIZED ALTITUDE

A **maximum authorized altitude (MAA)** is a published altitude representing the maximum usable altitude or flight level for an airspace structure or route segment. It is the highest altitude on a Federal airway, jet route, RNAV low or high route, or other direct route for which an MEA is designated at which adequate reception of navigation signals is assured. MAAs represent procedural limits determined by technical limitations or other factors such as limited airspace or frequency interference of ground based facilities.

IFR CRUISING ALTITUDE OR FLIGHT LEVEL

In controlled airspace, pilots must maintain the altitude or flight level assigned by ATC, although if the ATC clearance assigns *"VFR conditions on-top,"* an altitude or flight level as prescribed by Part 91.159 must be maintained. In uncontrolled airspace (except while in a holding pattern of 2 minutes or less or while turning) if operating an aircraft under IFR in level cruising flight, an appropriate altitude as depicted in the legend of NACO IFR en route high and low altitude charts must be maintained. [Figure 3-21]

When operating on an IFR flight plan below 18,000 feet MSL in accordance with a VFR-on-top clearance, any VFR cruising altitude appropriate to the direction of flight between the MEA and 18,000 feet MSL may be selected that allows the flight to remain in VFR conditions. Any change in altitude must be reported to ATC and pilots must comply with all other IFR reporting procedures. VFR-on-top is not authorized in Class A airspace. When cruising below 18,000 feet MSL, the altimeter must be adjusted to the current setting, as reported by a station within 100 NM of your position. In areas where weather-reporting stations are more than 100 NM from the route, the altimeter setting of a station that is closest may be used. During IFR flight, ATC advises flights periodically of the current altimeter setting, but it remains the responsibility of the pilot or flight crew to update altimeter settings in a timely manner. Altimeter settings and weather information are available from weather reporting facilities operated or approved by the U.S. National Weather Service, or a source approved by the FAA. Some commercial operators have the authority to act as a government-approved source of weather information, including altimeter settings, through certification under the FAA's Enhanced Weather Information System.

Flight level operations at or above 18,000 feet MSL require the altimeter to be set to 29.92. A **flight level (FL)** is defined as a level of constant atmospheric pressure related to a reference datum of 29.92 in. Hg. Each flight level is stated in three digits that represent hundreds of feet. For example, FL 250 represents an altimeter indication of 25,000 feet. Conflicts with traffic operating below 18,000 feet MSL may arise when actual altimeter settings along the route of flight are lower than 29.92. Therefore, Part 91.121 specifies the lowest usable flight levels for a given altimeter setting range.

LOWEST USABLE FLIGHT LEVEL

When the barometric pressure is 31.00 inches of mercury or less and pilots are flying below 18,000 feet MSL, use the current reported altimeter setting. This is important because the true altitude of an aircraft is lower than indicated when sea level pressure is lower than standard. When an aircraft is en route on an instrument flight plan, air traffic controllers furnish this information at least once while the aircraft is in the controller's area of jurisdiction. According to Part 91.144, when the barometric pressure exceeds 31.00 inches Hg., the following procedures are placed in effect by NOTAM defining the geographic area affected: Set 31.00 inches for en route operations below 18,000 feet MSL and maintain this setting until beyond the affected area. Air traffic control issues actual altimeter settings

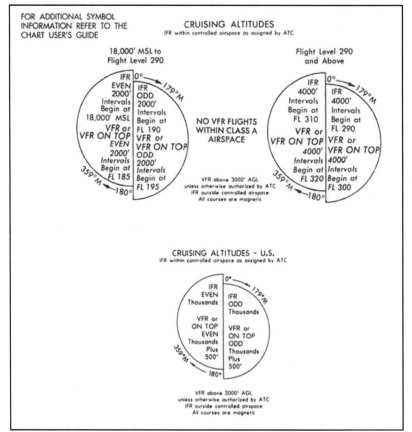

Figure 3-21. IFR Cruising Altitude or Flight Level.

and advises pilots to set 31.00 inches in their altimeter, for en route operations below 18,000 feet MSL in affected areas. If an aircraft has the capability of setting the current altimeter setting and operating into airports with the capability of measuring the current altimeter setting, no additional restrictions apply. At or above 18,000 feet MSL, altimeters should be set to 29.92 inches of mercury (standard setting). Additional procedures exist beyond the en route phase of flight.

The lowest usable flight level is determined by the atmospheric pressure in the area of operation. As local altimeter settings fall below 29.92, pilots operating in Class A airspace must cruise at progressively higher indicated altitudes to ensure separation from aircraft operating in the low altitude structure as follows:

Current Altimeter Setting	Lowest Usable Flight Level
• 29.92 or higher	180
• 29.91 to 29.42	185
• 29.41 to 28.92	190
• 28.91 to 28.42	195
• 28.41 to 27.92	200

When the minimum altitude, as prescribed in Parts 91.159 and 91.177, is above 18,000 feet MSL, the lowest usable flight level is the flight level equivalent of the minimum altitude plus the number of feet specified according to the lowest flight level correction factor as follows:

Altimeter Setting	Correction Factor
• 29.92 or higher	none
• 29.91 to 29.42	500 Feet
• 29.41 to 28.92	1000 Feet
• 28.91 to 28.42	1500 Feet
• 28.41 to 27.92	2000 Feet
• 27.91 to 27.42	2500 Feet

OPERATIONS IN OTHER COUNTRIES

When flight crews transition from the U.S. NAS to another country's airspace, they should be aware of differences not only in procedures but also airspace. For example, when flying into Canada regarding altimeter setting changes, as depicted in Figure 3-22 on page 3-18, notice the change from QNE to QNH when flying northbound into the Moncton **flight information region (FIR)**, an airspace of defined dimensions where flight

information service and alerting service are provided. **Transition altitude (QNH)** is the altitude in the vicinity of an airport at or below which the vertical position of the aircraft is controlled by reference to altitudes (MSL). The **transition level (QNE)** is the lowest flight level available for use above the transition altitude. **Transition height (QFE)** is the height in the vicinity of an airport at or below which the vertical position of the aircraft is expressed in height above the airport reference datum. The **transition layer** is the airspace between the transition altitude and the transition level. If descending through the transition layer, set the altimeter to local station pressure. When departing and climbing through the transition layer, use the standard altimeter setting (QNE) of 29.92 inches of Mercury, 1013.2 millibars, or 1013.2 hectopascals. Remember that most pressure altimeters are subject to mechanical, elastic, temperature, and installation errors. Extreme cold temperature differences also may require a correction factor.

REPORTING PROCEDURES

In addition to acknowledging a handoff to another Center en route controller, there are reports that should be made without a specific request from ATC. Certain reports should be made at all times regardless of whether a flight is in radar contact with ATC, while others are necessary only if radar contact has been lost or terminated. Refer to Figure 3-23 on page 3-19 for a review of these reports.

NONRADAR POSITION REPORTS

If radar contact has been lost or radar service terminated, the CFRs require pilots to provide ATC with position reports over designated VORs and intersections along their route of flight. These compulsory reporting points are depicted on NACO IFR en route charts by solid triangles. Position reports over fixes indicated by open triangles are noncompulsory reporting points, and are only necessary when requested by ATC. If on a direct course that is not on an established airway, report over the fixes used in the flight plan that define the route, since they automatically become compulsory reporting points. Compulsory reporting points also apply when conducting an IFR flight in accordance with a VFR-on-top clearance. Whether a route is on airways or direct, position reports are mandatory in a nonradar environment, and they must include specific information. A typical position report includes information pertaining to aircraft position, expected route, and estimated time of arrival (ETA). Time may be stated in minutes only when no misunderstanding is likely to occur. [Figure 3-24 on page 3-20]

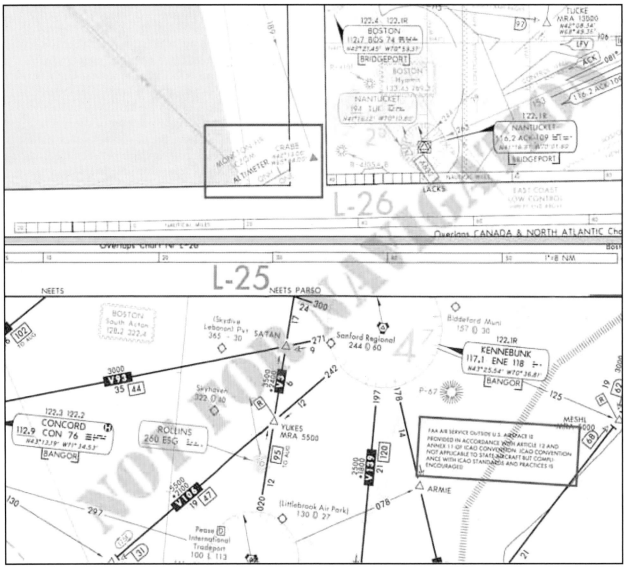

Figure 3-22. Altimeter Setting Changes.

COMMUNICATION FAILURE

Two-way radio communication failure procedures for IFR operations are outlined in Part 91.185. Unless otherwise authorized by ATC, pilots operating under IFR are expected to comply with this regulation. Expanded procedures for communication failures are found in the AIM. Pilots can use the transponder to alert ATC to a radio communication failure by squawking code 7600. [Figure 3-25 on page 3-20] If only the transmitter is inoperative, listen for ATC instructions on any operational receiver, including the navigation receivers. It is possible ATC may try to make contact with pilots over a VOR, VORTAC, NDB, or localizer frequency. In addition to monitoring NAVAID receivers, attempt to reestablish communications by contacting ATC on a previously assigned frequency, calling a FSS or Aeronautical Radio Incorporated (ARINC).

The primary objective of the regulations governing communication failures is to preclude extended IFR no-radio operations within the ATC system since these operations may adversely affect other users of the airspace. If the radio fails while operating on an IFR clearance, but in VFR conditions, or if encountering VFR conditions at any time after the failure, continue the flight under VFR conditions, if possible, and land as soon as practicable. The requirement to land as soon as practicable should not be construed to mean as soon as possible. Pilots retain the prerogative of exercising their best judgment and are not required to land at an unauthorized airport, at an airport unsuitable for the type of aircraft flown, or to land only minutes short of their intended destination. However, if IFR conditions prevail, pilots must comply with procedures designated in the CFRs to ensure aircraft separation.

If pilots must continue their flight under IFR after experiencing two-way radio communication failure, they should fly one of the following routes:

• The route assigned by ATC in the last clearance received.

RADAR/NONRADAR REPORTS	
These reports should be made at all times without a specific ATC request.	
REPORTS	EXAMPLE:
Leaving one assigned flight altitude or flight level for another	*"Marathon 564, leaving 8,000, climb to 10,000."*
VFR-on-top change in altitude	*"Marathon 564, VFR-on-top, climbing to 10,500."*
Leaving any assigned holding fix or point	*"Marathon 564, leaving FARGO Intersection."*
Missed approach	*"Marathon 564, missed approach, request clearance to Chicago."*
Unable to climb or descend at least 500 feet per minute	*"Marathon 564, maximum climb rate 400 feet per minute."*
TAS variation from filed speed of 5% or 10 knots, whichever is greater	*"Marathon 564, advises TAS decrease to 140 knots."*
Time and altitude or flight level upon reaching a holding fix or clearance limit	*"Marathon 564, FARGO Intersection at 05, 10,000, holding east."*
Loss of nav/comm capability (required by Part 91.187)	*"Marathon 564, ILS receiver inoperative."*
Unforecast weather conditions or other information relating to the safety of flight (required by Part 91.183)	*"Marathon 564, experiencing moderate turbulence at 10,000."*

NONRADAR REPORTS	
When you are not in radar contact, these reports should be made without a specific request from ATC.	
REPORTS	EXAMPLE:
Leaving FAF or OM inbound on final approach	*"Marathon 564, outer marker inbound, leaving 2,000."*
Revised ETA of more than three minutes	*"Marathon 564, revising SCURRY estimate to 55."*
Position reporting at compulsory reporting points (required by Part 91.183)	*See Figure 3-24 on page 3-20 for position report items.*

Figure 3-23. ATC Reporting Procedure Examples.

- If being radar vectored, the direct route from the point of radio failure to the fix, route, or airway specified in the radar vector clearance.

- In the absence of an assigned route, the route ATC has advised to expect in a further clearance.

- In the absence of an assigned or expected route, the route filed in the flight plan.

It is also important to fly a specific altitude should two-way radio communications be lost. The altitude to fly after a communication failure can be found in Part 91.185 and must be the highest of the following altitudes for each route segment flown.

- The altitude or flight level assigned in the last ATC clearance.

- The minimum altitude or flight level for IFR operations.

- The altitude or flight level ATC has advised to expect in a further clearance.

In some cases, the assigned or expected altitude may not be as high as the MEA on the next route segment. In this situation, pilots normally begin a climb to the higher MEA when they reach the fix where the MEA rises. If the fix also has a published minimum crossing altitude, they start the climb so they will be at or above the MCA when reaching the fix. If the next succeeding route segment has a lower MEA, descend to the applicable altitude — either the last assigned altitude or the altitude expected in a further clearance — when reaching the fix where the MEA decreases.

CLIMBING AND DESCENDING EN ROUTE

Before the days of nationwide radar coverage, en route aircraft were separated from each other primarily by specific altitude assignments and position reporting procedures. Much of the pilot's time was devoted to inflight calculations, revising ETAs, and relaying

Position Report Items	
Identification	*"Marathon 564,*
Position	*Sidney*
Time	*15, (minutes after the hour)*
Altitude/Flight Level	*9,000,*
IFR or VFR (in a report to an FSS only)	*IFR,*
ETA over the next reporting fix	*Akron 35, (minutes after the hour)*
Following reporting point	*Thurman next."*
Pertinent remarks	*(If necessary)*

Figure 3-24. Nonradar Position Reports.

position reports to ATC. Today, pilots and air traffic controllers have far more information and better tools to make inflight computations and, with the expansion of radar, including the use of an en route flight progress strip shown in Figure 3-26, position reports may only be necessary as a backup in case of radar failure or for RNAV random route navigation. Figure 3-26 also depicts the numerous en route data entries used on a flight progress strip, generated by the ARTCC computer. Climbing, level flight, and descending during the en route phase of IFR flight involves staying in communication with ATC, making necessary reports, responding to clearances,

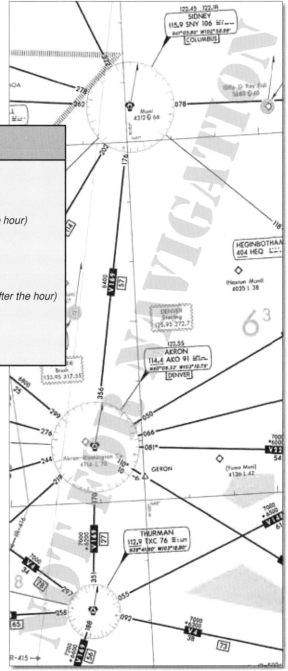

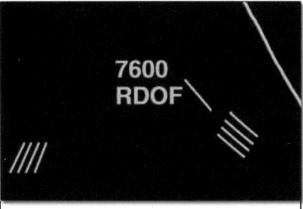

When an aircraft squawks code 7600 during a two-way radio communication failure, the information block on the radar screen flashes RDOF (radio failure) to alert the controller.

Figure 3-25. Two-Way Radio Communication Failure Transponder Code.

monitoring position, and staying abreast of any changes to the airplane's equipment status or weather.

PILOT/CONTROLLER EXPECTATIONS

When ATC issues a clearance or instruction, pilots are expected to execute its provisions upon receipt. In some cases, ATC includes words that modify their expectation. For example, the word "*immediately*" in a clearance or instruction is used to impress urgency to avoid an imminent situation, and expeditious compliance is expected and necessary for safety. The addition of a climb point or time restriction, for example, does not authorize pilots to deviate from the route of flight or any other provision of the ATC clearance. If you receive a term

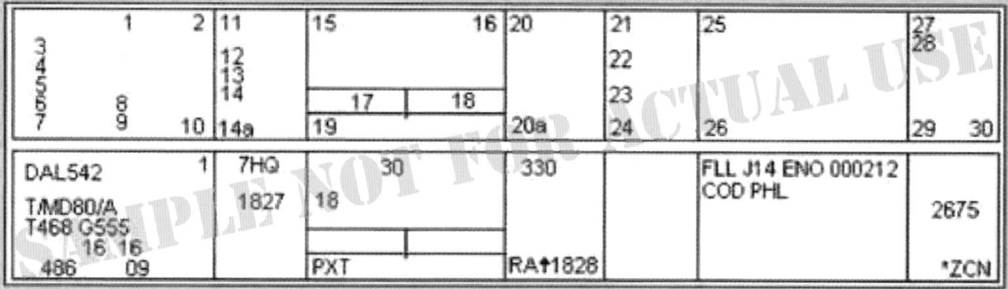
SAMPLE NOT FOR ACTUAL USE
DAL542
T/MD80/A
T468 G555
16 16
486 09
7HQ
1827
30
18
PXT
330
RA↑1828
FLL J14 ENO 000212
COD PHL
2675
*ZCN

Block	Information Recorded
1.	Verification symbol if required.
2.	Revision number. DSR - Not used.
3.	Number of aircraft if more than one, TCAS/ heavy aircraft indicator if appropriate, type of aircraft, and aircraft equipment suffix. The TCAS indicator is "T/" and the heavy aircraft indicator is "H/". For aircraft that are both TCAS and heavy, the indicator is "B/". For B757, the indicator is "F/" and for B757 with TCAS, the indicator is "L/".
4.	Number of aircraft if more than one, TCAS/ heavy aircraft indicator if appropriate, type of aircraft, and aircraft equipment suffix. The TCAS indicator is "T/" and the heavy aircraft indicator is "H/". For aircraft that are both TCAS and heavy, the indicator is "B/". For B757, the indicator is "F/" and for B757 with TCAS, the indicator is "L/".
5.	Filed true airspeed.
6.	Sector number.
7.	Computer identification number if required.
8.	Estimated ground speed.
9.	Revised ground speed or strip request (SR) originator.
10.	Strip number. DSR-Strip number/Revision number.
11.	Previous fix.
12.	Estimated time over previous fix.
13.	Revised estimated time over previous fix.
14.	Actual time over previous fix, or actual departure time entered on first fix posting after departure.
14a.	Plus time expressed in minutes from the previous fix to the posted fix.
15.	Center-estimated time over fix (in hours and minutes), or clearance information for departing aircraft.
16.	Arrows to indicate if aircraft is departing (↑)or arriving (↓).
17.	Pilot-estimated time over fix.
18.	Actual time over fix, time leaving holding fix, arrival time at nonapproach control airport, or symbol indicating cancellation of IFR flight plan for arriving aircraft, or departure time (actual or assumed).
19.	Fix. For departing aircraft, add proposed departure time.
20.	Altitude information (in hundreds of feet) or as noted below.
NOTE	*Altitude information may be written in thousands of feet provided the procedure is authorized by the facility manager, and is defined in a facility directive, i.e. FL 330 as 33, 5,000 feet as 5, and 2,800 as 2.8.*
20a.	**OPTIONAL USE,** when voice recorders are operational; **REQUIRED USE,** when the voice recorders are not operating and strips are being used at the facility. This space is used to record reported RA events. The letters RA followed by a climb or descent arrow (if the climb or descent action is reported) and the time (hhmm) the event is reported.
21.	Next posted fix or coordination fix.
22.	Pilot's estimated time over next fix.
23.	Arrows to indicate north (↑), south (↓), east (→), or west (←) direction of flight if required.
24.	Requested altitude.
NOTE	*Altitude information may be written in thousands of feet provided the procedure is authorized by the facility manager, and is defined in a facility directive, i.e.* *FL 330 as 33, 5000 feet as 5, and 2,800 as 2.8.*
25.	*Point of origin, route as required for control and data relay, and destination.*
26.	Pertinent remarks, minimum fuel, point out/radar vector/speed adjustment information or sector/position number (when applicable in accordance with para 2-2-1, RECORDING INFORMATION), or NRP.
27.	Mode 3/A beacon code if applicable.
28.	Miscellaneous control data (expected further clearance time, time cleared for approach, etc.).
29-30.	Transfer of control data and coordination indicators.

Figure 3-26. En Route Flight Progress Strip and Data Entries.

"*climb at pilot's discretion*" in the altitude information of an ATC clearance, it means that you have the option to start a climb when you wish, that you are authorized to climb at any rate, and to temporarily level off at any intermediate altitude as desired, although once you vacate an altitude, you may not return to that altitude. When ATC has not used the term "*at pilot's discretion*" nor imposed any climb restrictions, you should climb promptly on acknowledgment of the clearance. Climb at an optimum rate consistent with the operating characteristics of your aircraft to 1,000 feet below the assigned altitude, and then attempt to climb at a rate of between 500 and 1,500 feet per minute until you reach your assigned altitude. If at anytime you are unable to climb at a rate of at least 500 feet a minute, advise ATC. If it is necessary to level off at an intermediate altitude during climb, advise ATC.

"Expedite climb" normally indicates you should use the approximate best rate of climb without an exceptional change in aircraft handling characteristics. Normally controllers will inform you of the reason for an instruction to expedite. If you fly a turbojet airplane equipped with afterburner engines, such as a military aircraft, you should advise ATC prior to takeoff if you intend to use afterburning during your climb to the en route altitude. Often, the controller may be able to plan traffic to accommodate a high performance climb and allow you to climb to the planned altitude without restriction. If you receive an *"expedite"* clearance from ATC, and your altitude to maintain is subsequently changed or restated without an expedite instruction, the expedite instruction is canceled.

During en route climb, as in any other phase of flight, it is essential that you clearly communicate with ATC regarding clearances. In the following example, a flight crew experienced an apparent clearance readback/hearback error, that resulted in confusion about the clearance, and ultimately, to inadequate separation from another aircraft. "Departing IFR, clearance was to maintain 5,000 feet, expect 12,000 in ten minutes. After handoff to Center, we understood and read back, 'Leaving 5,000 turn left heading 240° for vector on course.' The First Officer turned to the assigned heading climbing through 5,000 feet. At 5,300 feet Center advised assigned altitude was 5,000 feet. We immediately descended to 5,000. Center then informed us we had traffic at 12 o'clock and a mile at 6,000. After passing traffic, a higher altitude was assigned and climb resumed. We now believe the clearance was probably 'reaching' 5,000, etc. Even our readback to the controller with 'leaving' didn't catch the different wording." "Reaching" and "leaving" are commonly used ATC terms having different usages. They may be used in clearances involving climbs, descents, turns, or speed changes. In the cockpit, the words "reaching" and "leaving" sound much alike.

For altitude awareness during climb, professional pilots often call out altitudes on the flight deck. The pilot monitoring may call 2,000 and 1,000 feet prior to reaching an assigned altitude. The callout may be, *"two to go"* and *"one to go."* Climbing through the transition altitude (QNH), both pilots set their altimeters to 29.92 inches of mercury and announce *"2992 inches"* (or *'standard,'* on some airplanes) and the flight level

passing. For example, *"2992 inches" ('standard'),* *flight level one eight zero."* The second officer on three pilot crews may ensure that both pilots have inserted the proper altimeter setting. On international flights, pilots must be prepared to differentiate, if necessary, between barometric pressure equivalents with inches of mercury, and millibars or hectopascals, to eliminate any potential for error, for example, 996 millibars erroneously being set as 2996.

For a typical IFR flight, the majority of inflight time often is flown in level flight at cruising altitude, from top of climb to **top of descent (TOD)**. Generally, TOD is used in airplanes with a flight management system (FMS), and represents the point at which descent is first initiated from cruise altitude. FMSs also assist in level flight by cruising at the most fuel saving speed, providing continuing guidance along the flight plan route, including great circle direct routes, and continuous evaluation and prediction of fuel consumption along with changing clearance data. Descent planning is discussed in more detail in the next chapter, "Arrivals."

AIRCRAFT SPEED AND ALTITUDE

During the en route descent phase of flight, an additional benefit of flight management systems is that the FMS provides fuel saving idle thrust descent to your destination airport. This allows an uninterrupted profile descent from level cruising altitude to an appropriate **minimum IFR altitude (MIA)**, except where level flight is required for speed adjustment. Controllers anticipate and plan that you may level off at 10,000 feet MSL on descent to comply with the Part 91 indicated airspeed limit of 250 knots. Leveling off at any other time on descent may seriously affect air traffic handling by ATC. It is imperative that you make every effort to fulfill ATC expected actions on descent to aid in safely handling and expediting air traffic.

ATC issues speed adjustments if you are being radar controlled to achieve or maintain required or desired spacing. They express speed adjustments in terms of knots based on indicated airspeed in 10 knot increments except that at or above FL 240 speeds may be expressed in terms of Mach numbers in 0.01 increments. The use of Mach numbers by ATC is restricted to turbojets. If complying with speed adjustments, pilots are expected to maintain that speed within plus or minus 10 knots or 0.02 Mach.

Speed and altitude restrictions in clearances are subject to misinterpretation, as evidenced in this case where a corporate flight crew treated instructions in a published procedure as a clearance. "…We were at FL 310 and had already programmed the 'expect-crossing altitude' of 17,000 feet at the VOR. When the altitude alerter sounded, I advised Center that we were leaving FL 310. ATC acknowledged with a 'Roger.' At FL 270, Center quizzed us about our descent. I told the controller we were descending so as to cross the VOR at 17,000 feet. ATC advised us that we did not have clearance to descend. What we thought was a clearance was in fact an 'expect' clearance. We are both experienced pilots…which just means that experience is no substitute for a direct question to Center when you are in doubt about a clearance. Also, the term 'Roger' only means that ATC received the transmission, not that they understood the transmission. The AIM indicates that 'expect' altitudes are published for planning purposes. 'Expect' altitudes are not considered crossing restrictions until verbally issued by ATC."

HOLDING PROCEDURES

The criteria for holding pattern airspace is developed both to provide separation of aircraft, as well as obstacle clearance The alignment of holding patterns typically coincides with the flight course you fly after leaving the holding fix. For level holding, a minimum of 1,000 feet obstacle clearance is provided throughout the primary area. In the secondary area 500 feet of obstacle clearance is provided at the inner edge, tapering to zero feet at the outer edge. Allowance for precipitous terrain is considered, and the altitudes selected for obstacle clearance may be rounded to the nearest 100 feet. When criteria for a climb in hold are applied, no obstacle penetrates the holding surface. [Figure 3-27]

There are many factors that affect aircraft during holding maneuvers, including navigational aid ground and airborne tolerance, effect of wind, flight procedures, application of air traffic control, outbound leg length, maximum holding airspeeds, fix to NAVAID distance, DME slant range effect, holding airspace size, and altitude holding levels. In order to allow for these factors when establishing holding patterns, procedure specialists must apply complex criteria contained in Order 7130.3, Holding Pattern Criteria.

ATC HOLDING INSTRUCTIONS

When controllers anticipate a delay at a clearance limit or fix, pilots will usually be issued a holding clearance at least five minutes before the ETA at the clearance limit or fix. If the holding pattern assigned by ATC is depicted on the appropriate aeronautical chart, pilots are expected to hold as published. In this situation, the controller will issue a holding clearance which includes the name of the fix, directs you to hold as published,

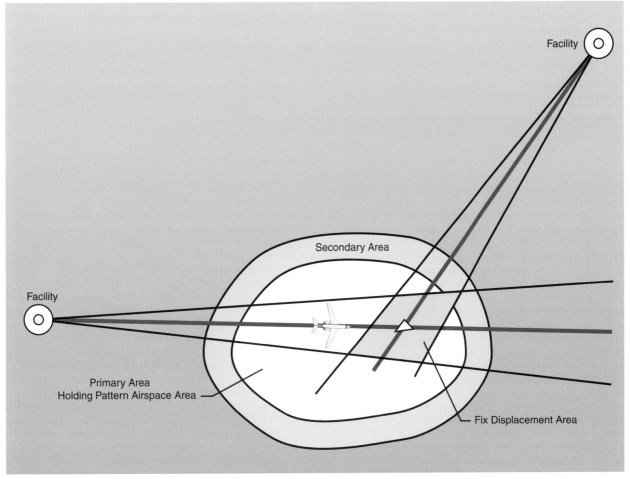

Figure 3-27. Typical Holding Pattern Design Criteria Template.

and includes an expect further clearance (EFC) time. An example of such a clearance is: *"Marathon five sixty four, hold east of MIKEY Intersection as published, expect further clearance at 1521."* When ATC issues a clearance requiring you to hold at a fix where a holding pattern is not charted, you will be issued complete holding instructions. This information includes the direction from the fix, name of the fix, course, leg length, if appropriate, direction of turns (if left turns are required), and the EFC time. You are required to maintain your last assigned altitude unless a new altitude is specifically included in the holding clearance, and you should fly right turns unless left turns are assigned. Note that all holding instructions should include an EFC time. If you lose two-way radio communication, the EFC allows you to depart the holding fix at a definite time. Plan the last lap of your holding pattern to leave the fix as close as possible to the exact time. [Figure 3-28]

If you are approaching your clearance limit and have not received holding instructions from ATC, you are expected to follow certain procedures. First, call ATC and request further clearance before you reach the fix. If you cannot obtain further clearance, you are expected to hold at the fix in compliance with the published holding pattern. If a holding pattern is not charted at

the fix, you are expected to hold on the inbound course using right turns. This procedure ensures that ATC will provide adequate separation. [Figure 3-29] Assume you are eastbound on V214 and the Cherrelyn VORTAC is your clearance limit. If you have not been able to obtain further clearance and have not received holding instructions, you should plan to hold southwest on the 221 degrees radial using left-hand turns, as depicted. If this holding pattern was not charted, you would hold west of the VOR on V214 using right-hand turns.

Where required for aircraft separation, ATC may request that you hold at any designated reporting point in a standard holding pattern at the MEA or the MRA, whichever altitude is the higher at locations where a minimum holding altitude has not been established. Unplanned holding at en route fixes may be expected on airway or route radials, bearings, or courses. If the fix is a facility, unplanned holding could be on any radial or bearing. There may be holding limitations required if standard holding cannot be accomplished at the MEA or MRA.

MAXIMUM HOLDING SPEED

As you have seen, the size of the holding pattern is directly proportional to the speed of the airplane. In order to limit the amount of airspace that must be protected by ATC, maximum holding speeds in knots

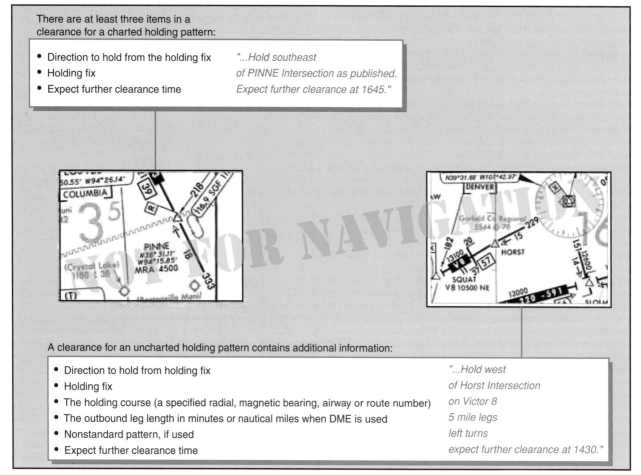

Figure 3-28. ATC Holding Instructions.

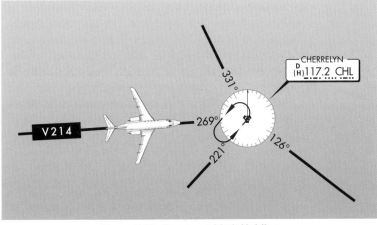

Figure 3-29. Clearance Limit Holding.

indicated airspeed (KIAS) have been designated for specific altitude ranges [Figure 3-30]. Even so, some holding patterns may have additional speed restrictions to keep faster airplanes from flying out of the protected area. If a holding pattern has a nonstandard speed restriction, it will be depicted by an icon with the limiting airspeed. If the holding speed limit is less than you feel is necessary, you should advise ATC of your revised holding speed. Also, if your indicated airspeed exceeds the applicable maximum holding speed, ATC expects you to slow to the speed limit within three minutes of your ETA at the holding fix. Often pilots can avoid flying a holding pattern, or reduce the length of time spent in the holding pattern, by slowing down on the way to the holding fix.

HIGH PERFORMANCE HOLDING

Certain limitations come into play when you operate at higher speeds; for instance, aircraft do not make standard rate turns in holding patterns if the bank angle will exceed 30 degrees. If your aircraft is using a flight director system, the bank angle is limited to 25 degrees. Since any aircraft must be traveling at over 210 knots TAS for the bank angle in a standard rate turn to exceed 30 degrees, this limit applies to relatively fast airplanes. An aircraft using a flight director would have to be holding at more than 170 knots TAS to come up against the 25 degrees limit. These true airspeeds correspond to indicated airspeeds of about 183 and 156 knots, respectively, at 6,000 feet in a standard atmosphere [Figure 3-31 on page 3-26]. Since some military airplanes

need to hold at higher speeds than the civilian limits, the maximum at military airfields is higher. For example, the maximum holding airspeed at USAF airfields is 310 KIAS.

FUEL STATE AWARENESS

In order to increase fuel state awareness, commercial operators and other professional flight crews are required to record the time and fuel remaining during IFR flight. For example, on a flight scheduled for one hour or less, the flight crew may record the time and fuel remaining at the **top of climb (TOC)** and at one additional waypoint listed in the flight plan. Generally, TOC is used in airplanes with a flight management system, and represents the point at which cruise altitude is first reached. TOC is calculated based on current airplane altitude, climb speed, and cruise altitude. The captain may elect to delete the additional waypoint recording requirement if the flight is so short that the record will not assist in the management of the flight. For flights scheduled for more than one hour, the flight crew may record the time and fuel remaining shortly after the top of climb and at selected waypoints listed in the flight plan, conveniently spaced approximately one hour apart. The flight crew compares actual fuel burn to planned fuel burn. Each fuel tank must be monitored to verify proper burn off and appropriate fuel remaining. On two pilot airplanes,

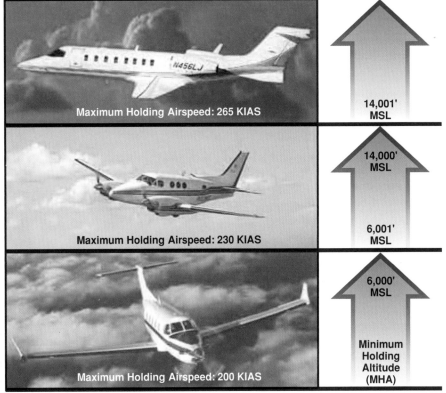

Figure 3-30. Maximum Holding Speed Examples.

Figure 3-31. High Performance Holding.

the pilot monitoring (PM) keeps the flight plan record. On three pilot airplanes, the second officer and PM coordinate recording and keeping the flight plan record. In all cases, the pilot making the recording communicates the information to the pilot flying.

DIVERSION PROCEDURES

Operations Specifications (OpsSpecs) for commercial operators include provisions for en route emergency diversion airport requirements. Operators are expected to develop a sufficient set of emergency diversion airports, such that one or more can be reasonably expected to be available in varying weather conditions. The flight must be able to make a safe landing, and the airplane maneuvered off of the runway at the selected diversion airport. In the event of a disabled airplane following landing, the capability to move the disabled airplane must exist so as not to block the operation of any recovery airplane. In addition, those airports designated for use must be capable of protecting the safety of all personnel by being able to:

- Offload the passengers and flight crew in a safe manner during possible adverse weather conditions.

- Provide for the physiological needs of the passengers and flight crew for the duration until safe evacuation.

- Be able to safely extract passengers and flight crew as soon as possible. Execution and completion of the recovery is expected within 12 to 48 hours following diversion.

Part 91 operators also need to be prepared for a diversion. Designation of an alternate in the IFR flight plan is a good first step; although, changing weather conditions or equipment issues may require pilots to consider other options.

EN ROUTE RNAV PROCEDURES

RNAV is a method of navigation that permits aircraft operations on any desired course within the coverage of station-referenced signals, or within the limits of self-contained system capability. The continued growth in aviation creates increasing demands on airspace capacity and emphasizes the need for optimum utilization of available airspace. These factors, allied with the requirement for NAS operational efficiency, along with the enhanced accuracy of current navigation systems, resulted in the required navigation performance (RNP) concept. RNAV is incorporated into RNP requirements.

OFF AIRWAY ROUTES

Part 95 prescribes altitudes governing the operation of your aircraft under IFR on Federal airways, jet routes, RNAV low or high altitude routes, and other direct routes for which an MEA is designated in this regulation. In addition, it designates mountainous areas and changeover points. **Off-airway routes** are established in the same manner, and in accordance with the same criteria as airways and jet routes. If you fly for a scheduled air carrier or operator for compensation or hire, any requests for the establishment of off-airway routes are initiated by your company through your **principal operations inspector (POI)** who works directly with your company and coordinates FAA approval. Air carrier authorized routes are contained in the company's OpsSpecs under the auspices of the air carrier operating certificate. [Figure 3-32]

Off-airway routes predicated on public navigation facilities and wholly contained within controlled airspace are published as direct Part 95 routes. Off-airway routes predicated on privately owned navigation facilities or not contained wholly within controlled airspace are published as off-airway non-Part 95 routes. In evaluating the adequacy of off-airway routes, the following items are considered; the type of aircraft and navigation systems used; proximity to military bases, training areas, low level military routes; and the adequacy of communications along the route. If you are a commercial operator, and you plan to fly off-airway routes, your OpsSpecs will likely address en route limitations and provisions regarding en route authorizations to use the global positioning system (GPS) or other RNAV systems in the NAS. Your POI must ensure that your long-range navigation program incorporates the required practices and procedures. These procedures must be in your manuals and in checklists, as appropriate. Training on the use of long range navigation equipment and procedures must be included in your training curriculums, and your minimum equipment lists (MELs) and maintenance programs must address the long range navigation equipment. Examples of other selected areas requiring specialized en route authorization include the following:

AUTHORIZED AREAS OF EN ROUTE OPERATION	LIMITATIONS, PROVISIONS, AND REFERENCE PARAGRAPHS
The 48 contiguous United States and the District of Columbia	Note 1
Canada, excluding Canadian MNPS airspace and the areas of magnetic unreliability as established in the Canadian AIP	Note 3

SPECIAL REQUIREMENTS:

Note 1 - B-737 Class II navigation operations with a single long-range system is authorized only within this area of en route operation.

Note 3 - Only B-747 and DC-10 operations authorized in these areas.

Figure 3-32. Excerpt of Authorized Areas of En Route Operation.

- Class I navigation in the U.S. Class A airspace using area or long range navigation systems.

- Class II navigation using multiple long range navigation systems.

- Operations in central East Pacific airspace.

- North Pacific operations.

- Operations within North Atlantic (NAT) minimum navigation performance specifications (MNPS) airspace.

- Operations in areas of magnetic unreliability.

- North Atlantic operation (NAT/OPS) with two engine airplanes under Part 121.

- Extended range operations (ER-OPS) with two engine airplanes under Part 121.

- Special fuel reserves in international operations.

- Planned inflight redispatch or rerelease en route.

- Extended over water operations using a single long-range communication system.

- Operations in reduced vertical separation minimum (RVSM) airspace.

DIRECT FLIGHTS

There are a number of ways to create shorter routes and fly off the airways. You can use NACO low and high altitude en route charts to create routes for direct flights, although many of the charts do not share the same scale as the adjacent chart, so a straight line is virtually impossible to use as a direct route for long distances. Generally speaking, NACO charts are plotted accurately enough to draw a direct route that can be flown. A straight line drawn on a NACO en route chart can be used to determine if a direct route will avoid airspace such as Class B airspace, restricted areas, prohibited areas, etc. Because NACO en route charts use the Lambert Conformal Conic projection, a straight line is as close as possible to a geodesic line (better than a great circle route). The closer that your route is to the two standard parallels of 33 degrees and 45 degrees on the chart, the better your straight line. There are cautions, however. Placing our round earth on a flat piece of paper causes distortions, particularly on long east-west routes. If your route is 180 degrees or 360 degrees, there is virtually no distortion in the course line.

About the only way you can confidently avoid protected airspace is by the use of some type of airborne database, including a graphic display of the airspace on the long-range navigation system moving map, for example. When not using an airborne database, leaving a few miles as a buffer helps ensure that you stay away from protected airspace.

In Figure 3-33 on page 3-28, a straight line on a magnetic course from SCRAN intersection of 270 degrees direct to the Fort Smith Regional Airport in Arkansas will pass just north of restricted area R-2401A and B, and R-2402. Since the airport and the restricted areas are precisely plotted, there is an assurance that you will stay north of the restricted areas. From a practical standpoint, it might be better to fly direct to the Wizer NDB. This route goes even further north of the restricted areas and places you over the final approach fix to Runway 25 at Fort Smith.

One of the most common means for you to fly direct routes is to use conventional navigation such as VORs. When flying direct off-airway routes, remember to apply the FAA distance limitations, based upon NAVAID service volume.

RANDOM RNAV ROUTES

Random RNAV routes may be an integral solution in meeting the worldwide demand for increased air traffic system capacity and safety. Random RNAV routes are direct routes, based on RNAV capability. They are typically flown between waypoints defined in terms of latitude and longitude coordinates, degree and distance fixes, or offsets from established routes and airways at a specified distance and direction. Radar monitoring by ATC is required on all random RNAV routes.

With IFR certified RNAV units (GPS or FMS), there are several questions to be answered, including "Should I fly airways or should I fly RNAV direct?" One of the considerations is the determination of the MIA. In most

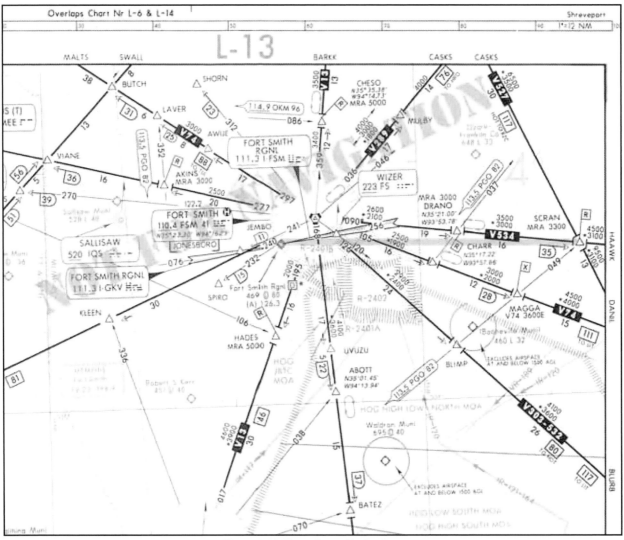

Figure 3-33. Direct Route Navigation.

places in the world at FL 180 and above, the MIA is not significant since you are well above any terrain or obstacles. On the other hand, a direct route at 18,000 feet from Salt Lake City, Utah to Denver, Colorado, means terrain and obstacles are very important. This RNAV direct route across the Rocky Mountains reduces your distance by about 17 NM, but radar coverage over the Rockies at lower altitudes is pretty spotty. This raises numerous questions. What will air traffic control allow on direct flights? What will they do if radar coverage is lost? What altitudes will they allow when they can't see you on radar? Do they have altitudes for direct routes? The easy answer is to file the airways, and then all the airway MIAs become usable. But with RNAV equipment, a direct route is more efficient. Even though on some routes the mileage difference may be negligible, there are many other cases where the difference in distance is significant. ATC is required to provide radar separation on random RNAV routes at FL 450 and below. It is logical to assume that ATC will clear you at an altitude that allows it to maintain radar contact along the entire route, which could mean spending additional time and fuel climbing to an altitude that gives full radar coverage.

All air route traffic control centers have MIAs for their areas of coverage. Although these altitudes are not published anywhere, they are available when airborne from ATC.

OFF ROUTE OBSTRUCTION CLEARANCE ALTITUDE

An **off-route obstruction clearance altitude (OROCA)** is an off-route altitude that provides obstruction clearance with a 1,000-foot buffer in non-mountainous terrain areas and a 2,000-foot buffer in designated mountainous areas within the U.S. This altitude may not provide signal coverage from ground-based navigational aids, air traffic control radar, or communications coverage. OROCAs are intended primarily as a pilot tool for emergencies and situational awareness. OROCAs depicted on NACO en route charts do not provide you with an acceptable altitude for terrain and obstruction clearance for the purposes of off-route, random RNAV direct flights in either controlled or uncontrolled airspace. OROCAs are not subject to the same scrutiny as MEAs, MVAs, MOCAs, and other minimum IFR altitudes. Since they do not undergo the same obstruction evaluation, airport airspace analysis procedures, or flight inspection, they cannot provide the same level of confidence as the other minimum IFR altitudes.

If you depart an airport VFR intending to or needing to obtain an IFR clearance en route, you must be aware of the position of your aircraft relative to terrain and obstructions. When accepting a clearance below the MEA, MIA, MVA, or the OROCA, you are responsible for your own terrain/obstruction clearance until reaching the MEA, MIA, or MVA. If you are unable to visually maintain terrain/obstruction clearance, you should advise the controller and state your intentions. [Figure 3-34]

For all random RNAV flights, there needs to be at least one waypoint in each ARTCC area through which you intend to fly. One of the biggest problems in creating an RNAV direct route is determining if the route goes through special use airspace. For most direct routes, the chances of going through prohibited, restricted,

or special use airspace are good. In the U.S., all direct routes should be planned to avoid prohibited or restricted airspace by at least 3 NM. If a bend in a direct route is required to avoid special use airspace, the turning point needs to be part of the flight plan. Two of the most prominent long range navigation systems today include FMS with integrated GPS and stand-alone GPS. The following example is a simplified overview showing how the RNAV systems might be used to fly a random RNAV route:

In Figure 3-35 on page 3-30, you are northeast of Tuba City VORTAC at FL 200 using RNAV (showing both GPS and FMS), RNAV direct on a southwesterly heading to Lindbergh Regional Airport in Winslow. As you monitor your position and cross check your avionics against the high altitude en route chart, you receive a company message instructing you to divert to Las Vegas, requiring a change in your flight plan as highlighted on the depicted chart excerpt.

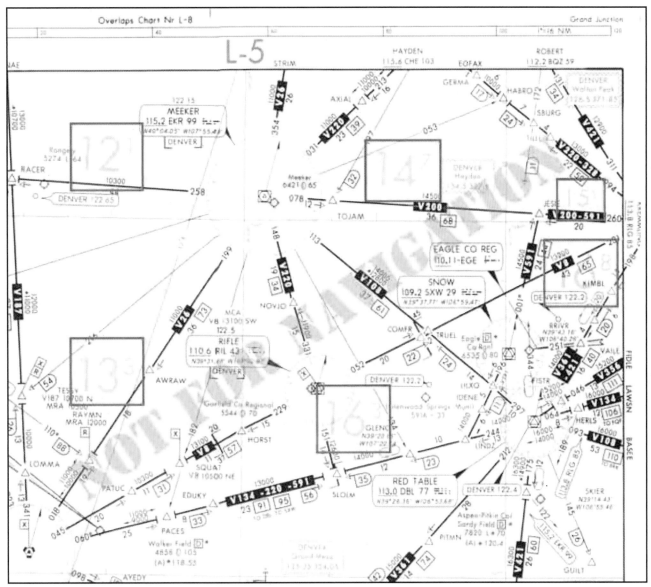

Figure 3-34. Off-Route Obstacle Clearance Altitude.

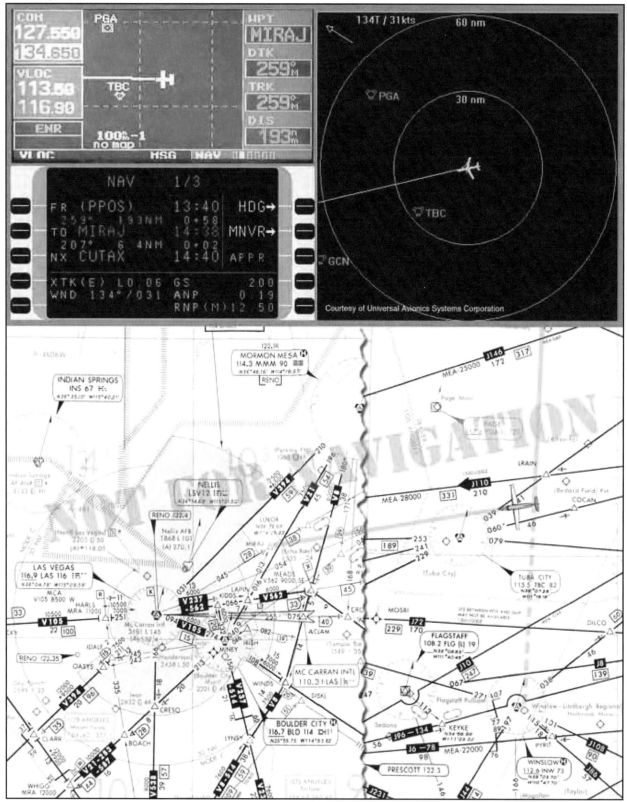

Figure 3-35. Random RNAV Route.

During your cockpit review of the high and low altitude en route charts, you determine that your best course of action is to fly direct to the MIRAJ waypoint, 28 DME northeast of the Las Vegas VORTAC on the 045° radial. This places you 193 NM out on a 259° magnetic course inbound, and may help you avoid diverting north,

allowing you to bypass the more distant originating and intermediate fixes feeding into Las Vegas. You request an RNAV random route clearance direct MIRAJ to expedite your flight. Denver Center comes back with the following amended flight plan and initial clearance into Las Vegas:

"Marathon five sixty four, turn right heading two six zero, descend and maintain one six thousand, cleared present position direct MIRAJ."

The latitude and longitude coordinates of your present position on the high altitude chart are N36 19.10, and W110 40.24 as you change your course. Notice your GPS moving map (upper left) and the FMS control display unit (below the GPS), and FMS map mode navigation displays (to the right of the GPS) as you reroute your flight to Las Vegas. For situational awareness, you note that your altitude is well above any of the OROCAs on your direct route as you arrive in the Las Vegas area using the low altitude chart.

PUBLISHED RNAV ROUTES

Although RNAV systems allow you to select any number of routes that may or may not be published on a chart, en route charts are still crucial and required for RNAV flight. They assist you with both flight planning and inflight navigation. NACO en route charts are very helpful in the context of your RNAV flights. Published RNAV routes are fixed, permanent routes that can be flight planned and flown by aircraft with RNAV capability. These are being expanded worldwide as new RNAV routes are developed, and existing charted, conventional routes are being designated for RNAV use. It is important to be alert to the rapidly changing application of RNAV techniques being applied to conventional en route airways. Published RNAV routes may potentially be found on any NACO en route chart. The published RNAV route designation may be obvious, or, on the other hand, RNAV route designations may be less obvious, as in the case where a published route shares a common flight track with a conventional airway. Note: Since the use of RNAV is dynamic and rapidly changing, NACO en route charts are continuously being updated for information changes and you may find some differences between charts.

According to the **International Civil Aviation Organization (ICAO)**, who develops standard principles and techniques for international air navigation, basic designators for **air traffic service (ATS)** routes and their use in voice communications have been established in Annex 11. ATS is a generic ICAO term for flight information service, alerting service, air traffic advisory service, and air traffic control service. One of the main purposes of a system of route designators is to allow both pilots and ATC to make unambiguous reference to RNAV airways and routes. Many countries have adopted ICAO recommendations with regard to ATS route designations. Basic designators for ATS routes consist of a maximum of five, and in no case exceed six, alpha/numeric characters in order to be usable by both ground and airborne automation systems. The designator indicates the type of the route such as high/low altitude, specific airborne navigation equipment requirements such as RNAV, and the aircraft type using the route primarily and exclusively. The basic route designator consists of one or two letter(s) followed by a number from 1 to 999.

COMPOSITION OF DESIGNATORS

The prefix letters that pertain specifically to RNAV designations are included in the following list:

1. The basic designator consists of one letter of the alphabet followed by a number from 1 to 999. The letters may be:

    a) A, B, G, R — for routes that form part of the regional networks of ATS routes and are not RNAV routes;

    b) L, M, N, P — for RNAV routes that form part of the regional networks of ATS routes;

    c) H, J, V, W — for routes that do not form part of the regional networks of ATS routes and are not RNAV routes;

    d) Q, T, Y, Z — for RNAV routes that do not form part of the regional networks of ATS routes.

2. Where applicable, one supplementary letter must be added as a prefix to the basic designator as follows:

    a) K — to indicate a low level route established for use primarily by helicopters.

    b) U — to indicate that the route or portion thereof is established in the upper airspace;

    c) S — to indicate a route established exclusively for use by supersonic airplanes during acceleration/deceleration and while in supersonic flight.

3. Where applicable, a supplementary letter may be added after the basic designator of the ATS route as a suffix as follows:

    a) F — to indicate that on the route or portion thereof advisory service only is provided;

    b) G — to indicate that on the route or portion thereof flight information service only is provided;

    c) Y — for RNP 1 routes at and above FL 200 to indicate that all turns on the route between 30° and 90° must be made within the tolerance of a tangential arc between the straight leg segments defined with a radius of 22.5 NM.

d) Z — for RNP 1 routes at and below FL 190 to indicate that all turns on the route between 30° and 90° shall be made within the tolerance of a tangential arc between the straight leg segments defined with a radius of 15 NM.

USE OF DESIGNATORS IN COMMUNICATIONS

In voice communications, the basic letter of a designator should be spoken in accordance with the ICAO spelling alphabet. Where the prefixes K, U or S, specified in 2., above, are used in voice communications, they should be pronounced as:

K = "*Kopter*"   U = "*Upper*"   S = "*Supersonic*"

as in the English language.

Where suffixes "F", "G", "Y" or "Z" specified in 3., above, are used, the flight crew should not be required to use them in voice communications.

Example:

A11   will be spoken *Alfa Eleven*

UR5   will be spoken *Upper Romeo Five*

KB34   will be spoken *Kopter Bravo Thirty Four*

UW456 F   will be spoken *Upper Whiskey Four Fifty Six*

Figure 3-36 depicts published RNAV routes in the Gulf of Mexico (black Q100, Q102, and Q105) that have been added to straighten out the flight segments and provide an alternative method of navigation to the LF airway (brown G26), that has since been terminated in this case. The "Q" designation is derived from the list of basic route designators previously covered, and correlates with the description for RNAV routes that do not form part of the regional networks of ATS routes. Notice the indirect reference to the RNAV requirement, with the note, "Navigational Equipment Other than LF or VHF Required."

Notice in Figure 3-37 that this en route chart excerpt depicts three published RNAV jet routes, J804R, J888R, and J996R. The "R" suffix is a supplementary route designator denoting an RNAV route. The overlapping symbols for the AMOTT intersection and waypoint indicate that AMOTT can be identified by conventional navigation or by latitude and longitude coordinates. Although coordinates were originally included for aircraft equipped with INS systems, they are now a good way to cross check between the coordinates on the chart and in the FMS or GPS databases to ensure you are tracking on your intended en route course. The AMOTT RNAV waypoint includes bearing and distance from the ANCHORAGE VORTAC. In an effort to simplify the conversion to RNAV, some controlling agencies

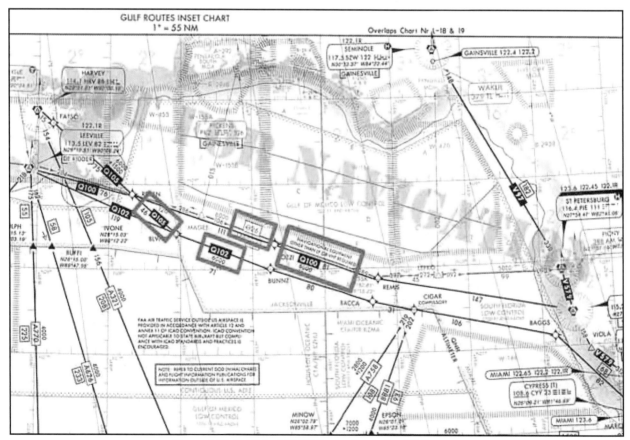

Figure 3-36. Published RNAV Routes Replacing LF Airways.

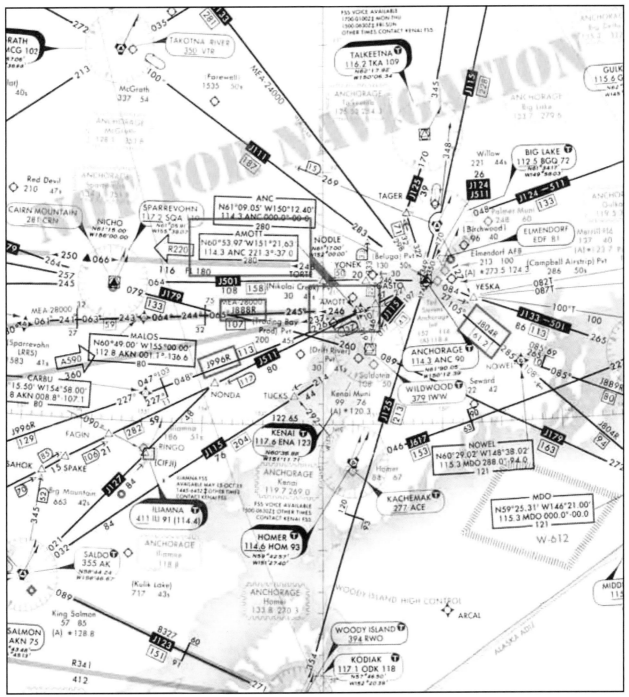

Figure 3-37. Published RNAV Jet Routes.

outside the U.S. have simply designated all conventional routes as RNAV routes at a certain flight level.

RNAV MINIMUM EN ROUTE ALTITUDE
RNAV MEAs are depicted on some NACO IFR en route charts, allowing both RNAV and non-RNAV pilots to use the same chart for instrument navigation.

MINIMUM IFR ALTITUDE
The **Minimum IFR altitude (MIA)** for operations is prescribed in Part 91. These MIAs are published on

NACO charts and prescribed in Part 95 for airways and routes, and in Part 97 for standard instrument approach procedures. If no applicable minimum altitude is prescribed in Parts 95 or 97, the following MIA applies: In designated mountainous areas, 2,000 feet above the highest obstacle within a horizontal distance of 4 NM from the course to be flown; or other than mountainous areas, 1,000 feet above the highest obstacle within a horizontal distance of 4 NM from the course to be flown; or as otherwise authorized by the Administrator or assigned by ATC. MIAs are not flight checked for communication.

3-33

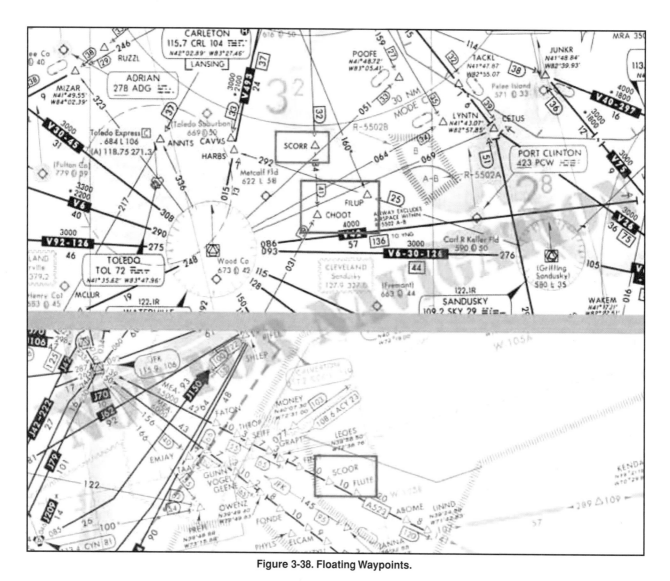

Figure 3-38. Floating Waypoints.

WAYPOINTS

Waypoints are specified geographical locations, or fixes, used to define an RNAV route or the flight path of an aircraft employing RNAV. Waypoints may be any of the following types: predefined, published waypoints, floating waypoints, or user-defined waypoints. Predefined, published waypoints are defined relative to VOR-DME or VORTAC stations or, as with GPS, in terms of latitude/longitude coordinates.

USER-DEFINED WAYPOINTS

Pilots typically create **user-defined waypoints** for use in their own random RNAV direct navigation. They are newly established, unpublished airspace fixes that are designated geographic locations/positions that help provide positive course guidance for navigation and a means of checking progress on a flight. They may or may not be actually plotted by the pilot on en route charts, but would normally be communicated to ATC in terms of bearing and distance or latitude/longitude. An example of user-defined waypoints typically includes those derived from database RNAV systems whereby latitude/longitude coordinate-based waypoints are gen-

erated by various means including keyboard input, and even electronic map mode functions used to establish waypoints with a cursor on the display. Another example is an offset phantom waypoint, which is a point-in-space formed by a bearing and distance from NAVAIDs, such as VORTACs and tactical air navigation (TACAN) stations, using a variety of navigation systems. When specifying unpublished waypoints in a flight plan, they can be communicated using the frequency/bearing/distance format or latitude and longitude, and they automatically become compulsory reporting points unless otherwise advised by ATC. All airplanes with latitude and longitude navigation systems flying above FL 390 must use latitude and longitude to define turning points.

FLOATING WAYPOINTS

Floating waypoints, or reporting points, represent airspace fixes at a point in space not directly associated with a conventional airway. In many cases, they may be established for such purposes as ATC metering fixes, holding points, RNAV-direct routing, gateway waypoints, STAR origination points leaving the en route structure, and SID terminating points joining the en route structure. In Figure 3-38, in the top example, a NACO low altitude en route chart depicts three floating waypoints that have been highlighted, SCORR, FILUP, and CHOOT. Notice that waypoints are named with five-letter identifiers that are unique and pronounceable. Pilots must be careful of similar waypoint names. Notice on the high altitude en route chart excerpt in the bottom example, the similar sounding and spelled floating waypoint named SCOOR, rather than SCORR. This emphasizes the importance of correctly entering waypoints into database-driven navigation systems. One waypoint character incorrectly entered into your navigation system could adversely affect your flight. The SCOOR floating reporting point also is depicted on a Severe Weather Avoidance Plan (SWAP) en route chart. These waypoints and SWAP routes assist pilots and controllers when severe weather affects the East Coast.

COMPUTER NAVIGATION FIXES
An integral part of RNAV using en route charts typically involves the use of airborne navigation databases. Database identifiers are depicted on NACO en route charts enclosed in parentheses, for example AWIZO waypoint, shown in Figure 3-39. These identifiers, sometimes referred to as computer navigation fixes (CNFs), have no ATC function and should not be used in filing flight plans nor should they be used when communicating with ATC. Database identifiers on en route charts are shown

only to enable you to maintain orientation as you use charts in conjunction with database navigation systems, including RNAV.

Many of the RNAV systems available today make it all too easy to forget that en route charts are still required and necessary for flight. As important as databases are, they really are onboard the airplane to provide navigation guidance and situational awareness; they are not intended as a substitute for paper charts. When flying with GPS, FMS, or planning a flight with a computer, it is important to understand the limitations of the system you are using, for example, incomplete information, uncodeable procedures, complex procedures, and database storage limitations. For more information on databases, refer to Appendix A, Airborne Navigation Database.

HIGH ALTITUDE AIRSPACE REDESIGN
Historically in the U.S., IFR flights have navigated along a system of Federal Airways that require pilots to fly directly toward or away from ground-based navigation aids. RNAV gives users the capability to fly direct routes between any two points, offering far more flexible and efficient en route operations in the high-altitude airspace environment. As part of the ongoing National Airspace Redesign (NAR), the FAA has implemented the High Altitude Redesign (HAR) program with the goal of obtaining maximum system efficiency by introducing advanced RNAV routes for suitably equipped aircraft to use.

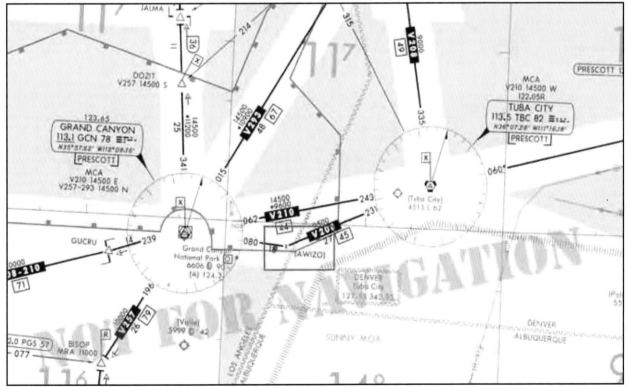

Figure 3-39. Computer Navigation Fix.

Q-ROUTES

Naturally, the routes between some points are very popular, so these paths are given route designators and published on charts. The U.S. and Canada use "Q" as a designator for RNAV routes. Q-Routes 1 through 499 are allocated to the U.S., while Canada is allocated Q-Routes numbered from 500 through 999. The first Q-Routes were published in 2003. One benefit of this system is that aircraft with RNAV or RNP capability can fly safely along closely spaced parallel flight paths on high-density routes, which eases airspace congestion. While the initial overall HAR implementation will be at FL390 and above, some of the features may be used at lower altitudes, and some Q-Routes may be used as low as FL180. A Q-Route is shown in figure 3-40.

NON-RESTRICTIVE ROUTING

HAR also includes provisions for pilots to choose their own routes, unconstrained by either conventional airways or Q-Routes. This non-restrictive routing (NRR) allows pilots of RNAV-equipped aircraft to plan the most advantageous route for the flight. There are two ways to designate an NRR route on your flight plan. One method, point-to-point (PTP), uses the traditional fixes in the aircraft equipment database and is shown by placing "PTP" in the first part of the "Remarks" block of the flight plan. For aircraft that have the additional waypoints of the Navigation Reference System (NRS) in their databases, "HAR" is placed in the first part of the "Remarks" block.

NAVIGATION REFERENCE SYSTEM

The NRS is a grid of waypoints overlying the U.S. that will be the basis for flight plan filing and operations in the redesigned high altitude environment. It will provide increased flexibility to aircraft operators and controllers. The NRS supports flight planning in a NRR environment and provides ATC with the ability to more efficiently manage tactical route changes for aircraft separation, traffic flow management, and weather avoidance. It provides navigation reference waypoints that pilots can use in requesting route deviations around weather areas, which will improve common understand-

ing between pilots and ATC of the desired flight path. The NRS will initially include waypoints every 30 minutes of latitude and every two degrees of longitude. In its final version, the NRS waypoints will have a grid resolution of 1-degree longitude by 10 minutes of latitude. As database capabilities for the preponderance of aircraft operating in the high altitude airspace environment becomes adequate to support more dense NRS resolution, additional NRS waypoints will be established.

T-ROUTES

T-Routes are being created for those who operate at lower altitudes. T-Routes have characteristics that are similar to Q-Routes, but they are depicted on low altitude en route charts and are intended for flights below FL180. The first T-Routes are being pioneered in Alaska.

IFR TRANSITION ROUTES

In order to expedite the handling of IFR overflight traffic through Charlotte Approach Control Airspace, several RNAV routes are published in the *Airport/Facility Directory* and available for you when filing your flight plan. Any RNAV capable aircraft filing flight plan equipment codes of /E, /F, or /G may file for these routes. Other aircraft may request vectors along these routes but should only expect vector routes

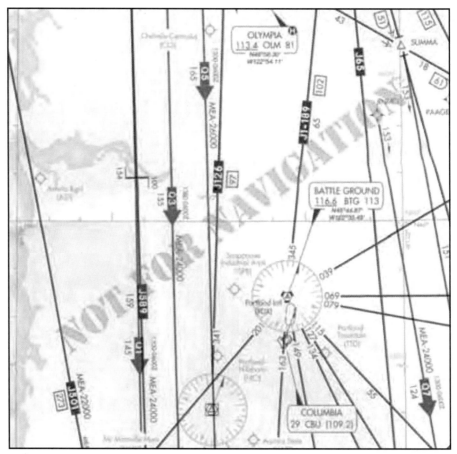

Figure 3-40. Q-Route

as workload permits. Altitudes are assigned by ATC based upon traffic. [Figure 3-41]

IFR transition routes through Class B airspace for general aviation aircraft en route to distant destinations are highly desirable. Since general aviation aircraft cruise at altitudes below the ceiling of most Class B airspace areas, access to that airspace for en route transition reduces cost and time, and is helpful to pilots in their flight planning. Establishing RNAV fixes could facilitate the implementation of IFR transition routes, although every effort should be made to design routes that can be flown with RNAV or VOR equipment. IFR transition routes are beneficial even if access is not available at certain times because of arriving or departing traffic saturation at the primary airport. For these locations, information can be published to advise pilots when IFR transition access is not available.

REQUIRED NAVIGATION PERFORMANCE

As RNAV systems grow in sophistication, high technology FMS and GPS avionics are gaining popularity as NDBs, VORs, and LORAN are being phased out. As a result, new procedures are being introduced, including RNP, RVSM, and **minimum navigation performance specifications (MNPS)**. ICAO defines an RNP "X" specification as requiring on-board performance monitoring and alerting. Even such terms as **gross navigation errors (GNEs)** are being introduced into the navigation equation. If you commit a GNE in the North Atlantic oceanic region of more than 25 NM laterally or 300 feet vertically, it has a detrimental effect on the overall targeted level of safety of the ATC airspace system in this region. This applies to commercial operators, as well as Part 91 operators, all of whom must be knowledgeable on procedures for operations in North Atlantic airspace, contained in the *North Atlantic MNPS Operations Manual*.

RNP types are identified by a single accuracy value. For example, RNP 1 refers to a required navigation performance accuracy within 1 NM of the desired flight path at least 95 percent of the flying time. Countries around the world are establishing required navigation performance values. For Federal Airways in the U.S. that extend 4 NM from either side of the airway centerline, the airway has an equivalent RNP of 2. Figure 3-42 on page 3-38 shows ICAO RNP containment parameters, including reference to lateral and longitudinal total system errors (TSEs).

RNP requires you to learn new procedures, communications, and limitations; and to learn new terminology that defines and describes navigation concepts. One of

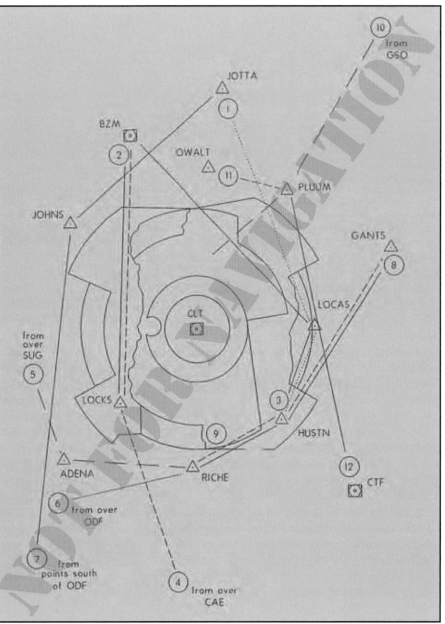

Figure 3-41. IFR Transition Routes in the *Airport/Facility Directory*.

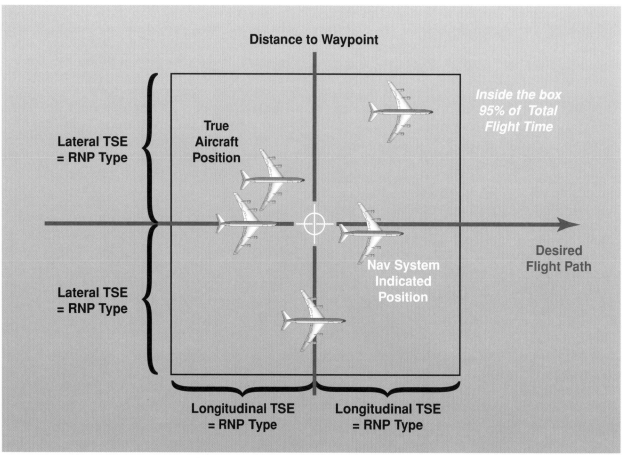

Figure 3-42. ICAO RNP Containment Parameters.

these terms is RNP Airspace, a generic term designating airspace, routes, legs, operations, or procedures where minimum RNP has been established. P-RNAV represents a 95 percent containment value of ±1 NM. B-RNAV provides a 95 percent containment value of ±5 NM. RNP is a function of RNAV equipment that calculates, displays, and provides lateral guidance to a profile or path. Estimated position error (EPE) is a measure of your current estimated navigational performance, also referred to as actual navigation performance (ANP).

RNP RNAV is an industry-expanded specification beyond ICAO-defined RNP. Some of the benefits of RNP RNAV includes being an aid in both separation and collision risk assessment. RNP RNAV can further reduce route separation. Figure 3-43 depicts route separation, that can now be reduced to four times the RNP value, which further increases route capacity within the same airspace. The containment limit quantifies the navigation performance where the probability of an unannunciated deviation greater than 2 x RNP is less than 1 x 10^{-5}. This means that the pilot will be alerted when the TSE can be greater than the containment limit. Figure 3-44 shows the U.S. RNP RNAV levels by airspace control regions, including RNP 2 for the en

route phase of flight, and Figure 3-45 on page 3-40 illustrates the U.S. standard RNP (95%) levels.

REDUCED VERTICAL SEPARATION MINIMUMS

In 1960, the minimum vertical separation between airplanes above FL 290 was officially increased to 2,000 feet. This was necessary because of the relatively large errors in barometric altimeters at high altitudes. Since that time, increased air traffic worldwide has begun to approach (and sometimes exceed) the capacity of the most popular high-altitude routes. Likewise, very accurate altitude determination by satellite positioning systems makes it possible to change the minimum vertical separation for properly equipped airplanes back to the pre-1960 standard of 1,000 feet. [Figure 3-46 on page 3-41] RVSM airspace is any airspace between FL 290 and FL 410 inclusive, where airplanes are separated by 1,000 feet vertically. In the early 1980's, programs were established to study the concept of **reduced vertical separation minimums (RVSM)**. RVSM was found to be technically feasible without imposing unreasonable requirements on equipment. RVSM is the most effective way to increase airspace capacity to cope with traffic growth. Most of the preferred international and domestic flight routes are under both RVSM and RNP RNAV rules.

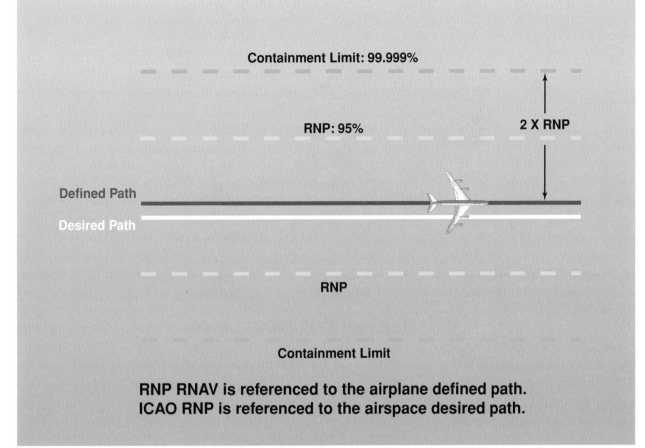

Figure 3-43. RNP RNAV Containment.

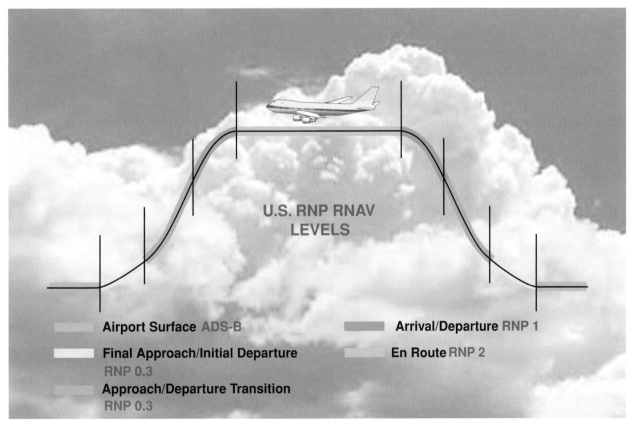

Figure 3-44. Airspace Control Regions.

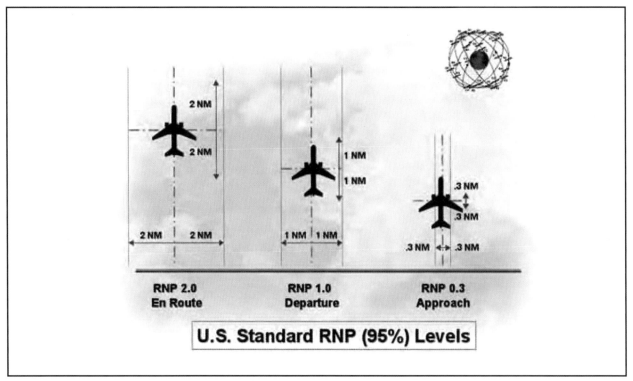

Figure 3-45. U.S. Standard RNP Levels.

In 1997, the first RVSM 1,000-foot separation was implemented between FL 330 and FL 370 over the North Atlantic. In 1998, RVSM was expanded to include altitudes from FL 310 to FL 390. Today States (governments) around the globe are implementing RVSM from FL 290 to FL 410. There are many requirements for operator approval of RVSM. Each aircraft must be in compliance with specific RVSM criteria. A program must be in place to assure continued airworthiness of all RVSM critical systems. Flight crews, dispatchers, and flight operations must be properly trained, and operational procedures, checklists, etc. must be established and published in the Ops Manual and AFM, plus operators must participate in a height monitoring program.

Using the appropriate suffix in Block 3 on the IFR flight plan lets ATC know that your flight conforms to the necessary standards and is capable of using RNP routes or flying in RVSM airspace. The equipment codes changed significantly in 2005 and are shown in Figure 3-47.

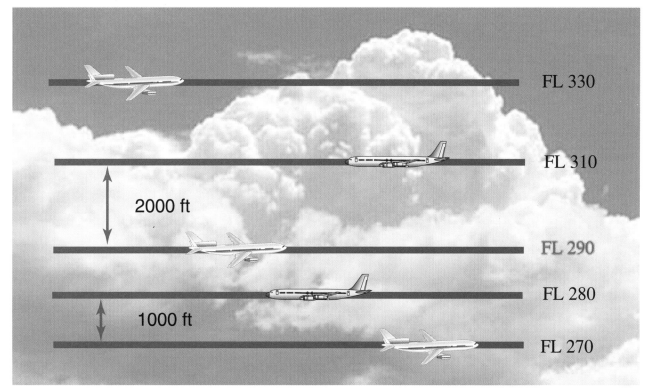

Figure 3-46. Prior to implementation of RVSM, all traffic above FL290 required vertical separation of 2,000 feet.

	No DME	DME	TACAN only	Area Navigation (RNAV) LORAN, VOR/DME, or INS
No transponder	/X	/D	/M	/Y
Transponder without Mode C	/T	/B	/N	/C
Transponder with Mode C	/U	/A	/P	/ I

Advanced RNAV with transponder and Mode C (If an aircraft is unable to operate with a transponder and/or Mode C, it will revert to the appropriate code listed above under Area Navigation.)	With RVSM
/J	/E
/K	/F
/L	/G
/Q	/R
/W	RVSM

Figure 3-47. When filed in your IFR flight plan, these codes inform ATC about your aircraft navigation capability.

Preparation for the arrival and approach begins long before the descent from the en route phase of flight. Planning early, while there are fewer demands on your attention, leaves you free to concentrate on precise control of the aircraft and better equipped to deal with problems that might arise during the last segment of the flight.

TRANSITION FROM EN ROUTE

This chapter focuses on the current procedures pilots and air traffic control (ATC) use for instrument flight rule (IFR) arrivals in the National Airspace System (NAS). The objective is to provide pilots with an understanding of ATC arrival procedures and pilot responsibilities as they relate to the transition between the en route and approach phases of flight. This chapter emphasizes standard terminal arrival routes (STARs), descent clearances, descent planning, and ATC procedures, while the scope of coverage focuses on transitioning from the en route phase of flight, typically the origination point of a STAR to the STAR termination fix. This chapter also differentiates between area navigation (RNAV) STARs and STARs based on conventional navigational aids (NAVAIDs).

Optimum IFR arrival options include flying directly from the en route structure to an approach gate or initial approach fix (IAF), a visual arrival, STARs, and radar vectors. Within controlled airspace, ATC routinely uses radar vectors for separation purposes, noise abatement considerations, when it is an operational advantage, or when requested by pilots. Vectors outside of controlled airspace are provided only on pilot request. The controller tells you the purpose of the vector when the vector is controller-initiated and takes the aircraft off a previously assigned nonradar route. Typically, when operating on RNAV routes, you are allowed to remain on your own navigation.

TOP OF DESCENT

Planning the descent from cruise is important because of the need to dissipate altitude and airspeed in order to arrive at the approach gate properly configured. Descending early results in more flight at low altitudes with increased fuel consumption, and starting down late results in problems controlling both airspeed and descent rates on the approach. Top of descent (TOD) from the en route phase of flight for high performance airplanes is often used in this process and is calculated manually or automatically through a flight management system (FMS) [Figure 4-1], based upon the altitude of

Top of
Descent

PROGRESS   2 / 3
SPD / ALT   CMD   VS@TOD
240 / 3000   2400↓
TOC   FUEL QTY
151.5NM / 00 + 23   20000
TOD   GROSS WT
1022NM / 02 + 17   62850
◁ AIR DATA   FLT SUM ▷

Figure 4-1. Top of Descent and FMS Display.

the approach gate. The approach gate is an imaginary point used by ATC to vector aircraft to the final approach course. The approach gate is established along the final approach course 1 nautical mile (NM) from the final approach fix (FAF) on the side away from the airport and is located no closer than 5 NM from the landing threshold. The altitude of the approach gate or initial approach fix is subtracted from the cruise altitude, and then the target rate of descent and groundspeed is applied, resulting in a time and distance for TOD, as depicted in Figure 4-1 on page 4-1.

Achieving an optimum stabilized, constant rate descent during the arrival phase requires different procedures for turbine-powered and reciprocating-engine airplanes. Controlling the airspeed and rate of descent is important for a stabilized arrival and approach, and it also results in minimum time and fuel consumption. Reciprocating-engine airplanes require engine performance and temperature management for maximum engine longevity, especially for turbocharged engines. Pilots of turbine-powered airplanes must not exceed the airplane's maximum operating limit speed above 10,000 feet, or exceed the 250-knot limit below 10,000 feet. Also, consideration must be given to turbulence that may be encountered at lower altitudes that may necessitate slowing to the turbulence penetration speed. If necessary, speed brakes should be used.

DESCENT PLANNING

Prior to flight, calculate the fuel, time, and distance required to descend from your cruising altitude to the approach gate altitude for the specific instrument approach of your destination airport. In order to plan your descent, you need to know your cruise altitude, approach gate altitude or initial approach fix altitude, descent groundspeed, and descent rate. Update this information while in flight for changes in altitude, weather, and wind. Your flight manual or operating handbook may also contain a fuel, time, and distance to descend chart that contains the same information. The calculations should be made before the flight and "rules of thumb" updates should be applied in flight. For example, from the charted STAR you might plan a descent based on an expected clearance to "cross 40 DME West of Brown VOR at 6,000" and then apply rules of thumb for slowing down from 250 knots. These might include planning your airspeed at 25 NM from the runway threshold to be 250 knots, 200 knots at 20 NM, and 150 knots at 15 NM until gear and flap speeds are reached, never to fall below approach speed.

The need to plan the IFR descent into the approach gate and airport environment during the preflight planning stage of flight is particularly important for turbojet powered airplanes. A general rule of thumb for initial IFR descent planning in jets is the 3 to 1 formula. This means that it takes 3 NM to descend 1,000 feet. If an airplane is

at flight level (FL) 310 and the approach gate or initial approach fix is at 6,000 feet, the initial descent requirement equals 25,000 feet (31,000 - 6,000). Multiplying 25 times 3 equals 75; therefore begin descent 75 NM from the approach gate, based on a normal jet airplane, idle thrust, speed Mach 0.74 to 0.78, and vertical speed of 1,800 - 2,200 feet per minute. For a tailwind adjustment, add 2 NM for each 10 knots of tailwind. For a headwind adjustment, subtract 2 NM for each 10 knots of headwind. During the descent planning stage, try to determine which runway is in use at the destination airport, either by reading the latest aviation routine weather report (METAR) or checking the automatic terminal information service (ATIS) information. There can be big differences in distances depending on the active runway and STAR. The objective is to determine the most economical point for descent.

An example of a typical jet descent-planning chart is depicted in Figure 4-2. Item 1 is the pressure altitude from which the descent begins; item 2 is the time required for the descent in minutes; item 3 is the amount of fuel consumed in pounds during descent to sea level; and item 4 is the distance covered in NM. Item 5 shows that the chart is based on a Mach .80 airspeed until 280 knots indicated airspeed (KIAS) is obtained. The 250-knot airspeed limitation below 10,000 feet mean sea level (MSL) is not included on the chart, since its effect is minimal. Also, the effect of temperature or weight variation is negligible and is therefore omitted.

Press Alt - 1000 Ft	Time - Min	Fuel - Lbs	Dist - NAM
39	20	850	124
37	19	800	112
35	18	700	101
33	17	650	92
31	16	600	86
29	15	600	80
27	14	550	74
25	13	550	68
23	12	500	63
21	11	500	58
19	10	450	52
17	10	450	46
15	9	400	41
10	6	300	26
5	3	150	13

Note: Subtract 30 lb. of fuel and 36 seconds for each 1,000 feet that the destination airport is above sea level.

Figure 4-2. Typical Air Carrier Descent Planning Chart

Due to the increased cockpit workload, you want to get as much done ahead of time as possible. As with the climb and cruise phases of flight, you should consult the proper performance charts to compute your fuel requirements as well as the time and distance needed for your descent. Figure 4-3 is an example of a descent-planning chart. If you are descending from 17,000 feet to a final (approach gate) altitude of 5,650, your time to descend is 11 minutes and distance to descend is 40 NM.

During the cruise and descent phases of flight, you need to monitor and manage the airplane according to the appropriate manufacturer's recommendations. The flight manuals and operating handbooks contain cruise and descent checklists, performance charts for specific cruise configurations, and descent charts that provide information regarding the fuel, time, and distance required to descend. Review this information prior to

the departure of every flight so you have an understanding of how your airplane is supposed to perform at cruise and during descent. A stabilized descent constitutes a pre-planned maneuver in which the power is properly set, and minimum control input is required to maintain the appropriate descent path. Excessive corrections or control inputs indicate the descent was improperly planned. Plan your IFR descent from cruising altitude so you arrive at the approach gate altitude or initial approach fix altitude prior to beginning the instrument approach. [Figure 4-4 on page 4-4]

Descending from cruise altitude and entering the approach environment can be a busy time during the flight. You are talking on the radio, changing radio frequencies, pulling out different charts, adjusting controls,

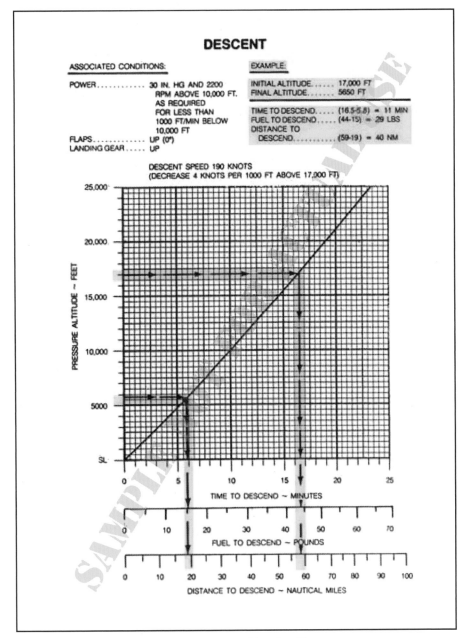

Figure 4-3. Descent Planning Chart.

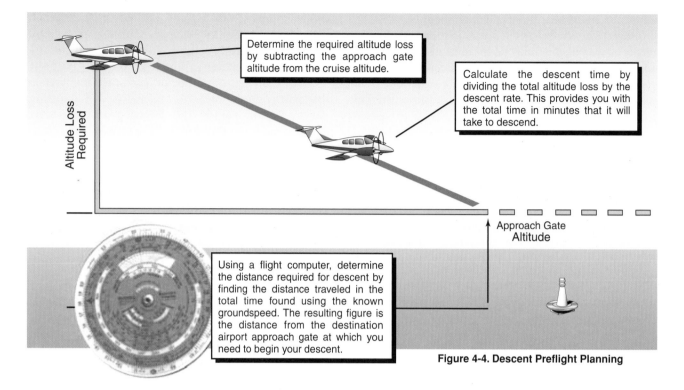

Determine the required altitude loss by subtracting the approach gate altitude from the cruise altitude.

Calculate the descent time by dividing the total altitude loss by the descent rate. This provides you with the total time in minutes that it will take to descend.

Altitude Loss Required

Approach Gate Altitude

Using a flight computer, determine the distance required for descent by finding the distance traveled in the total time found using the known groundspeed. The resulting figure is the distance from the destination airport approach gate at which you need to begin your descent.

Figure 4-4. Descent Preflight Planning

reading checklists, all of which can be distracting. By planning your descent in advance, you reduce the workload required during this phase of flight, which is smart workload management. Pilots often stay as high as they can as long as they can, so planning the descent prior to arriving at the approach gate is necessary to achieve a stabilized descent, and increases situational awareness. Using the information given, calculate the distance needed to descend to the approach gate.

- Cruise Altitude: 17,000 feet MSL

- Approach Gate Altitude: 2,100 feet MSL

- Descent Rate: 1,500 feet per minute

- Descent Groundspeed: 155 knots

Subtract 2,100 feet from 17,000 feet, which equals 14,900 feet. Divide this number by 1,500 feet per minute, which equals 9.9 minutes, round this off to 10 minutes. Using your flight computer, find the distance required for the descent by using the time of 10 minutes and the groundspeed of 155 knots. This gives you a distance of 25.8 NM. You need to begin your descent approximately 26 NM prior to arriving at your destination airport approach gate.

CRUISE CLEARANCE
The term "cruise" may be used instead of "maintain" to assign a block of airspace to an aircraft. The block extends from the minimum IFR altitude up to and including the altitude that is specified in the cruise clearance. On a cruise clearance, you may level off at any intermediate altitude within this block of airspace. You are allowed to climb or descend within the block at your own discretion. However, once you start descent and verbally report leaving an altitude in the block to ATC, you may not return to that altitude without an additional ATC clearance. A cruise clearance also authorizes you to execute an approach at the destination airport. When operating in uncontrolled airspace on a cruise clearance, you are responsible for determining the minimum IFR altitude. In addition, your descent and landing at an airport in uncontrolled airspace are governed by the applicable visual flight rules (VFR) and/or Operations Specifications (OpsSpecs), i.e., CFR, 91.126, 91.155, 91.175, 91.179, etc.

HOLDING PATTERNS
If you reach a clearance limit before receiving a further clearance from ATC, a holding pattern is required at your last assigned altitude. Controllers assign holds for a variety of reasons, including deteriorating weather or high traffic volume. Holding might also be required following a missed approach. Since flying outside the area set aside for a holding pattern could lead to an encounter with terrain or other aircraft, you need to understand the size of the protected airspace that a holding pattern provides.

Each holding pattern has a fix, a direction to hold from the fix, and an airway, bearing, course, radial, or route on which the aircraft is to hold. These elements, along with the direction of the turns, define the holding pattern.

Since the speed of the aircraft affects the size of a holding pattern, maximum holding airspeeds have been designated to limit the amount of airspace that must be protected. The three airspeed limits are shown in Figure 3-31 in Chapter 3 of this book. Some holding patterns have additional airspeed restrictions to keep faster airplanes from flying out of the protected area. These are depicted on charts by using an icon and the limiting airspeed.

Distance-measuring equipment (DME) and IFR-certified global positioning system (GPS) equipment offer some additional options for holding. Rather than being based on time, the leg lengths for DME/GPS holding patterns are based on distances in nautical miles. These patterns use the same entry and holding procedures as conventional holding patterns. The controller or the instrument approach procedure chart will specify the length of the outbound leg. The end of the outbound leg is determined by the DME or the along track distance (ATD) readout. The holding fix on conventional procedures, or controller-defined holding based on a conventional navigation aid with DME, is a specified course or radial and distances are from the DME station for both the inbound and outbound ends of the holding pattern. When flying published GPS overlay or standalone procedures with distance specified, the holding fix is a waypoint in the database and the end of the outbound leg is determined by the ATD. Instead of using the end of the outbound leg, some FMSs are programmed to cue the inbound turn so that the inbound leg length will match the charted outbound leg length.

Normally, the difference is negligible, but in high winds, this can enlarge the size of the holding pattern. Be sure you understand your aircraft's FMS holding program to ensure that the holding entry procedures and leg lengths match the holding pattern. Some situations may require pilot intervention in order to stay within protected airspace. [Figure 4-5]

DESCENDING FROM THE EN ROUTE ALTITUDE

As you near your destination, ATC issues a descent clearance so that you arrive in approach control's airspace at an appropriate altitude. In general, ATC issues either of two basic kinds of descent clearances.

- ATC may ask you to descend to and maintain a specific altitude. Generally, this clearance is for en route traffic separation purposes, and you need to respond to it promptly. Descend at the optimum rate for your aircraft until 1,000 feet above the assigned altitude, then descend at a rate between 500 and 1,500 feet per minute (FPM) to the assigned altitude. If at any time, other than when slowing to 250 KIAS at 10,000 feet MSL, you cannot descend at a rate of at least 500 FPM, advise ATC.

- The second type of clearance allows you to descend "... *at pilot's discretion.*" When ATC

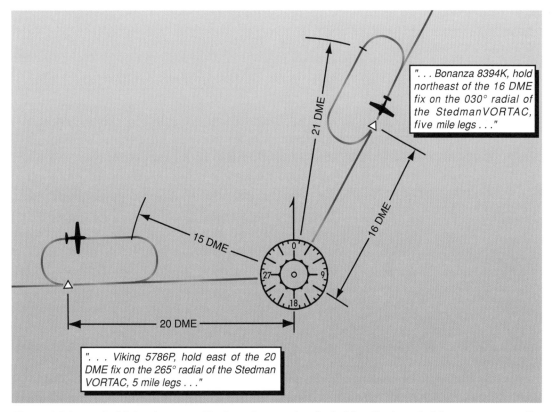

Figure 4-5. Instead of flying for a specific time after passing the holding fix, these holding patterns use distances to mark where the turns are made. The distances come from DME or IFR-certified GPS equipment.

issues a clearance to descend at pilot's discretion, you may begin the descent whenever you choose and at any rate you choose. You also are authorized to level off, temporarily, at any intermediate altitude during the descent. However, once you leave an altitude, you may not return to it.

A descent clearance may also include a segment where the descent is at your discretion—such as *"cross the Joliet VOR at or above 12,000, descend and maintain 5,000."* This clearance authorizes you to descend from your current altitude whenever you choose, as long as you cross the Joliet VOR at or above 12,000 feet MSL. After that, you should descend at a normal rate until you reach the assigned altitude of 5,000 feet MSL.

Clearances to descend at pilot's discretion are not just an option for ATC. You may also request this type of clearance so that you can operate more efficiently. For example, if you are en route above an overcast layer, you might ask for a descent at your discretion to allow you to remain above the clouds for as long as possible. This might be particularly important if the atmosphere is conducive to icing and your aircraft's icing protection is limited. Your request permits you to stay at your cruising altitude longer to conserve fuel or to avoid prolonged IFR flight in icing conditions. This type of descent can also help to minimize the time spent in turbulence by allowing you to level off at an altitude where the air is smoother.

APPROACH CLEARANCE

The approach clearance provides guidance to a position from where you can execute the approach, and it also clears you to fly that approach. If only one approach procedure exists, or if ATC authorizes you to execute the approach procedure of your choice, the clearance may be worded as simply as *"... cleared for approach."* If ATC wants to restrict you to a specific approach, the controller names the approach in the clearance—for example, *"...cleared ILS Runway 35 Right approach."*

When the landing will be made on a runway that is not aligned with the approach being flown, the controller may issue a circling approach clearance, such as *"...cleared for VOR Runway 17 approach, circle to land Runway 23."*

When cleared for an approach prior to reaching a holding fix, ATC expects the pilot to continue to the holding fix, along the feeder route associated with the fix, and then to the IAF. If a feeder route to an IAF begins at a fix located along the route of flight prior to reaching the holding fix, and clearance for an approach is issued, the pilot should commence the approach via the published feeder route. The pilot is expected to commence the approach in a similar manner at the IAF, if the IAF is located along the route to the holding fix.

ATC also may clear an aircraft directly to the IAF by using language such as *"direct"* or *"proceed direct."* Controllers normally identify an approach by its published name, even if some component of the approach aid (such as the glide slope of an ILS) is inoperative or unreliable. The controller uses the name of the approach as published but advises the aircraft when issuing the approach clearance that the component is unusable.

PRESENT POSITION DIRECT

In addition to using National Aeronautical Charting Office (NACO) high and low altitude en route charts as resources for your arrival, NACO area charts can be helpful as a planning aid for situational awareness. Many pilots find the area chart helpful in locating a depicted fix after ATC clears them to proceed to a fix and hold, especially at unfamiliar airports.

Looking at Figures 4-6, and 4-7 on page 4-8, assume you are V295 northbound en route to Palm Beach International Airport. You are en route on the airway when the controller clears you present position direct to the outer marker compass locator and for the instrument landing system (ILS) approach. There is no transition authorized or charted between your present position and the approach facility. There is no minimum altitude published for the route you are about to travel.

In Figure 4-6, you are just north of HEATT Intersection at 5,000 feet when the approach controller states, *"Citation 9724J, 2 miles from HEATT, cleared present position direct RUBIN, cleared for the Palm Beach ILS Runway 9L Approach, contact Palm Beach Tower on 119.1 established inbound."* With no minimum altitude published from that point to the RUBIN beacon, you should maintain the last assigned altitude until you reach the IAF (that's the fix, not the facility). Then, in Figure 4-7 on page 4-8, after passing the beacon outbound, commence your descent to 2,000 feet for the course reversal.

The ILS procedure relies heavily on the controller's recognition of the restriction upon you to maintain your last assigned altitude until "established" on a published segment of the approach. Refer to Appendix B, "Staying Within Protected Airspace," for a comprehensive discussion of "established." Prior to issuing a clearance for the approach, the controller usually assigns the pilot an altitude compatible with glide slope intercept.

RADAR VECTORS TO FINAL APPROACH COURSE

Arriving aircraft usually are vectored to intercept the final approach course, except with vectors for a visual approach, at least 2 NM outside the approach gate unless one of the following exists:

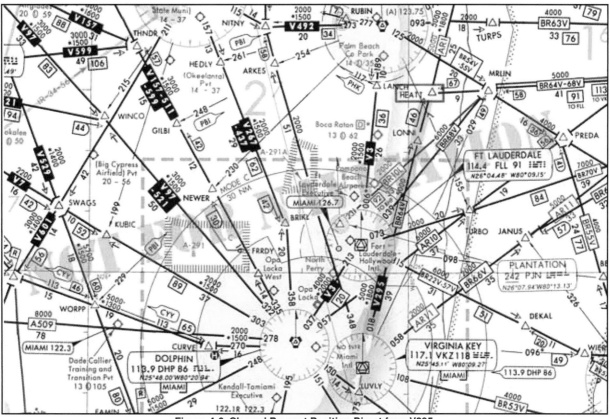

Figure 4-6. Cleared Present Position Direct from V295.

1. When the reported ceiling is at least 500 feet above the minimum vectoring altitude or minimum IFR altitude and the visibility is at least 3 NM (report may be a pilot report if no weather is reported for the airport), aircraft may be vectored to intercept the final approach course closer than 2 NM outside the approach gate but no closer than the approach gate.

2. If specifically requested by a pilot, ATC may vector aircraft to intercept the final approach course inside the approach gate but no closer than the FAF.

For a precision approach, aircraft are vectored at an altitude that is not above the glide slope/glidepath or below the minimum glide slope intercept altitude specified on the approach procedure chart. For a nonprecision approach, aircraft are vectored at an altitude that allows descent in accordance with the published procedure.

When a vector will take the aircraft across the final approach course, pilots are informed by ATC and the reason for the action is stated. In the event that ATC is not able to inform the aircraft, the pilot is not expected to turn inbound on the final approach course unless an approach clearance has been issued. An example of ATC phraseology in this case is, *"...expect vectors across final for spacing."*

The following ATC arrival instructions are issued to an IFR aircraft before it reaches the approach gate:

1. Position relative to a fix on the final approach course. If none is portrayed on the controller's radar display or if none is prescribed in the instrument approach procedure, ATC issues position information relative to the airport or relative to the navigation aid that provides final approach guidance.

2. Vector to intercept the final approach course if required.

3. Approach clearance except when conducting a radar approach. ATC issues the approach clearance only after the aircraft is established on a segment of a published route or instrument approach procedure, or in the following examples as depicted in Figure 4-8 on page 4-9.

Aircraft 1 was vectored to the final approach course but clearance was withheld. It is now at 4,000 feet and established on a segment of the instrument approach procedure. *"Seven miles from X-RAY. Cleared ILS runway three six approach."*

Aircraft 2 is being vectored to a published segment of the final approach course, 4 NM from LIMA at 2,000 feet. The minimum vectoring altitude for this area is 2,000 feet. *"Four miles from LIMA. Turn right heading three four zero. Maintain two thousand until established on the localizer. Cleared ILS runway three six approach."*

There are many times when it is desirable to position an aircraft onto the final approach course prior to a published, charted segment of an instrument approach procedure (IAP). Sometimes IAPs have no initial segment and require vectors. "RADAR REQUIRED" will be charted in the planview. Sometimes a route will intersect an extended final approach course making a long intercept desirable.

When ATC issues a vector or clearance to the final approach course beyond the published segment, controllers assign an altitude to maintain until the aircraft is established on a segment of a published route or IAP. This ensures that both the pilot and controller know precisely what altitude is to be flown and precisely where descent to appropriate minimum altitudes or step-down altitudes can begin.

Most aircraft are vectored onto a localizer or final approach course between an intermediate fix and the approach gate. These aircraft normally are told to maintain an altitude until established on a segment of the approach.

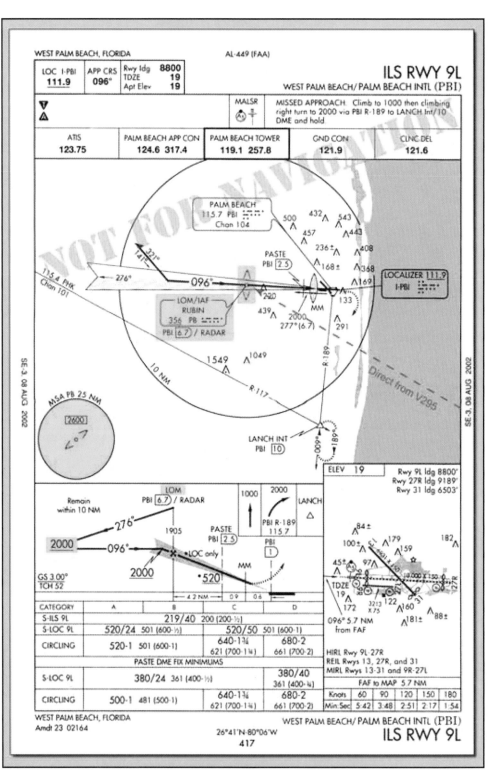

Figure 4-7. Cleared for the Palm Beach ILS Approach.

When an aircraft is assigned a route that will establish the aircraft on a published segment of an approach, the controller must issue an altitude to maintain until the aircraft is established on a published segment of the approach.

Aircraft 4 is established on the final approach course beyond the approach segments, 8 NM from Alpha at 6,000 feet. The minimum vectoring altitude for this area

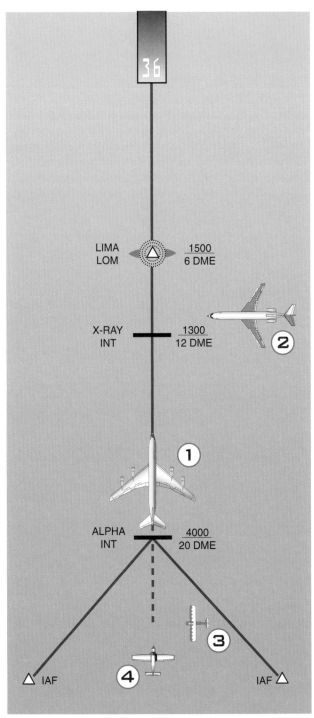

Figure 4-8. Arrival Instructions When Established.

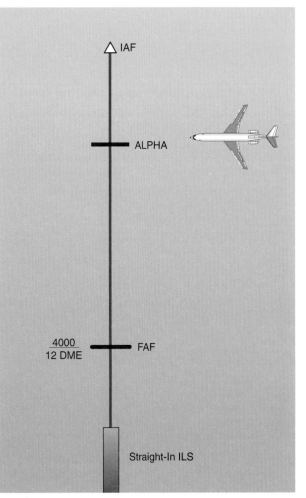

Figure 4-9. Arrival Instructions When Not Established.

is 4,000 feet. *"Eight miles from Alpha. Cross Alpha at or above four thousand. Cleared ILS runway three six approach."*

If an aircraft is not established on a segment of a published approach and is not conducting a radar approach, ATC will assign an altitude to maintain until the aircraft is established on a segment of a published route or instrument approach procedure, as depicted in Figure 4-9.

The aircraft is being vectored to a published segment of the ILS final approach course, 3 NM from Alpha at

4,000 feet. The minimum vectoring altitude for this area is 4,000 feet. *"Three miles from Alpha. Turn left heading two one zero. Maintain four thousand until established on the localizer. Cleared ILS runway one eight approach."*

The ATC assigned altitude ensures IFR obstruction clearance from the point at which the approach clearance is issued until established on a segment of a published route or instrument approach procedure.

ATC tries to make frequency changes prior to passing the FAF, although when radar is used to establish the FAF, ATC informs the pilot to contact the tower on the local control frequency after being advised that the aircraft is over the fix. For example, *"Three miles from final approach fix. Turn left heading zero one zero. Maintain two thousand until established on the localizer. Cleared ILS runway three six approach. I will advise when over the fix."*

"Over final approach fix. Contact tower one one eight point one."

Where a terminal arrival area (TAA) has been established to support RNAV approaches, as depicted in

Figure 4-10, ATC informs the aircraft of its position relative to the appropriate IAF and issues the approach clearance, as shown in the following examples:

Aircraft 1 is in the straight-in area of the TAA. *"Seven miles from CENTR, Cleared RNAV Runway One Eight Approach."*

Aircraft 2 is in the left base area of the TAA. *"Fifteen miles from LEFTT, Cleared RNAV Runway One Eight Approach."*

Aircraft 3 is in the right base area of the TAA. *"Four miles from WRITE, Cleared RNAV Runway One Eight Approach."*

IFR en route descent procedures should include a review of minimum, maximum, mandatory, and recommended altitudes that normally precede the fix or NAVAID facility to which they apply. The initial descent gradient for a low altitude instrument approach procedure does not exceed 500 feet per NM (approximately 5 degrees), and for a high altitude approach, the maximum

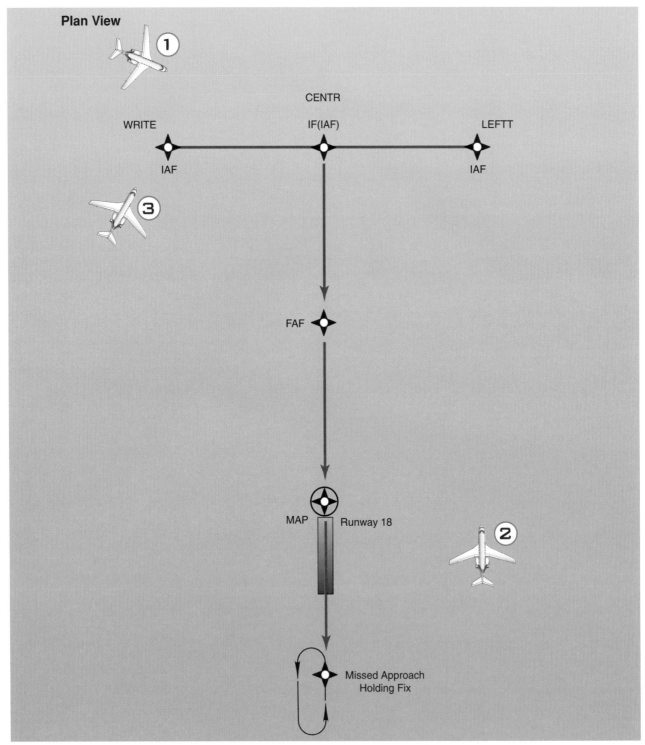

Figure 4-10. Basic "T" Design Terminal Arrival Area.

allowable initial gradient is 1,000 feet per NM (approximately 10 degrees).

Remember during arrivals, when cleared for an instrument approach, maintain the last assigned altitude until you are established on a published segment of the approach, or on a segment of a published route. If no altitude is assigned with the approach clearance and you are already on a published segment, you can descend to its minimum altitude.

HIGH PERFORMANCE AIRPLANE ARRIVALS

Procedures are established for the control of IFR high performance airplane arrivals, and are generally applied regardless of air traffic activity or time of day. This includes all turbojets and turboprops over 12,500 pounds. These procedures reduce fuel consumption and minimize the time spent at low altitudes. The primary objective is to ensure turbine-powered airplanes remain at the highest possible altitude as long as possible within reasonable operating limits and consistent with noise abatement policies.

AIRSPEED

During the arrival, expect to make adjustments in speed at the controller's request. When you fly a high-performance airplane on an IFR flight plan, ATC may ask you to adjust your airspeed to achieve proper traffic sequencing and separation. This also reduces the amount of radar vectoring required in the terminal area. When operating a reciprocating engine or turboprop airplane within 20 NM from your destination airport, 150 knots is usually the slowest airspeed you will be assigned. If your aircraft cannot maintain the assigned airspeed, you must advise ATC. Controllers may ask you to maintain the same speed as the aircraft ahead of or behind you on the approach. You are expected to maintain the specified airspeed ±10 knots. At other times, ATC may ask you to increase or decrease your speed by 10 knots, or multiples thereof. When the speed adjustment is no longer needed, ATC will advise you to *"...resume normal speed."* Keep in mind that the maximum speeds specified in Title 14 of the Code of Federal Regulations (14 CFR) Part 91.117 still apply during speed adjustments. It is your responsibility, as pilot in command, to advise ATC if an assigned speed adjustment would cause you to exceed these limits. For operations in Class C or D airspace at or below 2,500 feet above ground level (AGL), within 4 NM of the primary airport, ATC has the authority to request or approve a faster speed than those prescribed in Part 91.117.

Pilots operating at or above 10,000 feet MSL on an assigned speed adjustment that is greater than 250 KIAS are expected to reduce speed to 250 KIAS to comply with Part 91.117(a) when cleared below 10,000 feet MSL, within domestic airspace. This speed adjustment is made without notifying ATC. Pilots are expected to comply with the other provisions of Part 91.117 without notifying ATC. For example, it is normal for faster aircraft to level off at 10,000 feet MSL while slowing to the 250 KIAS limit that applies below that altitude, and to level off at 2,500 feet above airport elevation to slow to the 200 KIAS limit that applies within the surface limits of Class C or D airspace. Controllers anticipate this action and plan accordingly.

Speed restrictions of 250 knots do not apply to aircraft operating beyond 12 NM from the coastline within the United States (U.S.) Flight Information Region, in offshore Class E airspace below 10,000 feet MSL. In airspace underlying a Class B airspace area designated for an airport, pilots are expected to comply with the 200 KIAS limit specified in Part 91.117(c). (See Parts 91.117(c) and 91.703.)

Approach clearances cancel any previously assigned speed adjustment. Pilots are expected to make speed adjustments to complete the approach unless the adjustments are restated. Pilots complying with speed adjustment instructions should maintain a speed within plus or minus 10 knots or 0.02 Mach number of the specified speed.

Although standardization of these procedures for terminal locations is subject to local considerations, specific criteria apply in developing new or revised arrival procedures. Normally, high performance airplanes enter the terminal area at or above 10,000 feet above the airport elevation and begin their descent 30 to 40 NM from touchdown on the landing runway. Unless pilots indicate an operational need for a lower altitude, descent below 5,000 feet above the airport elevation is typically limited to the descent area where final descent and glide slope intercept can be made without exceeding specific obstacle clearance and other related arrival, approach, and landing criteria. Your descent should not be interrupted by controllers just to ensure that you cross the boundaries of the descent area at precisely 5,000 feet above the airport elevation. A typical descent area is shown in Figure 4-11 on page 4-12.

Arrival delays typically are absorbed at a metering fix. This fix is established on a route prior to the terminal airspace, 10,000 feet or more above the airport elevation. The metering fix facilitates profile descents, rather than controllers using delaying vectors or a holding pattern at low altitudes. Descent restrictions normally are applied prior to reaching the final approach phase to preclude relatively high descent rates close in to the destination airport. At least 10 NM from initial descent from 10,000 feet above the airport elevation, the controller issues an advisory that details when to expect to commence the descent. ATC typically uses the phraseology, *"Expect descent in (number) miles."* If cleared for a visual or contact approach, ATC usually restricts

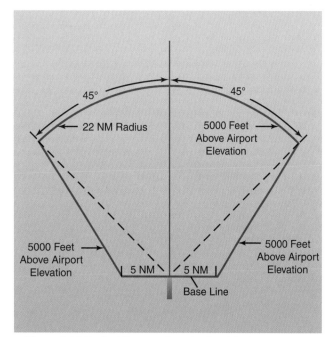

Figure 4-11. Typical Descent Area for Straight-In Approach.

you to at least 5,000 feet above the airport elevation until entering the descent area. Standard ATC phraseology is, *"Maintain (altitude) until (specified point; e.g., abeam landing runway end), cleared for visual approach or expect visual or contact approach clearance in (number of miles, minutes or specified point)."*

Once the determination is made regarding the instrument approach and landing runway you will use, with its associated descent area, ATC will not permit a change to another navigational aid that is not aligned with the landing runway. When altitude restrictions are required for separation purposes, ATC avoids assigning an altitude below 5,000 above the airport elevation.

There are numerous exceptions to the high performance airplane arrival procedures previously outlined. For example, in a nonradar environment, the controller may clear the flight to use an approach based on a NAVAID other than the one aligned with the landing runway, such as a circling approach. In this case, the descent to a lower altitude usually is limited to the descent area with the circle-to-land maneuver confined to the traffic pattern. Also in a nonradar environment, contact approaches may be approved from 5,000 above the airport elevation while the flight is within a descent area, regardless of landing direction.

Descent areas are established for all straight-in instrument approach procedures at an airport and may be established for runways not served by an instrument approach procedure to accommodate visual and contact approaches. More than one runway (descent area) may be used simultaneously for arriving high performance airplanes if there is an operational advantage for the pilot

or ATC, provided that the descent area serves the runway of intended landing.

CONTROLLED FLIGHT INTO TERRAIN

Inappropriate descent planning and execution during arrivals has been a contributing factor to many fatal aircraft accidents. Since the beginning of commercial jet operations, more than 9,000 people have died worldwide because of controlled flight into terrain (CFIT). CFIT is described as an event in which a normally functioning aircraft is inadvertently flown into the ground, water, or an obstacle. Of all CFIT accidents, 7.2 percent occurred during the descent phase of flight.

The basic causes of CFIT accidents involve poor flight crew situational awareness. One definition of situational awareness is an accurate perception by pilots of the factors and conditions currently affecting the safe operation of the aircraft and the crew. The causes of CFIT are the flight crews' lack of vertical position awareness or their lack of horizontal position awareness in relation to the ground, water, or an obstacle. More than two-thirds of all CFIT accidents are the result of an altitude error or lack of vertical situational awareness. CFIT accidents most often occur during reduced visibility associated with instrument meteorological conditions (IMC), darkness, or a combination of both.

The inability of controllers and pilots to properly communicate has been a factor in many CFIT accidents. Heavy workloads can lead to hurried communication and the use of abbreviated or non-standard phraseology. The importance of good communication during the arrival phase of flight was made evident in a report by an air traffic controller and the flight crew of an MD-80. The controller reported that he was scanning his radarscope for traffic and noticed that the MD-80 was descending through 6,400 feet. He immediately instructed a climb to at least 6,500 feet. The pilot responded that he had been cleared to 5,000 feet and then climbed to... The pilot reported that he had "heard" a clearance to 5,000 feet and read back 5,000 feet to the controller and received no correction from the controller. After almost simultaneous ground proximity warning system (GPWS) and controller warnings, the pilot climbed and avoided the terrain. The recording of the radio transmissions confirmed that the airplane was cleared to 7,000 feet and the pilot mistakenly read back 5,000 feet then attempted to descend to 5,000 feet. The pilot stated in the report: "I don't know how much clearance from the mountains we had, but it certainly makes clear the importance of good communications between the controller and pilot."

ATC is not always responsible for safe terrain clearance for the aircraft under its jurisdiction. Many times ATC will issue en route clearances for pilots to proceed off airway direct to a point. Pilots who accept this type of clearance also are accepting responsibility for maintaining safe terrain clearance. Know the height of the highest terrain and obstacles in the operating area. Know your position in relation to the surrounding high terrain.

The following are excerpts from CFIT accidents related to descending on arrival: "...delayed the initiation of the descent..."; "Aircraft prematurely descended too early..."; "...late getting down..."; "During a descent...incorrectly cleared down..."; "...aircraft prematurely let down..."; "...lost situational awareness..."; "Premature descent clearance..."; "Prematurely descended..."; "Premature descent clearance while on vector..."; "During initial descent..." [Figure 4-12]

Practicing good communication skills is not limited to just pilots and controllers. In its findings from a 1974 air carrier accident, the National Transportation Safety Board (NTSB) wrote, "...the extraneous conversation conducted by the flight crew during the descent was symptomatic of a lax atmosphere in the cockpit that continued throughout the approach." The NTSB listed the probable cause as "...the flight crew's lack of altitude awareness at critical points during the approach due to poor cockpit discipline in that the crew did not follow prescribed procedures." In 1981, the FAA issued Parts 121.542 and 135.100, Flight Crewmember Duties, commonly referred to as "sterile cockpit rules." The provisions in this rule can help pilots, operating under any regulations, to avoid altitude and course deviations during arrival. In part, it states: (a) No certificate holder shall require, nor may any flight crewmember perform, any duties during a critical phase of flight except those duties required for the safe operation of the aircraft. Duties such as company required calls made for such purposes as ordering galley supplies and confirming passenger connections, announcements made to passengers promoting the air carrier or pointing out sights of interest, and filling out company payroll and related records are not required for the safe operation of the aircraft. (b) No flight crewmember may engage in, nor

may any pilot in command permit, any activity during a critical phase of flight that could distract any flight crewmember from the performance of his or her duties or which could interfere in any way with the proper conduct of those duties. Activities such as eating meals, engaging in nonessential conversations within the cockpit and nonessential communications between the cabin and cockpit crews, and reading publications not related to the proper conduct of the flight are not required for the safe operation of the aircraft. (c) For the purposes of this section, critical phases of flight include all ground operations involving taxi, takeoff and landing, and all other flight operations conducted below 10,000 feet, except cruise flight.

ARRIVAL NAVIGATION CONCEPTS

Today, the most significant and demanding navigational requirement is the need to safely separate aircraft. In a nonradar environment, ATC does not have an independent means to separate air traffic and must depend entirely on information relayed from flight crews to determine the actual geographic position and altitude. In this situation, precise navigation is critical to ATC's ability to provide separation.

Even in a radar environment, precise navigation and position reports, when required, are still the primary means of providing separation. In most situations, ATC does not have the capability or the responsibility for navigating an aircraft. Because they rely on precise navigation by the flight crew, flight safety in all IFR operations depends directly on your ability to achieve and maintain certain levels of navigational performance. ATC uses radar to monitor navigational performance, detect possible navigational errors, and expedite traffic flow. In a nonradar environment, ATC has no independent knowledge of the actual position of your aircraft or its relationship to other aircraft in adjacent airspace. Therefore, ATC's ability to detect a navigational error and resolve collision hazards is seriously degraded when a deviation from a clearance occurs.

Figure 4-12. Altitude Management When Cleared Direct.

The concept of navigation performance, previously discussed in this book, involves the precision that must be maintained for both the assigned route and altitude. Required levels of navigation performance vary from area to area depending on traffic density and complexity of the routes flown. The level of navigation performance must be more precise in domestic airspace than in oceanic and remote land areas since air traffic density in domestic airspace is much greater. For example, there are two million flight operations conducted within Chicago Center's airspace each year. The minimum lateral distance permitted between co-altitude aircraft in Chicago Center's airspace is 8 NM (3 NM when radar is used). The route ATC assigns an aircraft has protected airspace on both sides of the centerline, equal to one-half of the lateral separation minimum standard. For example, the overall level of lateral navigation performance necessary for flight safety must be better than 4 NM in Center airspace. When STARs are reviewed subsequently in this chapter, you will see how the navigational requirements become more restrictive in the arrival phase of flight where air traffic density increases and procedural design and obstacle clearance become more limiting.

The concept of navigational performance is fundamental to the code of federal regulations, and is best defined in Parts 121.103 and 121.121, which state that each aircraft must be navigated to the degree of accuracy required for air traffic control. The requirements of Part 91.123 related to compliance with ATC clearances and instructions also reflect this fundamental concept. Commercial operators must comply with their Operations Specifications (OpsSpecs) and understand the categories of navigational operations and be able to navigate to the degree of accuracy required for the control of air traffic. In the broad concept of air navigation, there are two major categories of navigational operations consisting of Class I navigation and Class II navigation. Class I navigation is any en route flight operation conducted in controlled or uncontrolled airspace that is entirely within operational service volumes of ICAO standard NAVAIDs (VOR, VOR/DME, NDB). Class II navigation is any en route operation that is not categorized as Class I navigation and includes any operation or portion of an operation that takes place outside the operational service volumes of ICAO standard NAVAIDs. For example, your aircraft equipped only with VORs conducts Class II navigation when your flight operates in an area outside the operational service volumes of federal VORs. Class II navigation does not automatically require the use of long-range, specialized navigational systems if special navigational techniques are used to supplement

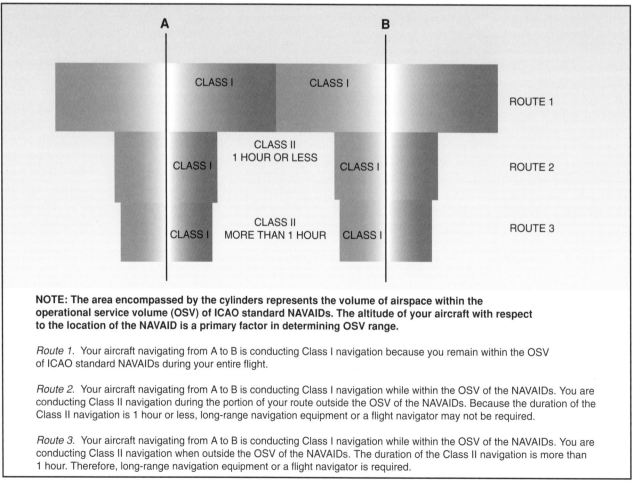

NOTE: The area encompassed by the cylinders represents the volume of airspace within the operational service volume (OSV) of ICAO standard NAVAIDs. The altitude of your aircraft with respect to the location of the NAVAID is a primary factor in determining OSV range.

Route 1. Your aircraft navigating from A to B is conducting Class I navigation because you remain within the OSV of ICAO standard NAVAIDs during your entire flight.

Route 2. Your aircraft navigating from A to B is conducting Class I navigation while within the OSV of the NAVAIDs. You are conducting Class II navigation during the portion of your route outside the OSV of the NAVAIDs. Because the duration of the Class II navigation is 1 hour or less, long-range navigation equipment or a flight navigator may not be required.

Route 3. Your aircraft navigating from A to B is conducting Class I navigation while within the OSV of the NAVAIDs. You are conducting Class II navigation when outside the OSV of the NAVAIDs. The duration of the Class II navigation is more than 1 hour. Therefore, long-range navigation equipment or a flight navigator is required.

Figure 4-13. Class I and II Navigation.

conventional NAVAIDs. Class II navigation includes transoceanic operations and operations in desolate and remote land areas such as the Arctic. The primary types of specialized navigational systems approved for Class II operations include inertial navigation system (INS), Doppler, and global positioning system (GPS). Figure 4-13 provides several examples of Class I and II navigation.

A typical limitations entry in a commercial operator's pilot handbook states, "The area navigation system used for IFR Class I navigation meets the performance/accuracy criteria of AC 20-130A for en route and terminal area navigation." The subject of AC 20-130A is *Airworthiness Approval of Navigation or Flight Management Systems Integrating Multiple Navigation Sensors.*

STANDARD TERMINAL ARRIVAL ROUTES

A standard terminal arrival route (STAR) provides a critical form of communication between pilots and ATC. Once a flight crew has accepted a clearance for a STAR, they have communicated with the controller what route, and in some cases what altitude and airspeed, they will fly during the arrival, depending on the type of clearance. The STAR provides a common method for leaving the en route structure and navigating to your destination. It is a preplanned instrument flight rule ATC arrival procedure published for pilot use in graphic and textual form that simplifies clearance delivery procedures.

When the repetitive complex departure clearances by controllers turned into standard instrument departures (SIDs) in the late 1970s, the idea caught on quickly. Eventually, most of the major airports in the U.S. developed standard departures with graphics for printed publication. The idea seemed so good that the standard arrival clearances also started being published in text and graphic form. The new procedures were named standard terminal arrival routes, or STARs.

The principal difference between SIDs or departure procedures (DPs) and STARs is that the DPs start at the airport pavement and connect to the en route structure. STARs on the other hand, start at the en route structure but don't make it down to the pavement; they end at a fix or NAVAID designated by ATC, where radar vectors commonly take over. This is primarily because STARs serve multiple airports. STARs greatly help to facilitate the transition between the en route and approach phases of flight. The objective when connecting a STAR to an instrument approach procedure is to ensure a seamless lateral and vertical transition. The STAR and approach procedure should connect to one another in such a way as to maintain the overall descent and deceleration profiles. This often results in a seamless transition between the en route, arrival, and approach phases of flight, and serves as a preferred route into high volume terminal areas. [Figure 4-14 on page 4-16]

STARs provide a transition from the en route structure to an approach gate, outer fix, instrument approach fix, or arrival waypoint in the terminal area, and they usually terminate with an instrument or visual approach procedure. STARs are included at the front of each *Terminal Procedures Publication* regional booklet.

For STARs based on conventional NAVAIDs, the procedure design and obstacle clearance criteria are essentially the same as that for en route criteria, covered in Chapter 3, *En Route Operations.* STAR procedures typically include a standardized descent gradient at and above 10,000 feet MSL of 318 feet per NM, or 3 degrees. Below 10,000 feet MSL the maximum descent rate is 330 feet per NM, or approximately 3.1 degrees. In addition to standardized descent gradients, STARs allow for deceleration segments at any waypoint that has a speed restriction. As a general guideline, deceleration considerations typically add 1 NM of distance for each ten knots of speed reduction required.

INTERPRETING THE STAR

STARs use much of the same symbology as departure and approach charts. In fact, a STAR may at first appear identical to a similar graphic DP, except the direction of flight is reversed and the procedure ends at an approach fix. The STAR officially begins at the common NAVAID, intersection, or fix where all the various transitions to the arrival come together. A STAR transition is a published segment used to connect one or more en route airways, jet routes, or RNAV routes to the basic STAR procedure. It is one of several routes that bring traffic from different directions into one STAR. This way, arrivals from several directions can be accommodated on the same chart, and traffic flow is routed appropriately within the congested airspace.

To illustrate how STARs can be used to simplify a complex clearance and reduce frequency congestion, consider the following arrival clearance issued to a pilot flying to Seattle, Washington, depicted in Figure 4-15 on page 4-17: *"Cessna 32G, cleared to the Seattle/Tacoma International Airport as filed. Maintain 12,000. At the Ephrata VOR intercept the 221° radial to CHINS Intersection. Intercept the 284° radial of the Yakima VOR to RADDY Intersection. Cross RADDY at 10,000. Continue via the Yakima 284° radial to AUBRN Intersection. Expect radar vectors to the final approach course."*

Now consider how this same clearance is issued when a STAR exists for this terminal area. *"Cessna 32G,*

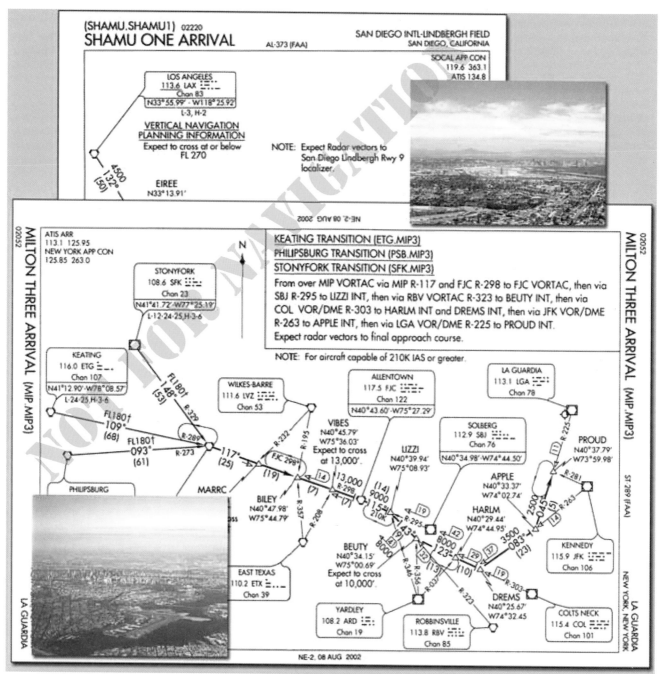

Figure 4-14. Arrival Charts.

cleared to Seattle/Tacoma International Airport as filed, then CHINS FOUR ARRIVAL, Ephrata Transition. Maintain 10,000 feet." A shorter transmission conveys the same information.

Safety is enhanced when both pilots and controllers know what to expect. Effective communication increases with the reduction of repetitive clearances, decreasing congestion on control frequencies. To accomplish this, STARs are developed according to the following criteria:

• STARs must be simple, easily understood and, if possible, limited to one page.

• A STAR transition should be able to accommodate as many different types of aircraft as possible.

• VORTACs are used wherever possible, with some exceptions on RNAV STARs, so that military and civilian aircraft can use the same arrival.

• DME arcs within a STAR should be avoided since not all aircraft in the IFR environment are so equipped.

• Altitude crossing and airspeed restrictions are included when they are assigned by ATC a majority of the time. [Figure 4-16 on page 4-18]

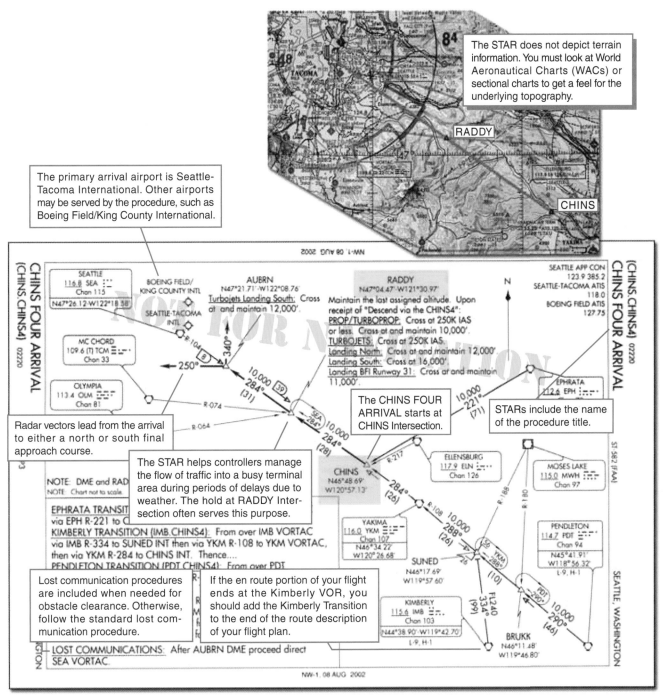

The STAR does not depict terrain information. You must look at World Aeronautical Charts (WACs) or sectional charts to get a feel for the underlying topography.

The primary arrival airport is Seattle-Tacoma International. Other airports may be served by the procedure, such as Boeing Field/King County International.

Radar vectors lead from the arrival to either a north or south final approach course.

The STAR helps controllers manage the flow of traffic into a busy terminal area during periods of delays due to weather. The hold at RADDY Intersection often serves this purpose.

The CHINS FOUR ARRIVAL starts at CHINS Intersection.

STARs include the name of the procedure title.

Lost communication procedures are included when needed for obstacle clearance. Otherwise, follow the standard lost communication procedure.

If the en route portion of your flight ends at the Kimberly VOR, you should add the Kimberly Transition to the end of the route description of your flight plan.

Figure 4-15. STAR Interpretation.

STARs usually are named according to the point at which the procedure begins. In the U.S., typically there are en route transitions before the STAR itself. So the STAR name is usually the same as the last fix on the en route transitions where they come together to begin the basic STAR procedure. A STAR that commences at the CHINS Intersection becomes the CHINS ONE ARRIVAL. When a significant portion of the arrival is revised, such as an altitude, a route, or data concerning the NAVAID, the number of the arrival changes. For example, the CHINS ONE ARRIVAL is now the CHINS FOUR ARRIVAL due to modifications in the procedure.

Studying the STARs for an airport may allow you to perceive the specific topography of the area. Note the initial fixes and where they correspond to fixes on the NACO en route or area chart. Arrivals may incorporate step-down fixes when necessary to keep aircraft within airspace boundaries, or for obstacle clearance. Routes between fixes contain courses, distances, and minimum altitudes, alerting you to possible obstructions or terrain under your arrival path. Airspeed restrictions also appear where they aid in managing the traffic flow. In addition, some STARs require that you use DME and/or ATC radar. You can decode the symbology on the PAWLING TWO ARRIVAL depicted in Figure 4-17 on page 4-18 by referring to the legend at the beginning of the NACO *Terminal Procedures Publication*.

Figure 4-16. Reducing Pilot/Controller Workload.

VERTICAL NAVIGATION PLANNING

Included within certain STARs is information on vertical navigation planning. This information is provided to reduce the amount of low altitude flying time for high performance airplanes, like jets and turboprops. An expected altitude is given for a key fix along the route. By knowing an intermediate altitude in advance when flying a high performance airplane, you can plan the power or thrust settings and airplane configurations that result in the most efficient descent in terms of time and fuel requirements. Pilots of high performance airplanes use the vertical navigation planning information from the RAMMS FIVE ARRIVAL at Denver, Colorado, to plan their descents. [Figure 4-18]

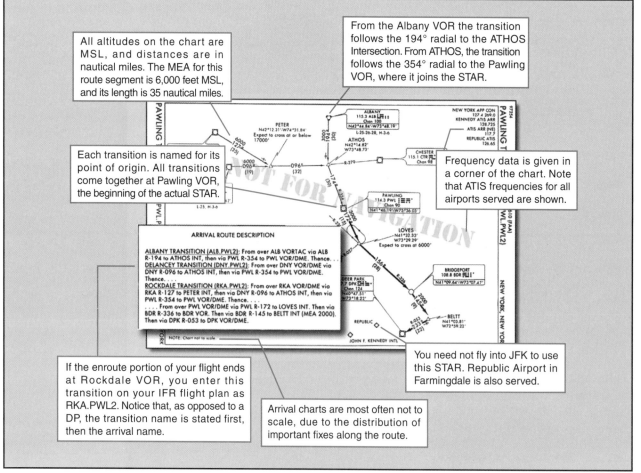

Figure 4-17. STAR Symbology.

All altitudes on the chart are MSL, and distances are in nautical miles. The MEA for this route segment is 6,000 feet MSL, and its length is 35 nautical miles.

From the Albany VOR the transition follows the 194° radial to the ATHOS Intersection. From ATHOS, the transition follows the 354° radial to the Pawling VOR, where it joins the STAR.

Each transition is named for its point of origin. All transitions come together at Pawling VOR, the beginning of the actual STAR.

Frequency data is given in a corner of the chart. Note that ATIS frequencies for all airports served are shown.

If the enroute portion of your flight ends at Rockdale VOR, you enter this transition on your IFR flight plan as RKA.PWL2. Notice that, as opposed to a DP, the transition name is stated first, then the arrival name.

Arrival charts are most often not to scale, due to the distribution of important fixes along the route.

You need not fly into JFK to use this STAR. Republic Airport in Farmingdale is also served.

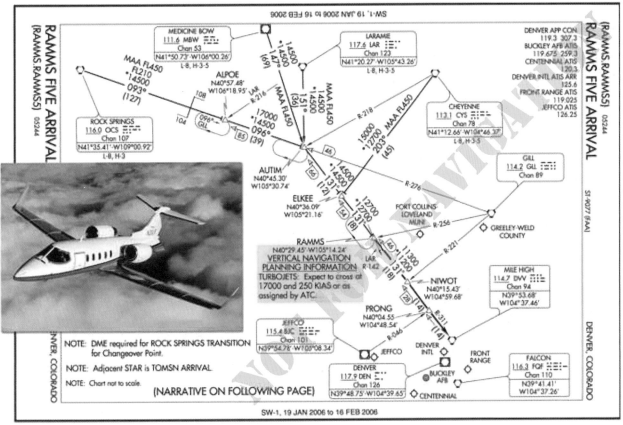

Figure 4-18. Vertical Navigation Planning.

ARRIVAL PROCEDURES

You may accept a STAR within a clearance or you may file for one in your flight plan. As you near your destination airport, ATC may add a STAR procedure to your original clearance. Keep in mind that ATC can assign a STAR even if you have not requested one. If you accept the clearance, you must have at least a textual description of the procedure in your possession. If you do not want to use a STAR, you must specify "No STAR" in the remarks section of your flight plan. You may also refuse the STAR when it is given to you verbally by ATC, but the system works better if you advise ATC ahead of time.

PREPARING FOR THE ARRIVAL

As mentioned before, STARs include navigation fixes that are used to provide transition and arrival routes from the en route structure to the final approach course. They also may lead to a fix where radar vectors will be provided to intercept the final approach course. You may have noticed that minimum crossing altitudes and airspeed restrictions appear on some STARs. These expected altitudes and airspeeds are not part of your clearance until ATC includes them verbally. A STAR is simply a published routing; it does not have the force of a clearance until issued specifically by ATC. For example, MEAs printed on STARs are not valid unless stated within an ATC clearance or in cases of

lost communication. After receiving your arrival clearance, you should review the assigned STAR procedure.

Obtain the airport and weather information as early as practical. It is recommended that you have this information prior to flying the STAR. If you are landing at an airport with approach control services that has two or more published instrument approach procedures, you will receive advance notice of which instrument approaches to expect. This information is broadcast either by ATIS or by a controller. It may not be provided when the visibility is 3 statute miles (SM) or better and the ceiling is at or above the highest initial approach altitude established for any instrument approach procedure for the airport. [Figure 4-19 on page 4-20]

For STAR procedures charted with radar vectors to the final approach, look for routes from the STAR terminating fixes to the IAF. If no route is depicted, you should have a predetermined plan of action to fly from the STAR terminating fix to the IAF in the event of a communication failure.

REVIEWING THE APPROACH

Once you have determined which approach to expect, review the approach chart thoroughly before you enter the terminal area. Check your fuel level and make sure

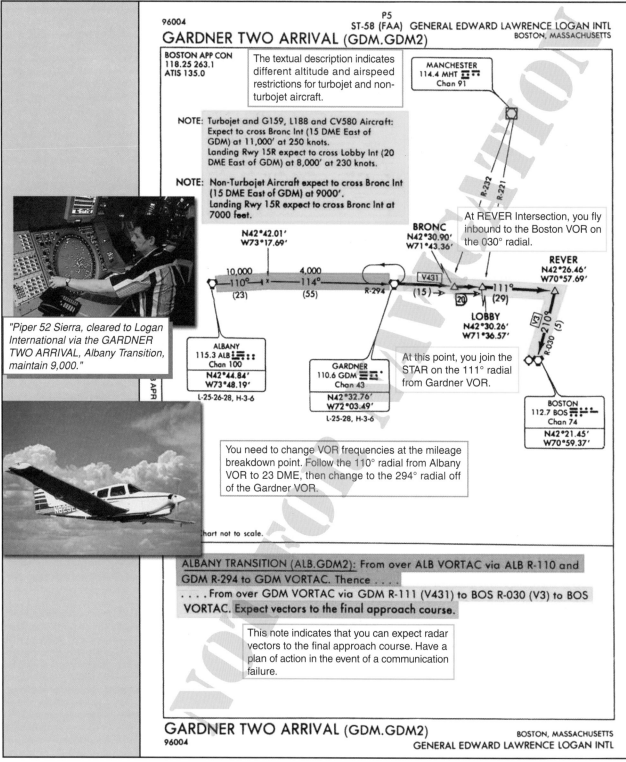

GARDNER TWO ARRIVAL (GDM.GDM2)

96004
P5
ST-58 (FAA) GENERAL EDWARD LAWRENCE LOGAN INTL
BOSTON, MASSACHUSETTS

BOSTON APP CON
118.25 263.1
ATIS 135.0

The textual description indicates different altitude and airspeed restrictions for turbojet and non-turbojet aircraft.

MANCHESTER
114.4 MHT
Chan 91

NOTE: Turbojet and G159, L188 and CV580 Aircraft: Expect to cross Bronc Int (15 DME East of GDM) at 11,000' at 250 knots. Landing Rwy 15R expect to cross Lobby Int (20 DME East of GDM) at 8,000' at 230 knots.

NOTE: Non-Turbojet Aircraft expect to cross Bronc Int (15 DME East of GDM) at 9000'. Landing Rwy 15R expect to cross Bronc Int at 7000 feet.

R-232 R-221

BRONC
N42°30.90'
W71°43.36'

At REVER Intersection, you fly inbound to the Boston VOR on the 030° radial.

REVER
N42°26.46'
W70°57.69'

N42°42.01'
W73°17.69'

10,000   4,000
110° x 114°
(23)   (55)   R-294

V431

111°
(15)   20   (29)

"Piper 52 Sierra, cleared to Logan International via the GARDNER TWO ARRIVAL, Albany Transition, maintain 9,000."

LOBBY
N42°30.26'
W71°36.57'

V3
R-030 210°
(5)

ALBANY
115.3 ALB
Chan 100

N42°44.84'
W73°48.19'

L-25-26-28, H-3-6

GARDNER
110.6 GDM
Chan 43

N42°32.76'
W72°03.49'

L-25-28, H-3-6

At this point, you join the STAR on the 111° radial from Gardner VOR.

BOSTON
112.7 BOS
Chan 74

N42°21.45'
W70°59.37'

You need to change VOR frequencies at the mileage breakdown point. Follow the 110° radial from Albany VOR to 23 DME, then change to the 294° radial off of the Gardner VOR.

Chart not to scale.

ALBANY TRANSITION (ALB.GDM2): From over ALB VORTAC via ALB R-110 and GDM R-294 to GDM VORTAC. Thence . . . .
. . . . From over GDM VORTAC via GDM R-111 (V431) to BOS R-030 (V3) to BOS VORTAC. Expect vectors to the final approach course.

This note indicates that you can expect radar vectors to the final approach course. Have a plan of action in the event of a communication failure.

GARDNER TWO ARRIVAL (GDM.GDM2)
96004

BOSTON, MASSACHUSETTS
GENERAL EDWARD LAWRENCE LOGAN INTL

Figure 4-19. Arrival Clearance.

a prolonged hold or increased headwinds have not cut into your fuel reserves because there is always a chance you will have to make a missed approach or go to an alternate. By completing landing checklists early, you free yourself to concentrate on the approach.

In setting up for the expected approach procedure when using an RNAV, GPS, or FMS system, it is important to understand how multiple approaches to

the same runway are coded in the database. When more than one RNAV procedure is issued for the same runway, there must be a way to differentiate between them within the equipment's database, as well as to select which procedure you want to use. (Multiple procedures may exist to accommodate GPS receivers and FMSs, both with and without VNAV capability.) Each procedure name incorporates a letter of the alphabet, starting with Z and working backward through Y, X, W, and so

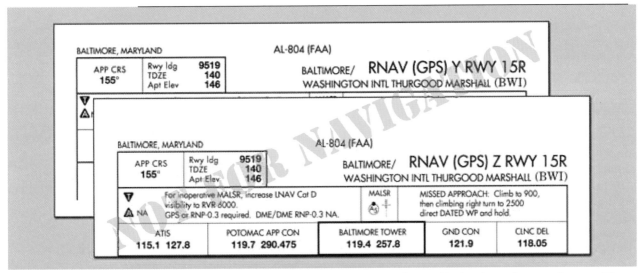

Figure 4-20. Here are two RNAV (GPS) approaches to Runway 15R at Baltimore. A controller issuing a clearance for one of these approaches would speak the identifying letter—for example, "...cleared for the RNAV (GPS) Yankee approach, Runway 15R..."

on. (Naming conventions for approaches are covered in more depth in the next chapter.) [Figure 4-20]

ALTITUDE
Upon your arrival in the terminal area, ATC either clears you to a specific altitude, or they give you a "descend via" clearance that instructs you to follow the altitudes published on the STAR. [Figure 4-21] You are not authorized to leave your last assigned altitude unless specifically cleared to do so. If ATC amends the altitude or route to one that is different from the published procedure, the rest of the charted

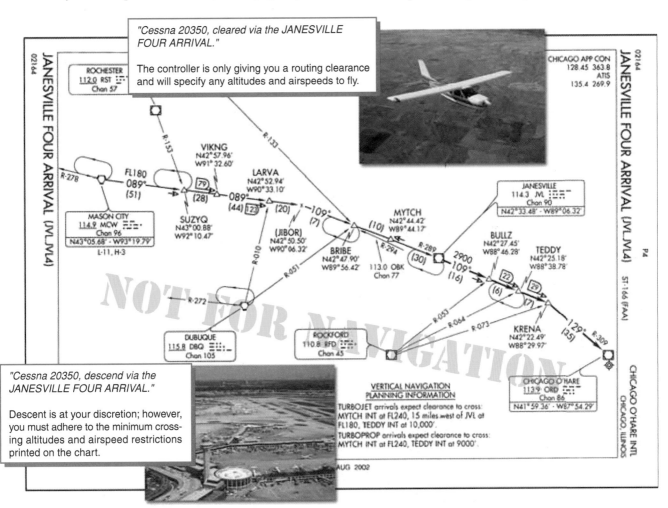

Figure 4-21. Assigned Altitudes.

descent procedure is canceled. ATC will assign you any further route, altitude, or airspeed clearances, as necessary. Notice the JANESVILLE FOUR ARRIVAL depicts only one published arrival route, with no named transition routes leading to the basic STAR procedure beginning at the Janesville VOR/DME. Vertical navigation planning information is included for turbojet and turboprop airplanes at the bottom of the chart. Additionally, note that there are several ways to identify the BRIBE reporting point using alternate formation radials, some of which are from off-chart NAVAIDs. ATC may issue a descent clearance that includes a crossing altitude restriction. In the PENNS ONE ARRIVAL, the ATC clearance authorizes you to descend at your discretion, as long as you cross the PENNS Intersection at 6,000 feet MSL. [Figure 4-22]

In the United States, Canada, and many other countries, the common altitude for changing to the standard altimeter setting of 29.92 inches of mercury (or 1013.2 hectopascals or millibars) when climbing to the high altitude structure is 18,000 feet. When descending from high altitude, the altimeter should be changed to the local altimeter setting when passing through FL 180, although in most countries throughout the world the change to or from the standard altimeter setting is not done at the same altitude for each instance.

For example, the flight level where you change your altimeter setting to the local altimeter setting is specified by ATC each time you arrive at a specific airport. This information is shown on STAR charts outside the U.S. with the words: TRANS LEVEL: BY ATC. When departing from that same airport (also depicted typically on the STAR chart), the altimeter should be set to the standard setting when passing through 5,000 feet, as an example. This means that altimeter readings when flying above 5,000 feet will actually be flight levels, not feet. This is common for Europe, but very different for pilots experienced with flying in the United States and Canada.

RNAV STARS OR STAR TRANSITIONS

STARs designated RNAV serve the same purpose as conventional STARs, but are only used by aircraft equipped with FMS or GPS. An RNAV STAR or STAR transition typically includes flyby waypoints, with flyover waypoints used only when operationally required. These waypoints may be assigned crossing altitudes and speeds to optimize the descent and deceleration profiles. RNAV STARs often are designed, coordinated, and approved by a joint effort between air carriers, commercial operators, and the ATC facilities that have jurisdiction for the affected airspace.

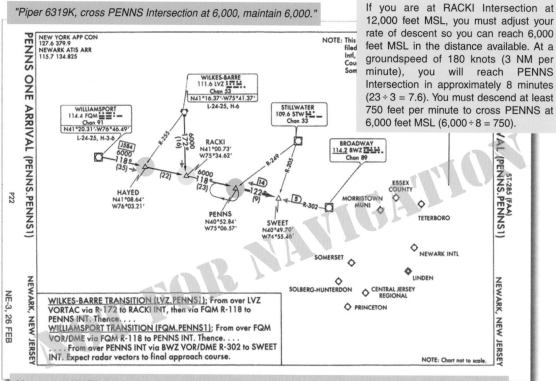

Figure 4-22. Altitude Restrictions.

RNAV STAR procedure design, such as minimum leg length, maximum turn angles, obstacle assessment criteria, including widths of the primary and secondary areas, use the same design criteria as RNAV DPs. Likewise, RNAV STAR procedures are designated as either Type A or Type B, based on the aircraft navigation equipment required, flight crew procedures, and the process and criteria used to develop the STAR. The Type A or Type B designation appears in the notes on the chart. Type B STARs have higher equipment requirements and, often, tighter RNP tolerances than Type A. For Type B STARS, pilots are required to use a CDI/flight director, and/or autopilot in LNAV mode while operating on RNAV courses. (These requirements are detailed in Chapter 2 of this book, under "RNAV Departure Procedures.") Type B STARs are generally designated for high-traffic areas. Controllers may clear you to use an RNAV STAR in various ways.

If your clearance simply states, *"cleared Hadly One arrival,"* you are to use the arrival for lateral routing only.

- A clearance such as *"cleared Hadly One arrival, descend and maintain flight level two four zero,"* clears you to descend only to the assigned altitude, and you should maintain that altitude until cleared for further vertical navigation.

- If you are cleared using the phrase *"descend via,"* the controller expects you to use the equipment for both lateral and vertical navigation, as published on the chart.

- The controller may also clear you to use the arrival with specific exceptions—for example, *"Descend via the Haris One arrival, except cross Bruno at one three thousand then maintain one zero thousand."* In this case, the pilot should track the arrival both laterally and vertically, descending so as to comply with all altitude and airspeed restrictions until reaching Bruno, and then maintain 10,000 feet until cleared by ATC to continue to descend.

- Pilots might also be given direct routing to intercept a STAR and then use it for vertical navigation. For example, *"proceed direct Mahem, descend via the Mahem Two arrival."*

[Figure 4-23 on page 4-24]

Figure 4-24 on page 4-25 depicts typical RNAV STAR leg (segment) types you can expect to see when flying these procedures.

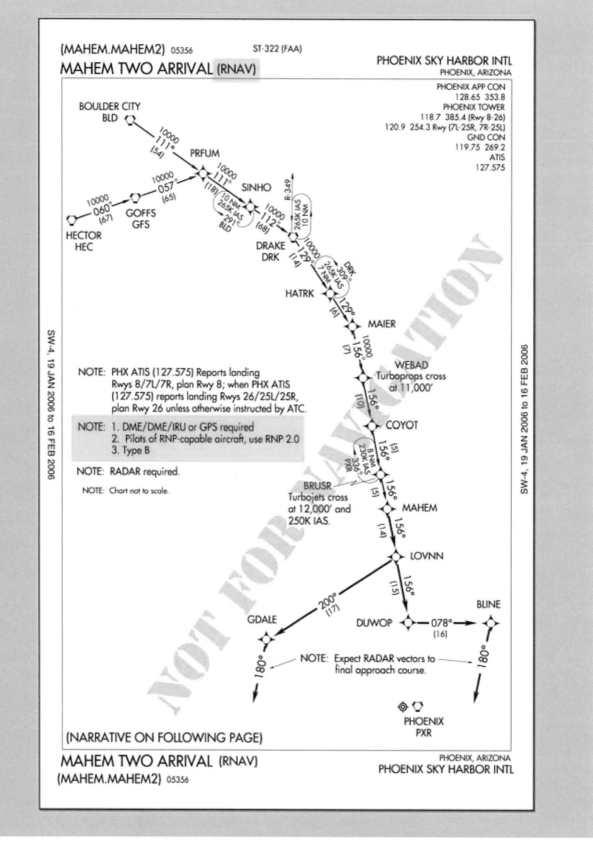

Figure 4-23. The notes show that this is a Type B RNAV STAR.

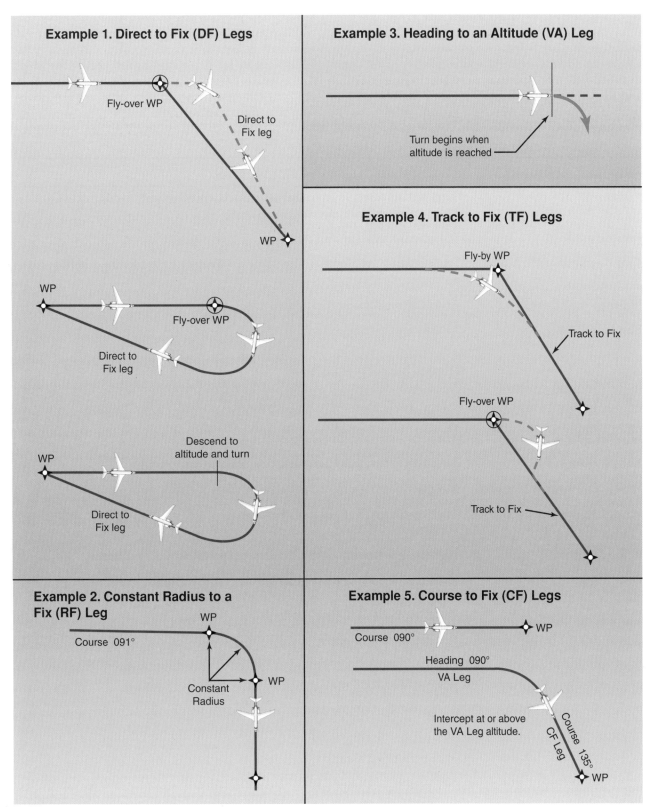

Figure 4-24. RNAV STAR Leg (Segment) Types.

SPECIAL AIRPORT QUALIFICATION

It is important to note an example of additional resources that are helpful for arrivals, especially into unfamiliar airports requiring special pilot or navigation qualifications. The operating rules governing domestic and flag air carriers require pilots in command to be qualified over the routes and into airports where scheduled operations are conducted, including areas, routes, and airports in which special pilot qualifications or special navigation qualifications are needed. For Part 119 certificate holders who conduct operations under Parts 121.443, there are provisions in OpsSpecs under which operators can comply with this regulation. The following are examples of special airports in the U.S, along with associated comments:

SPECIAL AIRPORTS	COMMENTS
Kodiak, AK	Airport is surrounded by mountainous terrain. Any go-around beyond ILS or GCA MAP will not provide obstruction clearance.
Petersburg, AK	Mountainous terrain in immediate vicinity of airport, all quadrants.
Cape Newenham AFS, AK	Runway located on mountain slope with high gradient factor; nonstandard instrument approach.
Washington, DC (National)	Special arrival/departure procedures.
Shenandoah Valley, VA (Stanton-Waynesboro-Harrisonburg)	Mountainous terrain.
Aspen, CO	High terrain; special procedures.
Gunnison, CO	VOR only; uncontrolled; numerous obstructions in airport area; complex departure procedures.
Missoula, MT	Mountainous terrain; special procedures.
Jackson Hole, WY	Mountainous terrain; all quadrants; complex departure procedures.
Hailey, ID (Friedman Memorial)	Mountainous terrain; special arrival/departure procedures.
Hayden, Yampa Valley, CO	Mountainous terrain; no control tower; special engine-out procedures for certain large airplanes.
Lihue, Kauai, HI	High terrain; mountainous to 2,300 feet within 3 miles of the localizer.
Ontario, CA	Mountainous terrain and extremely limited visibility in haze conditions.

CHAPTER 5
APPROACH

This chapter discusses general planning and conduct of instrument approaches by professional pilots operating under Title 14 of the Code of Federal Regulations (14 CFR) Parts 91, 121, 125, and 135. Operations specific to helicopters are covered in Chapter 7. The operations specifications (OpsSpecs), standard operating procedures (SOPs), and any other Federal Aviation Administration (FAA) approved documents for each commercial operator are the final authorities for individual authorizations and limitations as they relate to instrument approaches. While coverage of the various authorizations and approach limitations for all operators is beyond the scope of this chapter, an attempt is made to give examples from generic manuals where it is appropriate.

APPROACH PLANNING

Depending on speed of the aircraft, availability of weather information, and the complexity of the approach procedure or special terrain avoidance procedures for the airport of intended landing, the inflight planning phase of an instrument approach can begin as far as 100-200 NM from the destination. Some of the approach planning should be accomplished during preflight. In general, there are five steps that most operators incorporate into their Flight Standards manuals for the inflight planning phase of an instrument approach:

- Gathering weather information, field conditions, and Notices to Airmen (NOTAMs) for the runway of intended landing.

- Calculation of performance data, approach speeds, and thrust/power settings.

- Flight deck navigation/communication and automation setup.

- Instrument approach procedure (IAP) review and, for flight crews, IAP briefing.

- Operational review and, for flight crews, operational briefing.

Although often modified to suit each individual operator, these five steps form the basic framework for the inflight-planning phase of an instrument approach. The extent of detail that a given operator includes in their SOPs varies from one operator to another; some may designate which pilot performs each of the above actions, the sequence, and the manner in which each action is performed. Others may leave much of the detail up to individual flight crews and only designate which tasks should be performed prior to commencing an approach. Flight crews of all levels, from single-pilot to multi-crewmember Part 91 operators, can benefit from the experience of commercial operators in developing techniques to fly standard instrument approach procedures (SIAPs).

Determining the suitability of a specific IAP can be a very complex task, since there are many factors that can limit the usability of a particular approach. There are several questions that pilots need to answer during preflight planning and prior to commencing an approach. Is the approach procedure authorized for the company, if Part 91K, 121, 125, or 135? Is the weather appropriate for the approach? Is the aircraft currently at a weight that will allow it the necessary performance for the approach and landing or go around/missed approach? Is the aircraft properly equipped for the approach? Is the flight crew qualified and current for the approach? Many of these types of issues must be considered during preflight planning and within the framework of each specific air carrier's OpsSpecs, or Part 91.

WEATHER CONSIDERATIONS

Weather conditions at the field of intended landing dictate whether flight crews need to plan for an instrument approach and, in many cases, determine which approaches can be used, or if an approach can even be attempted. The gathering of weather information should be one of the first steps taken during the approach-planning phase. Although there are many possible types of weather information, the primary concerns for approach decision-making are wind speed, wind direction, ceiling, visibility, altimeter setting, temperature, and field conditions. It is also a good idea to check NOTAMs at this time in case there were any changes since preflight planning.

Wind speed and direction are factors because they often limit the type of approach that can be flown at a specific location. This typically is not a factor at airports with multiple precision approaches, but at airports with only a few or one approach procedure the wrong combination of wind and visibility can make all instrument approaches at an airport unavailable. As an example, consider the available approaches at the Chippewa Valley Regional Airport (KEAU) in Eau Claire, Wisconsin, shown in Figure 5-1. In the event that the visibility is reported as less than one mile, the only useable approach for Category C airplanes is the Instrument Landing System (ILS) to Runway 22. This leaves very few options for flight crews if the wind does not favor Runway 22; and, in cases where the wind restricts a landing on that runway altogether, even a circling approach cannot be flown because of the visibility.

WEATHER SOURCES

Most of the weather information that flight crews receive is issued to them prior to the start of each flight segment, but the weather used for inflight planning and execution of an instrument approach is normally obtained en route via government sources, company frequency, or Aircraft Communications Addressing and Reporting System (ACARS).

Air carriers and operators certificated under the provisions of Part 119 (Certification: Air Carriers and Commercial Operators) are required to use the aeronautical weather information systems defined in the OpsSpecs issued to that certificate holder by the FAA. These systems may use basic FAA/National Weather Service (NWS) weather services, contractor or operator-proprietary weather services and/or Enhanced Weather Information System (EWINS) when approved in the OpsSpecs. As an integral part of EWINS approval, the procedures for collecting, producing, and disseminating aeronautical weather information, as well as the crewmember and dispatcher training to support the use of system weather products, must be accepted or approved.

Operators not certificated under the provisions of Part 119 are encouraged to use FAA/NWS products through Automated Flight Service Stations (AFSSs), Direct User Access Terminal System (DUATS), and/or Flight Information Services Data Link (FISDL). Refer to the *Aeronautical Information Manual* (AIM) for more information regarding AFSSs, DUATS, and FISDL.

The suite of available aviation weather product types is expanding with the development of new sensor systems, algorithms, and forecast models. The FAA and NWS, supported by the National Center for Atmospheric Research and the Forecast Systems Laboratory, develop and implement new aviation

weather product types through a comprehensive process known as the Aviation Weather Technology Transfer process. This process ensures that user needs and technical and operational readiness requirements are met as experimental product types mature to operational application.

The development of enhanced communications capabilities, most notably the Internet, has allowed pilots access to an ever-increasing range of weather service providers and proprietary products. It is not the intent of the FAA to limit operator use of this weather information. However, pilots and operators should be aware that weather services provided by entities other than the FAA, NWS, or their contractors (such as the DUATS and FISDL providers) may not meet FAA/NWS quality control standards. Therefore, operators and pilots contemplating use of such services should consider the following in determining the suitability of that service or product. In many cases, this may be accomplished by provider disclosure or a description of services or products:

Is the service or product applicable for aviation use?

- Does the weather product or service provide information that is usable in aeronautical decision-making?

- Does the product or service fail to provide data necessary to make critical aeronautical weather decisions?

Does the service provide data/products produced by approved aviation weather information sources?

- Are these data or this product modified?

- If so, is the modification process described, and is the final product in a configuration that supports aeronautical weather decision-making?

Are the weather products professionally developed and produced and/or quality-controlled by a qualified aviation meteorologist?

Does the provider's quality assurance plan include the capability to monitor generated products and contain a procedure to correct deficiencies as they are discovered?

Is the product output consistent with original data sources?

Are education and training materials sufficient to enable users to use the new product effectively?

Are the following key elements of the product intuitive and easy for the user to interpret?

- Type of data/product.

- Currency or age of data/product.

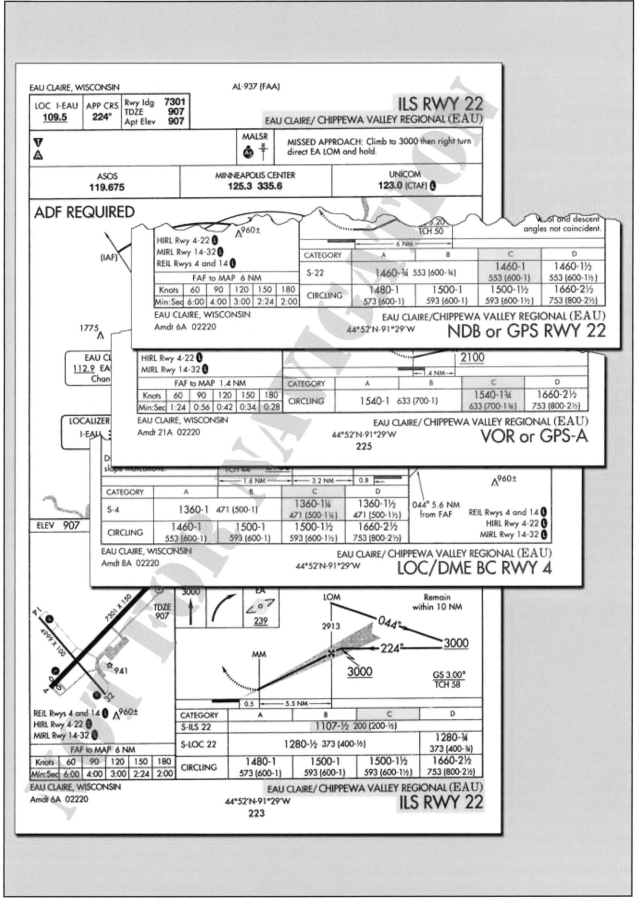

Figure 5-1. Chippewa Regional Airport (KEAU), Eau Claire, Wisconsin.

- Method for displaying and decoding the data/product.
- Location/mapping of the data.

Is the product suitable for use? Consider potential pilot misunderstandings due to:

- Complexity of the product.
- Nonstandard display (colors, labels).
- Incorrect mapping/display of data.
- Incorrect overlay of weather data with other data (terrain, navigational aids (NAVAIDs), waypoints, etc.).
- Inappropriate display of missing data.
- Missing or inaccurate time/date stamp on product.

Pilots and operators should be cautious when using unfamiliar products, or products not supported by technical specifications that satisfy the considerations noted above.

NOTE: When in doubt, use FAA/NWS products with the consultation of an FAA AFSS specialist.

BROADCAST WEATHER
The most common method used by flight crews to obtain specific inflight weather information is to use a source that broadcasts weather for the specific airport. Information about ceilings, visibility, wind, temperature, barometric pressure, and field conditions can be obtained from most types of broadcast weather services. Broadcast weather can be transmitted to the aircraft in radio voice format or digital format, if it is available, via an ACARS system.

AUTOMATIC TERMINAL INFORMATION SERVICE
The weather broadcast system found most often at airports with air traffic control towers in the National Airspace System (NAS) is the automatic terminal information service (ATIS). The AIM defines ATIS as the continuous broadcast of recorded non-control information in selected high activity terminal areas. The main purpose of ATIS is the reduction of frequency congestion and controller workload. It is broadcast over very high frequency (VHF) radio frequencies, and is designed to be receivable up to 60 NM from the transmitter at altitudes up to 25,000 feet above ground level (AGL). ATIS is typically derived from an automated weather observation system or a human weather observer's report.

AUTOMATED WEATHER OBSERVING PROGRAMS
Automated surface observation systems can provide pilots with weather information over discrete VHF frequencies or over the voice portion of local NAVAIDs.

The automated weather observing system (AWOS) and automated surface observing system (ASOS) provide real-time weather information that can be used by flight crews to make approach decisions, and by the NWS to generate aviation routine weather reports (METARs). Flight crews planning approaches to airports where ATIS is not available may be able to obtain current airport conditions from an AWOS/ASOS facility.

FAA-owned and operated AWOS-2 and AWOS-3 systems are approved sources of weather for Part 121 and 135 operations. Also, NWS-operated ASOSs are approved sources of weather for Part 121 and 135 operations. An AWOS/ASOS cannot be used as an authorized weather source for Part 121 or 135 instrument flight rules (IFR) operations if the visibility or altimeter setting is reported missing from the report. Refer to the AIM for the most current information on automated weather observation systems.

CENTER WEATHER
In the event that an airport has weather observation capability, but lacks the appropriate equipment to transmit that information over a radio frequency, air route traffic control centers (ARTCCs) can provide flight crews with hourly METAR or non-routine (special) aviation weather report (SPECI) information for those airports. For example, as an aircraft approaches an airport, the center controller can voluntarily or upon request provide the pilot with the most recent weather observation. Terminal Radar Approach Control (TRACON) facilities also provide weather observation information on a workload-permitting basis. Another option to obtain a current METAR or SPECI is to contact an En Route Flight Advisory Service facility (Flight Watch).

REGULATORY REQUIREMENTS
There are many practical reasons for reviewing weather information prior to initiating an instrument approach. Pilots must familiarize themselves with the condition of individual airports and runways so that they may make informed decisions regarding fuel management, diversions, and alternate planning. Because this information is critical, CFRs require pilots to comply with specific weather minimums for planning and execution of instrument flights and approaches.

PART 91 OPERATORS
According to Part 91.103, the pilot in command must become familiar with all available information concerning a flight prior to departure. Included in this directive is the fundamental basis for pilots to review NOTAMs and pertinent weather reports and forecasts for the intended route of flight. This review should include current weather reports and terminal forecasts for all intended points of landing and alternate airports. In addition, a thorough review of an airport's current weather conditions should always be conducted prior to initiating an instrument approach. Pilots should also

consider weather information as a planning tool for fuel management.

For flight planning purposes, weather information must be reviewed in order to determine the necessity and suitability of alternate airports. For Part 91 operations, the 600-2 and 800-2 rule applies to airports with precision and nonprecision approaches, respectively. Approaches with vertical guidance (APV) are considered semi-precision and nonprecision since they do not meet the International Civil Aviation Organization (ICAO) Annex 10 standards for a precision approach. (See Final Approach Segment section later in this chapter for more information regarding APV approaches.) Exceptions to the 600-2 and 800-2 alternate minimums are listed in the front of the National Aeronautical Charting Office (NACO) U.S. Terminal Procedures Publication (TPP) and are indicated by an " Ⓐ " symbol on the approach charts for the airport. This does not preclude flight crews from initiating instrument approaches at alternate airports when the weather conditions are below these minimums. The 600-2 and 800-2 rules, or any exceptions, only apply to flight planning purposes, while published landing minimums apply to the actual approach at the alternate.

PART 135 OPERATORS

Unlike Part 91 operators, Part 135 operators may not depart for a destination unless the forecast weather there will allow an instrument approach and landing. According to Part 135.219, flight crews and dispatchers may only designate an airport as a destination if the latest weather reports or forecasts, or any combination of them, indicate that the weather conditions will be at or above IFR landing minimums at the estimated time of arrival (ETA). This ensures that Part 135 flight crews consider weather forecasts when determining the suitability of destinations. Departures for airports can be made when the forecast weather shows the airport will be at or above IFR minimums at the ETA, even if current conditions indicate the airport to be below minimums. Conversely, Part 135.219 prevents departures when the first airport of intended landing is currently above IFR landing minimums, but the forecast weather is below those minimums at the ETA.

Another very important difference between Part 91 and Part 135 operations is the Part 135 requirement for airports of intended landing to meet specific weather criteria once the flight has been initiated. For Part 135, not only is the weather required to be forecast at or above IFR landing minimums for planning a departure, but it also must be above minimums for initiation of an instrument approach and, once the approach is initiated, to begin the final approach segment of an approach. Part 135.225 states that pilots may not begin an instrument approach unless the latest weather report indicates that the weather conditions are at or above the authorized IFR landing minimums for that procedure. Part 135.225 provides relief from this rule if the aircraft has already passed the FAF when the weather report is received. It should be noted that the controlling factor for determining whether or not the aircraft can proceed is reported visibility. Runway visual range (RVR), if available, is the controlling visibility report for determining that the requirements of this section are met. The runway visibility value (RVV), reported in statute miles (SM), takes precedent over prevailing visibility. There is no required timeframe for receiving current weather prior to initiating the approach.

PART 121 OPERATORS

Like Part 135 operators, flight crews and dispatchers operating under Part 121 must ensure that the appropriate weather reports or forecasts, or any combination thereof, indicate that the weather will be at or above the authorized minimums at the ETA at the airport to which the flight is dispatched (Part 121.613). This regulation attempts to ensure that flight crews will always be able to execute an instrument approach at the destination airport. Of course, weather forecasts are occasionally inaccurate; therefore, a thorough review of current weather is required prior to conducting an approach. Like Part 135 operators, Part 121 operators are restricted from proceeding past the FAF of an instrument approach unless the appropriate IFR landing minimums exist for the procedure. In addition, descent below the **minimum descent altitude (MDA)**, **decision altitude (DA)**, or **decision height (DH)** is governed, with one exception, by the same rules that apply to Part 91 operators. The exception is that during Part 121 and 135 operations, the airplane is also required to land within the touchdown zone (TDZ). Refer to the section titled Minimum Descent Altitude, Decision Altitude, and Decision Height later in this chapter for more information regarding MDA, DA, and DH.

PERFORMANCE CONSIDERATIONS

All operators are required to comply with specific airplane performance limitations that govern approach and landing. Many of these requirements must be considered prior to the origination of flight. The primary goal of these performance considerations is to ensure that the aircraft can remain clear of obstructions throughout the approach, landing, and go-around phase of flight, as well as land within the distance required by the FAA. Although the majority of in-depth performance planning for an instrument flight is normally done prior to the aircraft's departure, a general review of performance considerations is usually conducted prior to commencing an instrument approach.

AIRPLANE PERFORMANCE OPERATING LIMITATIONS

Generally speaking, air carriers must have in place an approved method of complying with Subpart I of Parts 121 and 135 (Airplane Performance Operating Limitations), thereby proving the airplane's performance capability for every flight that it intends to make. Flight crews must have an approved method of complying with the approach and landing performance criteria in the applicable regulations prior to departing for their intended destination. The primary source of information for performance calculations for all operators, including Part 91, is the approved Aircraft Flight Manual (AFM) or Pilot's Operating Handbook (POH) for the make and model of aircraft that is being operated. It is required to contain the manufacturer determined performance capabilities of the aircraft at each weight, altitude, and ambient temperature that are within the airplane's listed limitations. Typically, the AFM for a large turbine powered airplane should contain information that allows flight crews to determine that the airplane will be capable of performing the following actions, considering the airplane's landing weight and other pertinent environmental factors:

- Land within the distance required by the regulations.

- Climb from the missed approach point (MAP) and maintain a specified climb gradient with one engine inoperative.

- Perform a go-around from the final stage of landing and maintain a specified climb gradient with all engines operating and the airplane in the landing configuration.

Many airplanes have more than one allowable flap configuration for normal landing. Often, a reduced flap setting for landing will allow the airplane to operate at a higher landing weight into a field that has restrictive obstacles in the missed approach or rejected landing climb path. On these occasions, the full-flap landing speed may not allow the airplane enough energy to successfully complete a go-around and avoid any high terrain that might exist on the climb out. Therefore, all-engine and engine-out missed approaches, as well as rejected landings, must be taken into consideration in compliance with the regulations. [Figure 5-2]

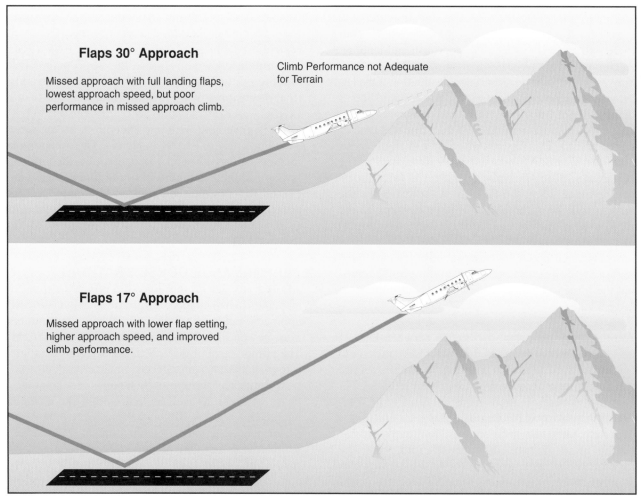

Figure 5-2. Reduced Flap Settings Effect on Go-Around.

APPROACH SPEED AND CATEGORY

Two other critical performance factors that should be considered during the planning phase of an instrument approach are aircraft approach category and planned approach speed. According to the December 26, 2002 amendment of Part 97.3 (b), **aircraft approach category** means a grouping of aircraft based on **reference landing speed (V_{REF})**, if specified, or if V_{REF} is not specified, 1.3 V_{SO} (the stalling speed or minimum steady flight speed in the landing configuration) at the maximum certificated landing weight. V_{REF} refers to the speed used in establishing the approved landing distance under the airworthiness regulations constituting the type certification basis of the airplane, regardless of whether that speed for a particular airplane is 1.3 V_{SO}, 1.23 V_{SR}, or some higher speed required for airplane controllability such as when operating with a failed engine. The categories are as follows:

- Category A: Speed less than 91 knots.

- Category B: Speed 91 knots or more but less than 121 knots.

- Category C: Speed 121 knots or more but less than 141 knots.

- Category D: Speed 141 knots or more but less than 166 knots.

- Category E: Speed 166 knots or more.

- NOTE: Helicopter pilots may use the Category A line of minimums provided the helicopter is operated at Category A airspeeds.

An airplane is certified in only one approach category, and although a faster approach may require higher category minimums to be used, an airplane cannot be flown to the minimums of a slower approach category. The certified approach category is permanent, and independent of the changing conditions of day-to-day operations. From a TERPS viewpoint, the importance of a pilot not operating an airplane at a category line of minimums lower than the airplane is certified for is primarily the margin of protection provided for containment of the airplane within the procedure design for a slower airplane. This includes height loss at the decision altitude, missed approach climb surface, and turn containment in the missed approach at the higher category speeds. Pilots are responsible for determining if a higher approach category applies. If a faster approach speed is used that places the aircraft in a higher approach category, the minimums for the appropriate higher category must be used. Emergency returns at weights in excess of maximum certificated landing weight, approaches made with inoperative flaps, and approaches made in icing conditions for some airplanes are examples of situations that can necessitate the use of a higher approach category minima.

Circling approaches conducted at faster-than-normal straight-in approach speeds also require a pilot to consider the larger circling approach area, since published circling minimums provide obstacle clearance only within the appropriate area of protection, and is based on the approach category speed. [Figure 5-3] The circling approach area is the obstacle clearance area for airplanes maneuvering to land on a runway that does not meet the criteria for a straight-in approach. The size of the circling area varies with the approach category of the airplane, as shown in Figure 5-3. A minimum of 300 feet of obstacle clearance is provided in the circling segment. Pilots should remain at or above the circling altitude until the airplane is continuously in a position from which a descent to a landing on the intended runway can be made at a normal rate of descent and using normal maneuvers. Since an approach category can make a difference in the approach and weather minimums and, in some cases, prohibit flight crews from initiating an approach, the approach speed should be calculated and the effects on the approach determined and briefed in the preflight planning phase, as well as reviewed prior to commencing an approach.

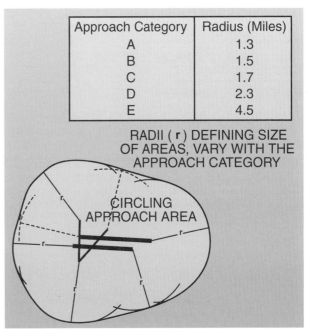

Approach Category	Radius (Miles)
A	1.3
B	1.5
C	1.7
D	2.3
E	4.5

RADII (**r**) DEFINING SIZE OF AREAS, VARY WITH THE APPROACH CATEGORY

CIRCLING APPROACH AREA

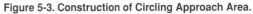

Figure 5-3. Construction of Circling Approach Area.

OPERATIONAL CONSIDERATIONS

Most commercial operators dictate standard procedures for conducting instrument approaches in their FAA approved manuals. These standards designate company callouts, flight profiles, configurations, and other specific duties for each cockpit crewmember during the conduct of an instrument approach.

APPROACH CHART FORMATS

Beginning in February 2000, NACO began issuing the current format for IAPs. This chart was developed by the Department of Transportation, Volpe National Transportation Systems Center and is commonly referred to as the **Pilot Briefing Information** format. The NACO

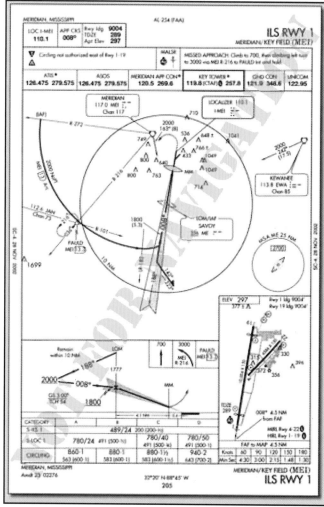

Figure 5-4. Pilot Briefing Information NACO Chart Format.

chart format is presented in a logical order, facilitating pilot briefing of the procedures. [Figure 5-4]

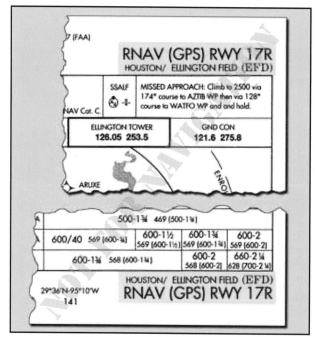

Figure 5-5. Chart Identification.

APPROACH CHART NAMING CONVENTIONS

Individual NACO charts are identified on both the top and the bottom of the page by their procedure name (based on the NAVAIDs required for the final approach), runway served, and airport location. The identifier for the airport is also listed immediately after the airport name, as shown in Figure 5-5.

There are several types of approach procedures that may cause some confusion for flight crews unfamiliar with the naming conventions. Although specific information about each type of approach will be covered later in this chapter, here are a few procedure names that can cause confusion.

STRAIGHT-IN PROCEDURES

When two or more straight-in approaches with the same type of guidance exist for a runway, a letter suffix is added to the title of the approach so that it can be more easily identified. These approach charts start with the letter Z and continue in reverse alphabetical order. For example, consider the RNAV (GPS) Z RWY 13C and RNAV (RNP) Y RWY 13C approaches at Chicago Midway International Airport. [Figure 5-6] Although these two approaches can both be flown with GPS to the same runway they are significantly different, e.g., one is a "SPECIAL AIRCRAFT & AIRCREW AUTHORIZATION REQUIRED (SAAAR); one has circling minimums and the other does not; the minimums are different; and the missed approaches are not the same. The approach procedure labeled Z will have lower landing minimums than Y (some older charts may not reflect this). In this example, the LNAV MDA for the RNAV (GPS) Z RWY 13C has the lowest minimums of either approach due to the differences in the final approach ROC evaluation. This convention also eliminates any confusion with approach procedures labeled A and B, where only circling minimums are published. The designation of two area navigation (RNAV) procedures to the same runway can occur when it is desirable to accommodate panel mounted global positioning system (GPS) receivers and flight management systems (FMSs), both with and without VNAV. It is also important to note that only one of each type of approach for a runway, including ILS, VHF omnidirectional range (VOR), non-directional beacon (NDB), etc., can be coded into a database.

CIRCLING ONLY PROCEDURES

Approaches that do not have straight-in landing minimums are identified by the type of approach followed by a letter. Examples in Figure 5-7 show four procedure titles at the same airport that have only circling minimums.

As can be seen from the example, the first approach of this type created at the airport will be labeled with the letter A, and the lettering will continue in alphabetical

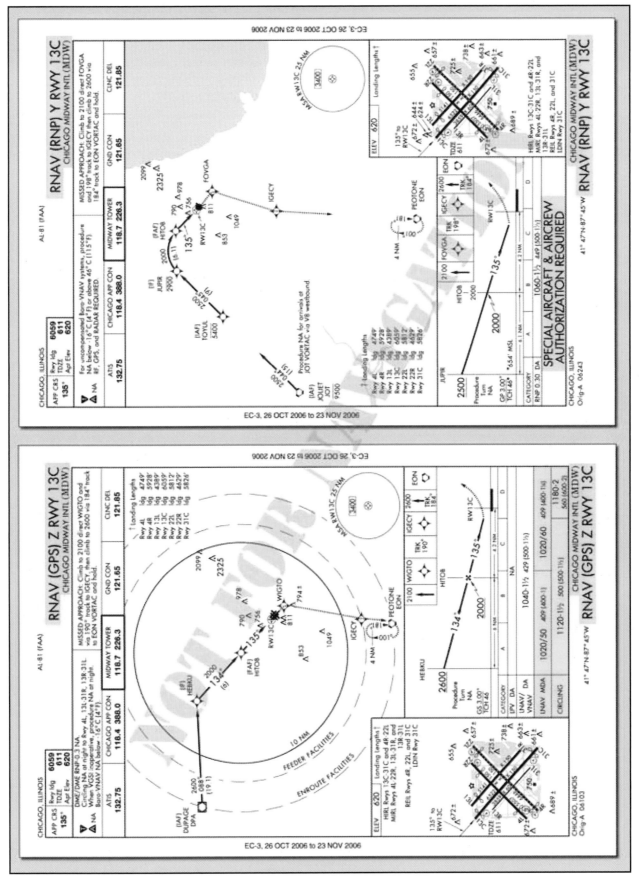

Figure 5-6. Multiple Approaches.

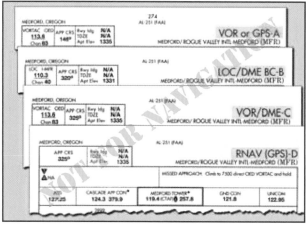

Figure 5-7. Procedures without Straight-in Landing Minimums.

order. Circling-only approaches are normally designed for one of the following reasons:

- The final approach course alignment with the runway centerline exceeds 30 degrees.

- The descent gradient is greater than 400 feet per NM from the FAF to the threshold crossing height (TCH). When this maximum gradient is

exceeded, the circling only approach procedure may be designed to meet the gradient criteria limits. This does not preclude a straight-in landing if a normal descent and landing can be made in accordance with the applicable CFRs.

AREA NAVIGATION APPROACHES

VOR distance-measuring equipment (DME) RNAV approach procedures that use collocated VOR and DME information to construct RNAV approaches are named "VOR/DME RNAV RWY XX," where XX stands for the runway number for which the approach provides guidance. Sometimes referred to as "station mover" approaches, these procedures were the first RNAV approaches issued by the FAA. They enable specific VOR/DME RNAV equipment to create waypoints on the final approach path by virtually "moving" the VOR a specific DME distance along a charted radial. [Figure 5-8]

GPS overlay procedures that are based on pre-existing nonprecision approaches contain the wording "or GPS" in the title. For instance, the title "VOR/DME or GPS A" denotes that throughout the GPS approach, the underlying ground-based NAVAIDs are not required to

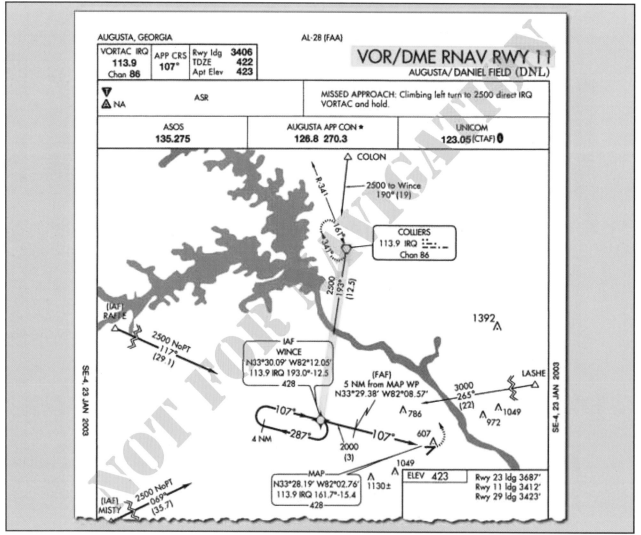

Figure 5-8. VOR/DME RNAV Approach Chart.

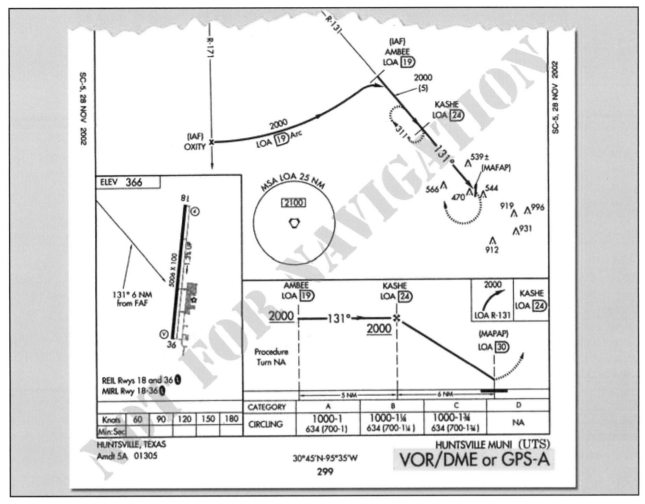

Figure 5-9. VOR/DME or GPS A Approach.

be operational and associated aircraft avionics need not be installed, operational, turned on, or monitored. [Figure 5-9] Monitoring of the underlying approach is suggested when equipment is available and functional. The procedure can be used as a GPS approach or as a traditional VOR/DME approach and may be requested using "GPS" or "VOR/DME," such as "GPS A" for the VOR/DME or GPS A. As previously mentioned, the "A" in the title shows that this is a circling approach without straight-in minimums. Many GPS overlay procedures have been replaced by stand-alone GPS or RNAV (GPS) procedures.

Stand-alone GPS procedures are not based on any other procedures, but they may replace other procedures. The naming convention used for stand-alone GPS approaches is "GPS RWY XX." The coding for the approach in the database does not accommodate multi-sensor FMSs because these procedures are designed only to accommodate aircraft using GPS equipment. These procedures will eventually be converted to RNAV (GPS) approaches. [Figure 5-10 on page 5-12]

RNAV (GPS) approach procedures have been developed in an effort to accommodate all RNAV systems, including multi-sensor FMSs used by airlines and

corporate operators. RNAV (GPS) IAPs are authorized as stand-alone approaches for aircraft equipped with RNAV systems that contain an airborne navigation database and are certified for instrument approaches. GPS systems require that the coding for a GPS approach activate the receiver autonomous integrity monitoring (RAIM) function, which is not a requirement for other RNAV equipment. The RNAV procedures are coded with both the identifier for a GPS approach and the identifier for an RNAV approach so that both systems can be used. In addition, so that the chart name, air traffic control (ATC) clearance, and database record all match, the charted title of these procedures uses both "RNAV" and "(GPS)," with GPS in parentheses. "GPS" is not included in the ATC approach clearance for these procedures.

RNP, a refinement of RNAV, is part of a collaborative effort by the FAA and the aviation industry to develop performance-based procedures. RNP is a statement of the navigation performance necessary for operation within defined airspace. RNP includes both performance and functional requirements, and is indicated by the RNP value. The RNP value designates the lateral performance requirement associated with a procedure. A key feature of

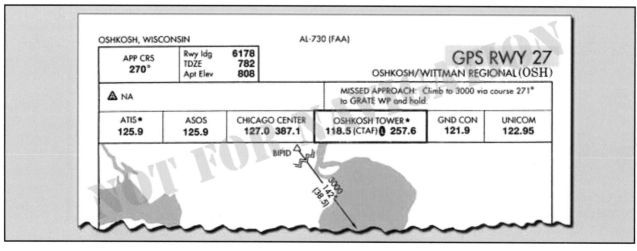

Figure 5-10. GPS Stand-alone Approach.

RNP is the concept of on-board monitoring and alerting. This means the navigation equipment is accurate enough to keep the aircraft in a specific volume of airspace, which moves along with the aircraft. The aircraft is expected to remain within this volume of airspace for at least 95 percent of the flight time, and the integrity of the system ensures the aircraft will do so. The aircraft avionics also continuously monitor sensor inputs, and through complex filtering, generate an indication in the level of confidence in the navigation performance sometimes referred to as actual navigation performance (ANP). An essential function required for RNP operations is the ability of the system to alert the pilot when the ANP exceeds the requisite RNP value.

Navigation performance for a particular RNP type is expressed numerically. Depending on the capability of each aircraft's system, RNP values can be as low as 0.1 of a nautical mile. A performance value of RNP 0.3, for example assures that the aircraft has the capability of remaining within 0.3 of a nautical mile to the right or left side of the centerline 95 percent of the time.

COMMUNICATIONS

The communication strip provided near the top of NACO approach charts gives flight crews the frequencies that they can expect to be assigned during the approach. The frequencies are listed in the logical order of use from arrival to touchdown. Having this information immediately available during the approach reduces the chances of a loss of contact between ATC and flight crews during this critical phase of flight.

It is important for flight crews to understand their responsibilities with regard to communications in the various approach environments. There are numerous differences in communication responsibilities when operating into and out of airports without air traffic control towers as compared to airports with control towers. Today's professional pilots face an ever-increasing range of ATC environments and conflicting traffic dangers, making approach

briefing and preplanning even more critical. Individual company operating manuals and SOPs dictate the duties for each crewmember.

Advisory Circular 120-71, *Standard Operating Procedures for Flight Deck Crewmembers*, contains the following concerning ATC communications:

ATC Communications: SOPs should state who handles the radios for each phase of flight (pilot flying [PF], pilot monitoring [PM], flight engineer/second officer (FE/SO), as follows:

PF makes input to aircraft/autopilot and/or verbally states clearances while PM confirms input is what he/she read back to ATC.

Any confusion in the flight deck is immediately cleared up by requesting ATC confirmation.

If any crewmember is off the flight deck, all ATC instructions are briefed upon his/her return. Or if any crewmember is off the flight deck all ATC instructions are written down until his/her return and then passed to that crewmember upon return. Similarly, if a crewmember is off ATC frequency (e.g., when making a PA announcement or when talking on company frequency), all ATC instructions are briefed upon his/her return.

Company policy should address use of speakers, headsets, boom mike and/or hand-held mikes.

APPROACH CONTROL

Approach control is responsible for controlling all instrument flights operating within its area of responsibility. Approach control may serve one or more airports. Control is exercised primarily through direct pilot and controller communication and airport surveillance radar (ASR). Prior to arriving at the IAF, instructions will be received from ARTCC to contact approach control on a specified frequency. Where radar is approved for approach control service, it is used not only for radar approaches, but also for vectors in conjunction with published non-radar approaches using conventional NAVAIDs or RNAV/GPS.

When radar handoffs are initiated between the ARTCC and approach control, or between two approach control facilities, aircraft are cleared (with vertical separation) to an outer fix most appropriate to the route being flown and, if required, given holding instructions. Or, aircraft are cleared to the airport or to a fix so located that the handoff will be completed prior to the time the aircraft reaches the fix. When radar handoffs are used, successive arriving flights may be handed off to approach control with radar separation in lieu of vertical separation.

After release to approach control, aircraft are vectored to the final approach course. ATC will occasionally vector the aircraft across the final approach course for spacing requirements. The pilot is not expected to turn inbound on the final approach course unless an approach clearance has been issued. This clearance will normally be issued with the final vector for interception of the final approach course, and the vector will enable the pilot to establish the aircraft on the final approach course prior to reaching the FAF.

AIR ROUTE TRAFFIC CONTROL CENTER

ARTCCs are approved for and may provide approach control services to specific airports. The radar systems used by these Centers do not provide the same precision as an ASR or precision approach radar (PAR) used by approach control facilities and control towers, and the update rate is not as fast. Therefore, pilots may be requested to report established on the final approach course. Whether aircraft are vectored to the appropriate final approach course or provide their own navigation on published routes to it, radar service is automatically terminated when the landing is completed; or when instructed to change to advisory frequency at airports without an operating air traffic control tower, whichever occurs first. When arriving on an IFR flight plan at an airport with an operating control tower, the flight plan will be closed automatically upon landing.

The extent of services provided by approach control varies greatly from location to location. The majority of Part 121 operations in the NAS use airports that have radar service and approach control facilities to assist in the safe arrival and departure of large numbers of aircraft. Many airports do not have approach control facilities. It is important for pilots to understand the differences between approaches with and without an approach control facility. For example, consider the Durango, Colorado, ILS DME RWY 2 and low altitude en route chart excerpt, shown in figure 5-11.

- High or lack of minimum vectoring altitudes (MVAs) – Considering the fact that most modern commercial and corporate aircraft are capable of direct, point-to-point flight, it is increasingly important for pilots to understand the limitations of ARTCC capabilities with regard to minimum altitudes. There are many airports that are below the coverage area of Center radar, and, therefore, off-route transitions into the approach environment may require that the aircraft be flown at a higher altitude than would be required for an on-route transition. In the Durango example, an airplane approaching from the northeast on a direct route to the Durango VOR may be restricted to a minimum IFR altitude (MIA) of 17,000 feet mean sea level (MSL) due to unavailability of Center radar coverage in that area at lower altitudes. An arrival on V95 from the northeast would be able to descend to a minimum en route altitude (MEA) of 12,000 feet, allowing a shallower transition to the approach environment. An off-route arrival may necessitate a descent in the published holding pattern over the DRO VOR to avoid an unstable approach into Durango.

- Lack of approach control terrain advisories – Flight crews must understand that terrain clearance cannot be assured by ATC when aircraft are operating at altitudes that are not served by Center or approach radar. Strict adherence to published routes and minimum altitudes is necessary to avoid a controlled flight into terrain (CFIT) accident. Flight crews should always familiarize themselves with terrain features and obstacles depicted on approach charts prior to initiating the approach. Approaches outside of radar surveillance require enhanced awareness of this information.

- Lack of approach control traffic advisories – If radar service is not available for the approach, the ability of ATC to give flight crews accurate traffic advisories is greatly diminished. In some cases, the common traffic advisory frequency (CTAF) may be the only tool available to enhance an IFR flight's awareness of traffic at the destination airport. Additionally, ATC will not clear an IFR flight for an approach until the preceding aircraft on the approach has cancelled IFR, either on the ground, or airborne once in visual meteorological conditions (VMC).

AIRPORTS WITH AN AIR TRAFFIC CONTROL TOWER

Towers are responsible for the safe, orderly, and expeditious flow of all traffic that is landing, taking off, operating on and in the vicinity of an airport and, when the responsibility has been delegated, towers

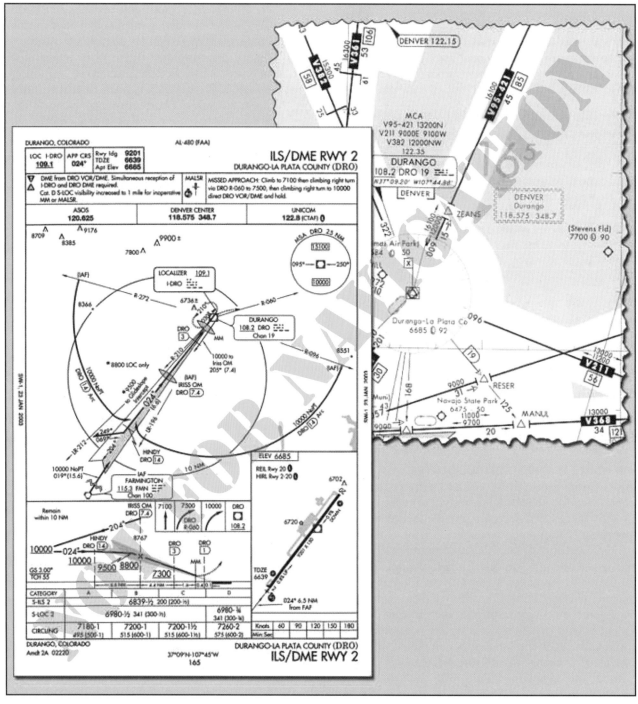

Figure 5-11. Durango Approach and Low Altitude En Route Chart Excerpt.

also provide for the separation of IFR aircraft in terminal areas. Aircraft that are departing IFR are integrated into the departure sequence by the tower. Prior to takeoff, the tower controller coordinates with departure control to assure adequate aircraft spacing.

AIRPORTS WITHOUT AN AIR TRAFFIC CONTROL TOWER

From a communications standpoint, executing an instrument approach to an airport that is not served by an ATC tower requires more attention and care than

making a visual approach to that airport. Pilots are expected to self-announce their arrival into the vicinity of the airport no later than 10 NM from the field. Depending on the weather, as well as the amount and type of conflicting traffic that exists in the area, an approach to an airport without an operating ATC tower will increase the difficulty of the transition to visual flight. In many cases, a flight arriving via an instrument approach will need to mix in with visual flight rules (VFR) traffic operating in the vicinity of the field. For this reason, many companies require that flight crews make contact with the arrival airport CTAF or company operations personnel via a secondary radio over 25 NM from the field in order to receive traffic advisories. In addition, pilots should attempt to listen to the CTAF well in advance of their arrival in order to determine the VFR traffic situation.

Since separation cannot be provided by ATC between IFR and VFR traffic when operating in areas where there is no radar coverage, pilots are expected to make radio announcements on the CTAF. These announcements allow other aircraft operating in the vicinity to plan their departures and arrivals with a minimum of conflicts. In addition, it is very important for crews to maintain a listening watch on the CTAF to increase their awareness of the current traffic situation. Flights inbound on an instrument approach to a field without a control tower should make several self-announced radio calls during the approach:

- Initial call within 5-10 minutes of the aircraft's arrival at the IAF. This call should give the aircraft's location as well as the crew's approach intentions.

- Departing the IAF, stating the approach that is being initiated.

- Procedure turn (or equivalent) inbound.

- FAF inbound, stating intended landing runway and maneuvering direction if circling.

- Short final, giving traffic on the surface notification of imminent landing.

When operating on an IFR flight plan at an airport without a functioning control tower, pilots must initiate cancellation of the IFR flight plan with ATC or an AFSS. Remote communications outlets (RCOs) or ground communications outlets (GCOs), if available, can be used to contact an ARTCC or an AFSS after landing. If a frequency is not available on the ground, the pilot has the option to cancel IFR while in flight if VFR conditions can be maintained while in contact with ARTCC, as long as those conditions can be maintained until landing. Additionally, pilots can relay a message through another aircraft or contact flight service via telephone.

PRIMARY NAVAID

Most conventional approach procedures are built around a primary final approach NAVAID; others, such as RNAV (GPS) approaches, are not. If a primary NAVAID exists for an approach, it should be included in the IAP briefing, set into the appropriate backup or active navigation radio, and positively identified at some point prior to being used for course guidance. Adequate thought should be given to the appropriate transition point for changing from FMS or other en route navigation over to the conventional navigation to be used on the approach. Specific company standards

and procedures normally dictate when this changeover occurs; some carriers are authorized to use FMS course guidance throughout the approach, provided that an indication of the conventional navigation guidance is available and displayed. Many carriers, or specific carrier fleets, are required to change over from RNAV to conventional navigation prior to the FAF of an instrument approach.

Depending on the complexity of the approach procedure, pilots may have to brief the transition from an initial NAVAID to the primary and missed approach NAVAIDs. Figure 5-12 shows the Cheyenne, Wyoming, ILS Runway 27 approach procedure, which requires additional consideration during an IAP briefing.

If the 15 DME arc of the CYS VOR is to be used as the transition to this ILS approach procedure, caution must be paid to the transition from en route navigation to the initial NAVAID and then to the primary NAVAID for the ILS approach. Planning when the transition to each of these NAVAIDs occurs may prevent the use of the incorrect NAVAID for course guidance during approaches where high pilot workloads already exist.

APPROACH CHART NOTES

The navigation equipment that is required to join and fly an instrument approach procedure is indicated by the title of the procedure and notes on the chart. Straight-in IAPs are identified by the navigation system by providing the final approach guidance and the runway with which the approach is aligned (for example, VOR RWY 13). Circling-only approaches are identified by the navigation system by providing final approach guidance and a letter (for example, VOR A). More than one navigation system separated by a slant indicates that more than one type of equipment must be used to execute the final approach (for example, VOR/DME RWY 31). More than one navigation system separated by the word "or" indicates either type of equipment can be used to execute the final approach (for example, VOR or GPS RWY 15).

In some cases, other types of navigation systems, including radar, are required to execute other portions of the approach or to navigate to the IAF (for example, an NDB procedure turn to an ILS, or an NDB in the missed approach, or radar required to join the procedure or identify a fix). When ATC radar or other equipment is required for procedure entry from the en route environment, a note is charted in the planview of the approach procedure chart (for example, RADAR REQUIRED or ADF REQUIRED). When radar or other equipment is required on portions of the procedure outside the final approach segment, including the missed approach, a note is charted in the notes box of the pilot briefing portion of the approach chart (for example, RADAR REQUIRED or DME REQUIRED). Notes are not charted when VOR is

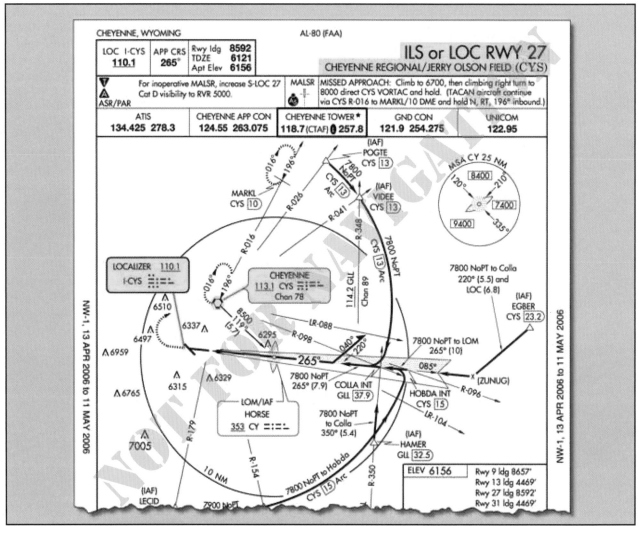

Figure 5-12. Cheyenne (KCYS), Cheyenne, Wyoming, ILS or LOC RWY 27.

required outside the final approach segment. Pilots should ensure that the aircraft is equipped with the required NAVAIDs to execute the approach, including the missed approach.

COURSES

An aircraft that has been cleared to a holding fix and subsequently "*cleared...approach,*" normally does not receive new routing. Even though clearance for the approach may have been issued prior to the aircraft reaching the holding fix, ATC would expect the pilot to proceed via the holding fix which was the last assigned route, and the feeder route associated with that fix, if a feeder route is published on the approach chart, to the IAF to commence the approach. When cleared for the approach, the published off-airway (feeder) routes that lead from the en route structure to the IAF are part of the approach clearance.

If a feeder route to an IAF begins at a fix located along the route of flight prior to reaching the holding fix, and clearance for an approach is issued, a pilot should commence the approach via the published feeder route; for

example, the aircraft would not be expected to overfly the feeder route and return to it. The pilot is expected to commence the approach in a similar manner at the IAF, if the IAF for the procedure is located along the route of flight to the holding fix.

If a route of flight directly to the IAF is desired, it should be so stated by the controller with phraseology to include the words *"direct," "proceed direct,"* or a similar phrase that the pilot can interpret without question. When a pilot is uncertain of the clearance, ATC should be queried immediately as to what route of flight is preferred.

The name of an instrument approach, as published, is used to identify the approach, even if a component of the approach aid is inoperative or unreliable. The controller will use the name of the approach as published, but must advise the aircraft at the time an approach clearance is issued that the inoperative or unreliable approach aid component is unusable. (Example: *"Cleared ILS RWY 4, glide slope unusable."*)

AREA NAVIGATION COURSES

RNAV (GPS) approach procedures introduce their own tracking issues because they are flown using an onboard navigation database. They may be flown as coupled approaches or flown manually. In either case, navigation system coding is based on procedure design, including **waypoint (WP)** sequencing for an approach and missed approach. The procedure design will indicate whether the WP is a fly-over or fly-by, and will provide appropriate guidance for each. A **fly-by (FB) waypoint** requires the use of turn anticipation to avoid overshooting the next flight segment. A **fly-over (FO) waypoint** precludes any turn until the waypoint is over-flown, and is followed by either an intercept maneuver of the next flight segment or direct flight to the next waypoint.

Approach waypoints, except for the **missed approach waypoint (MAWP)** and the **missed approach holding waypoint (MAHWP)**, are normally fly-by waypoints. Notice that in the planview for figure 5-13 there are five

fly-by waypoints, but only the circled waypoint symbols at RWY 13 and SMITS are fly-over waypoints. If flying manually to a selected RNAV waypoint, pilots should anticipate the turn at a fly-by waypoint to ensure a smooth transition and avoid overshooting the next flight segment. Alternatively, for a fly-over waypoint, no turn is accomplished until the aircraft passes the waypoint.

There are circumstances when a waypoint may be coded into the database as both a FB WP and a FO WP, depending on how the waypoints are sequenced during the approach procedure. For example, a waypoint that serves as an IAF may be coded as a FB WP for the approach and as a FO WP when it also serves as the MAHWP for the missed approach procedure.

ALTITUDES

Prescribed altitudes may be depicted in four different configurations: minimum, maximum, recommended, and mandatory. The U.S. Government distributes

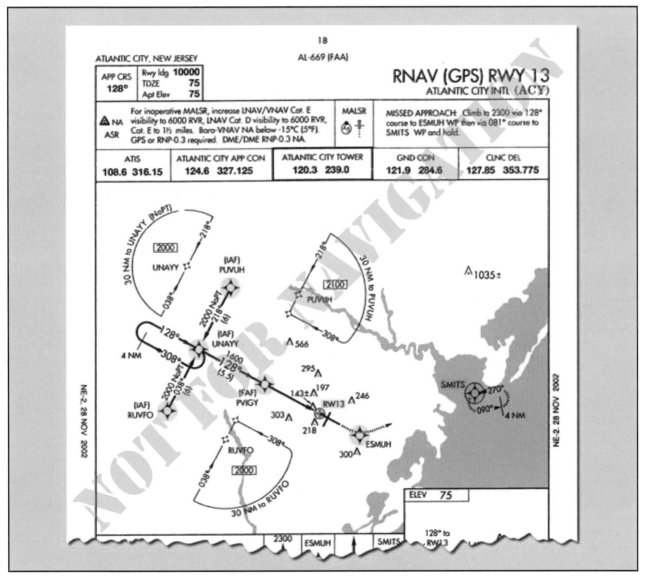

Figure 5-13. Fly-over and Fly-by Waypoints.

approach charts produced by the National Geospatial-Intelligence Agency (NGA) and NACO. Altitudes are depicted on these charts in the profile view with underscore, overscore, or both to identify them as minimum, maximum, or mandatory, respectively.

- Minimum altitudes are depicted with the altitude value underscored. Aircraft are required to maintain altitude at or above the depicted value.

- Maximum altitudes are depicted with the altitude value overscored. Aircraft are required to maintain altitude at or below the depicted value.

- Mandatory altitudes are depicted with the altitude value both underscored and overscored. Aircraft are required to maintain altitude at the depicted value.

- Recommended altitudes are depicted without an underscore or overscore.

NOTE: The underscore and overscore used to identify mandatory altitudes and overscore to identify maximum altitudes are used almost exclusively by the NGA for military charts. Pilots are cautioned to adhere to altitudes as prescribed because, in certain instances, they may be used as the basis for vertical separation of aircraft by ATC. When a depicted altitude is specified in the ATC clearance, that altitude becomes mandatory as defined above.

MINIMUM SAFE ALTITUDE

Minimum safe altitudes (MSAs) are published for emergency use on IAP charts. For conventional navigation systems, the MSA is normally based on the primary omnidirectional facility on which the IAP is predicated. The MSA depiction on the approach chart contains the facility identifier of the NAVAID used to determine the MSA. For RNAV approaches, the MSA is based on either the runway waypoint (RWY WP) or the missed approach waypoint (MAWP) for straight-in approaches, or the airport waypoint (APT WP) for circling only approaches. For RNAV (GPS) approaches with a terminal arrival area (TAA) the MSA is based on the IAF waypoint.

MSAs are expressed in feet above MSL and normally have a 25 NM radius. This radius may be expanded to 30 NM if necessary to encompass the airport landing surfaces. Ideally, a single sector altitude is established and depicted on the planview of approach charts. When necessary to maintain clearance from obstructions, the area may be further sectored and as many as four MSAs established. When established, sectors may be no less than 90°in spread. MSAs provide 1,000 feet clearance over all obstructions but do not necessarily assure acceptable navigation signal coverage.

FINAL APPROACH FIX ALTITUDE

Another important altitude that should be briefed during an IAP briefing is the FAF altitude, designated by the cross on a nonprecision approach, and the lightning bolt symbol designating the glide slope intercept altitude on a precision approach. Adherence to and crosscheck of this altitude can have a direct effect on the success of an approach.

Proper airspeed, altitude, and configuration, when crossing the FAF of a nonprecision approach, are extremely important no matter what type of aircraft is being flown. The stabilized approach concept, implemented by the FAA within the SOPs of each air carrier, suggests that crossing the FAF at the published altitude is often a critical component of a successful nonprecision approach, especially in a large turbojet aircraft.

The glide slope intercept altitude of a precision approach should also be included in the IAP briefing. Awareness of this altitude when intercepting the glide slope can ensure the flight crew that a "false glide slope" or other erroneous indication is not inadvertently followed. Many air carriers include a standard callout when the aircraft passes over the FAF of the nonprecision approach underlying the ILS. The pilot monitoring (PM) states the name of the fix and the charted glide slope altitude, thus allowing both pilots to crosscheck their respective altimeters and verify the correct indications.

MINIMUM DESCENT ALTITUDE, DECISION ALTITUDE, AND DECISION HEIGHT

MDA and DA are referenced to MSL and measured with a barometric altimeter. CAT II and III approach DHs are referenced to AGL and measured with a radio altimeter.

The height above touchdown (HAT) for a CAT I precision approach is normally 200 feet above touchdown zone elevation (TDZE). When a HAT of 250 feet or higher is published, it may be the result of the signal-in-space coverage, or there may be penetrations of either the final or missed approach obstacle clearance surfaces (OCSs). If there are OCS penetrations, the pilot will have no indication on the approach chart where the obstacles are located. It is important for pilots to brief the MDA, DA, or DH so that there is no ambiguity as to what minimums are being used. These altitudes can be restricted by many factors. Approach category, inoperative equipment in the aircraft or on the ground, crew qualifications, and company authorizations are all examples of issues that may limit or change the height of a published MDA, DA, or DH.

The primary authorization for the use of specific approach minimums by an individual air carrier can be found in Part C–Airplane Terminal Instrument Procedures, Airport Authorizations and Limitations, of its FAA approved OpsSpecs. This document lists the lowest authorized landing minimums that the carrier can use while conducting instrument approaches. Figure 5-14 shows an example of a carrier's OpsSpecs that lists minimum authorized MDAs and visibilities for nonprecision approaches.

OPERATIONS SPECIFICATIONS   30 MAR __

Straight-In Category I Approach Procedures Other Than ILS, MLS, or GPS and IFR Landing Minimums - All Airports
paragraph C53

The certificate holder shall not use any IFR Category I landing minimum lower than that prescribed by the applicable published instrument approach procedure. The IFR landing minimums prescribed in this paragraph are the lowest Category I minimums authorized for use at any airport.

a. Category Approach Procedures Other Than ILS, MLS, or GLS. The certificate holder shall not use an IFR landing minimum for straight-in nonprecision approach procedures, lower than that specified in the following table. Touchdown Zone (TDZ) RVR reports, when available for a particular runway, are controlling for all approaches to and landings on that runway. (See note 6.)

Straight-In Category I Approach (Approaches other than ILS, MLS, or GPS Landing System (GLS)					
Approach Light Configuration	HAT (See notes 1, 2, & 3)	Aircraft Category A, B, and C		Aircraft Category D	
		Visibility In Statute Miles	TDZ RVR In Feet	Visibility In Statute Miles	TDZ RVR In Feet
No Lights	250	1	5000	1	5000
ODALS or	250	3/4	4000	1	5000
MALS or	250	5/8	3000	1	5000
SALS				(See NOTE 5)	(See NOTES 5 & 6)
MALSR or SSALR or ALSF-1 or ALSF-2	250	1/2 (See NOTE 4)	2400 (See NOTES 4 & 6)	1 (See NOTE 5)	5000 (See NOTES 5 & 6)
DME ARC, Final Approach Segment, any light configuration	500	1	5000	1	5000

Note 1: For NDB approaches with a FAF, add 50 ft. to the HAT.

Note 2: For NDB approaches without a FAF, add 100 ft. to the HAT.

Note 3: For VOR approaches without a FAF, add 50 ft. to the HAT.

Note 4: For NDB approaches, the lowest authorized visibility is 3/4 and the lowest RVR is RVR 4000.

Note 5: For LOC approaches, the lowest authorized visibility is 3/4 and the lowest RVR is RVR 4000.

Note 6: The Mid RVR and Rollout RVR reports (if available) provide advisory information to pilots. The Mid RVR report may be substituted for the TDZ RVR report if the TDZ RVR report is not available.

Figure 5-14. Authorized Landing Minimums for Nonprecision Approaches.

As can be seen from the previous example, the OpsSpecs of this company rarely restrict it from using the published MDA for a nonprecision approach. In other words, most, if not all, nonprecision approaches that pilots for this company fly have published MDAs that meet or exceed its lowest authorized minimums. Therefore the published minimums are the limiting factor in these cases.

For many air carriers, OpsSpecs may be the limiting factor for some types of approaches. NDB and circling approaches are two common examples where the OpsSpecs minimum listed altitudes may be more restrictive than the published minimums. Many Part 121 and 135 operators are restricted from conducting circling approaches below 1,000-feet MDA and 3 SM visibility by Part C of their OpsSpecs, and many have specific visibility criteria listed for NDB approaches that exceed visibilities published for the approach (commonly 2 SM). In these cases, flight crews must determine which is the more restrictive of the two and comply with those minimums.

In some cases, flight crew qualifications can be the limiting factor for the MDA, DA, or DH for an instrument approach. There are many CAT II and III approach procedures authorized at airports throughout the U.S., but Special Aircraft and Aircrew Authorization Requirements (SAAAR) restrict their use to pilots who have received specific training, and aircraft that are equipped and authorized to conduct those approaches. Other rules pertaining to flight crew qualifications can also determine the lowest usable MDA, DA, or DH for a specific approach. Parts 121.652, 125.379, and 135.225 require that some pilots-in-command, with limited experience in the aircraft they are operating, increase the approach minimums and visibility by 100 feet and one-half mile respectively. Rules for these "high-minimums" pilots are usually derived from a combination of federal regulations and the company's OpsSpecs. There are many factors that can determine the actual minimums that can be used for a specific approach. All of them must be considered by pilots during the preflight and approach planning phases, discussed, and briefed appropriately.

VERTICAL NAVIGATION
One of the advantages of some GPS and multi-sensor FMS RNAV avionics is the advisory VNAV capability. Traditionally, the only way to get vertical path information during an approach was to use a ground-based precision NAVAID. Modern RNAV avionics can display an electronic vertical path that provides a constant-rate descent to minimums.

Since these systems are advisory and not primary guidance, the pilot must continuously ensure the aircraft remains at or above any published altitude constraint, including step-down fix altitudes, using the primary barometric altimeter. The pilots, airplane, and operator must be approved to use advisory VNAV inside the FAF on an instrument approach.

VNAV information appears on selected conventional nonprecision, GPS, and RNAV approaches (see Types of Approaches later in this chapter). It normally consists of two fixes (the FAF and the landing runway threshold), a FAF crossing altitude, a vertical descent angle (VDA), and may provide a visual descent point (VDP). [Figure 5-15] The published VDA is for information only, advisory in nature, and provides no additional obstacle protection below the MDA. Operators can be approved to add a height loss value to the MDA, and use this derived decision altitude (DDA) to ensure staying above the MDA. Operators authorized to use a VNAV DA in lieu of the MDA must commence a missed approach immediately upon reaching the VNAV DA if the required visual references to continue the approach have not been established.

A constant-rate descent has many safety advantages over nonprecision approaches that require multiple level-offs at stepdown fixes or manually calculating rates of descent. A stabilized approach can be maintained from the FAF to the landing when a constant-rate descent is used. Additionally, the use of an electronic vertical path produced by onboard avionics can serve to reduce CFIT, and minimize the effects of visual illusions on approach and landing.

WIDE AREA AUGMENTATION SYSTEM
In addition to the benefits that VNAV information provides for conventional nonprecision approaches, VNAV has a significant effect on approaches that are designed specifically for RNAV systems. Using an FMS or GPS that can provide both lateral navigation (LNAV) and VNAV, some RNAV approaches allow descents to lower MDAs or DAs than when using LNAV alone. The introduction of the Wide Area Augmentation System (WAAS), which became operational on July 10, 2003, provides even lower minimums for RNAV approaches that use GPS by providing electronic vertical guidance and increased accuracy.

The Wide Area Augmentation System, as its name implies, augments the basic GPS satellite constellation with additional ground stations and enhanced

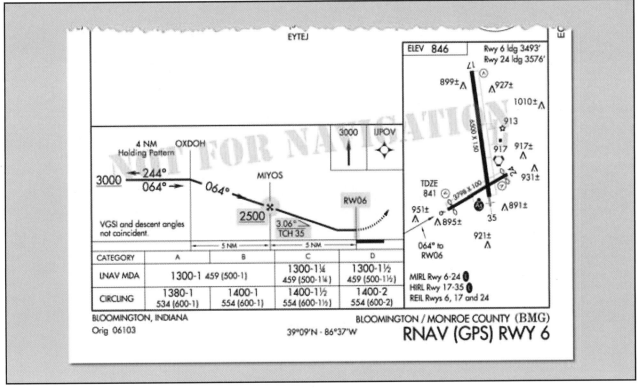

Figure 5-15. VNAV Information.

position integrity information transmitted from geostationary satellites. This capability of augmentation enhances both the accuracy and integrity of basic GPS, and may support electronic vertical guidance approach minimums as low as 200 feet HAT and 1/2 SM visibility. In order to achieve the lowest minimums, the requirements of an entire electronic vertical guidance system, including satellite availability; clear obstruction surfaces; *AC 150/5300-13, Airport Design*; and electronic vertical guidance runway and airport requirements, must be satisfied. The minimums are shown as DAs since electronically computed glidepath guidance is provided to the pilot. The electronically computed guidance eliminates errors that can be introduced when using barometric altimetry.

RNAV (GPS) approach charts presently can have up to four lines of approach minimums: LPV, LNAV/VNAV, LNAV, and Circling. Figure 5-16 shows how these minimums might be presented on an approach chart, with the exception of GLS.

• GLS — The acronym GLS stands for The Global Navigation Satellite System [GNSS] Landing System (GLS). GLS is a satellite based navigation system that provides course and glidepath information meeting the precision standards of ICAO Annex 10. Procedures based on the local area augmentation system (LAAS) will be charted separately under the GLS title as these systems are implemented.

NOTE: On RNAV approach charts the GLS minima line has been used as a placeholder only. As WAAS procedures are developed, LPV lines of minima will replace the "GLS DA-NA" lines of minima.

• LPV — APV minimums that take advantage of WAAS to provide electronic lateral and vertical guidance capability. The term **"LPV"** (localizer performance with vertical guidance) is used for approaches constructed with WAAS criteria where the value for the vertical alarm limit is more than 12 meters and less than 50 meters. WAAS avionics equipment approved for LPV approaches is required for this type of approach. The lateral guidance is equivalent to localizer accuracy, and the protected area is considerably smaller than the protected area for the present LNAV and LNAV/VNAV lateral protection. Aircraft can fly this minima line with a statement in the Aircraft Flight Manual that the installed equipment supports LPV approaches. Notice the WAAS information shown in the top left corner of the pilot briefing information on the chart depicted. Below the term WAAS is the WAAS channel number (CH 50102), and the WAAS approach identifier (W17A), indicating Runway 17R in this case, and then a letter to designate the first in a series of procedures to that runway.

• LNAV/VNAV — APV minimums used by aircraft with RNAV equipment that provides both

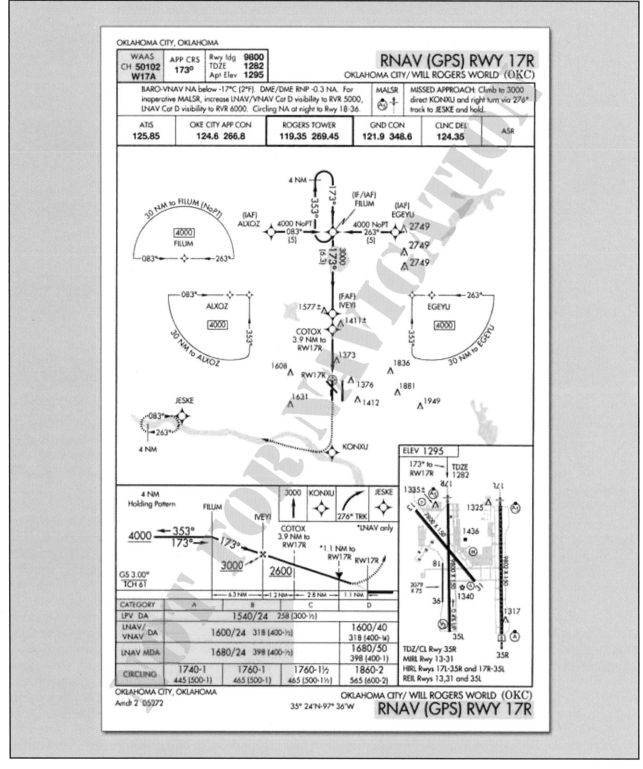

Figure 5-16. RNAV (GPS) Electronic Vertical Guidance Approach Minima.

lateral and vertical information in the approach environment, including WAAS avionics approved for LNAV/VNAV approaches, certified barometric-VNAV (Baro-VNAV) systems with an IFR approach approved GPS, or certified Baro-VNAV systems with an IFR approach approved WAAS system (See RNAV APPROACH AUTHORIZATION section for temperature limits on Baro-

VNAV). Many RNAV systems that have RNP 0.3 or less approach capability are specifically approved in the Aircraft Flight Manual. Airplanes that are commonly approved in these types of operations include Boeing 737NG, 767, and 777, as well as the Airbus A300 series. Landing minimums are shown as DAs because the approaches are flown using an electronic glidepath. Other

RNAV systems require special approval. In some cases, the visibility minimums for LNAV/VNAV might be greater than those for LNAV only. This situation occurs because DA on the LNAV/VNAV vertical descent path is farther away from the runway threshold than the LNAV MDA missed approach point.

- LNAV — minimums provided for RNAV systems that do not produce any VNAV information. IFR approach approved GPS, WAAS, or RNP 0.3 systems are required. Because vertical guidance is not provided, the procedure minimum altitude is published as an MDA. These minimums are used in the same manner as conventional nonprecision approach minimums. Other RNAV systems require special approval.

- Circling — minimums that may be used with any type of approach approved RNAV equipment when publication of straight-in approach minimums is not possible.

REQUIRED NAVIGATION PERFORMANCE

The operational advantages of RNP include accuracy and integrity monitoring, which provide more precision and lower minimums than conventional RNAV. RNP DAs can be as low as 250 feet with visibilities as low as 3/4 SM. Besides lower minimums, the benefits of RNP include improved obstacle clearance limits, as well as reduced pilot workload. When RNP-capable aircraft fly an accurate, repeatable path, ATC can be confident that these aircraft will be at a specific position, thus maximizing safety and increasing capacity.

To attain the benefits of RNP approach procedures, a key component is curved flight tracks. Constant radius turns around a fix are called "radius-to-fix legs," or RF legs. These turns, which are encoded into the navigation database, allow the aircraft to avoid critical areas of terrain or conflicting airspace while preserving positional accuracy by maintaining precise, positive course guidance along the curved track. The introduction of RF legs into the design of terminal RNAV procedures results in improved use of airspace and allows procedures to be developed to and from runways that are otherwise limited to traditional linear flight paths or, in some cases, not served by an IFR procedure at all. Navigation systems with RF capability are a prerequisite to flying a procedure that includes an RF leg. Refer to the notes box of the pilot briefing portion of the approach chart in figure 5-17.

In the United States, all RNP procedures are in the category of Special Aircraft and Aircrew Authorization Required (SAAAR). Operators who seek to take advan-

tage of RNP approach procedures must meet the special RNP requirements outlined in FAA AC 90-101, Approval Guidance for RNP Procedures with SAAAR. Currently, most new transport category airplanes receive an airworthiness approval for RNP operations. However, differences can exist in the level of precision that each system is qualified to meet. Each individual operator is responsible for obtaining the necessary approval and authorization to use these instrument flight procedures with navigation databases.

RNAV APPROACH AUTHORIZATION

Like any other authorization given to air carriers and Part 91 operators, the authorization to use VNAV on a conventional nonprecision approach, RNAV approaches, or LNAV/VNAV approaches is found in that operator's OpsSpecs, AFM, or other FAA-approved documents. There are many different levels of authorizations when it comes to the use of RNAV approach systems. The type of equipment installed in the aircraft, the redundancy of that equipment, its operational status, the level of flight crew training, and the level of the operator's FAA authorization are all factors that can affect a pilot's ability to use VNAV information on an approach.

Because most Part 121, 125, 135, and 91 flight departments include RNAV approach information in their pilot training programs, a flight crew considering an approach to North Platte, Nebraska, using the RNAV (GPS) RWY 30 approach shown in figure 5-18, would already know which minimums they were authorized to use. The company's OpsSpecs, Flight Operations Manual, and the AFM for the pilot's aircraft would dictate the specific operational conditions and procedures by which this type of approach could be flown.

There are several items of note that are specific to this type of approach that should be considered and briefed. One is the **terminal arrival area (TAA)** that is displayed in the approach planview. TAAs, discussed later in this chapter, depict the boundaries of specific arrival areas, and the MIA for those areas. The TAAs should be included in an IAP briefing in the same manner as any other IFR transition altitude. It is also important to note that the altitudes listed in the TAAs should be referenced in place of the MSAs on the approach chart for use in emergency situations.

In addition to the obvious differences contained in the planview of the previous RNAV (GPS) approach procedure example, pilots should be aware of the issues related to Baro-VNAV and RNP. The notes section of the procedure in the example contains restrictions relating to these topics.

RNP procedures are sequenced in the same manner as RNAV (GPS) procedures.

Procedure title "RNAV" includes parenthetical "(RNP)" terminology.

RNP-required sensors, FMS capabilities, and relevant procedure notes are included in the Pilot Briefing Information procedure notes section.

RF legs can be used in any segment of the procedure (transition, intermediate, final, or missed approach). RF leg turn directions (left or right) are not noted in the planview because the graphic depiction of the flight tracks is intuitive. Likewise, the arc center points, arc radius, and associated RF leg performance limits—such as bank angles and speeds—are not depicted because these aircraft performance characteristics are encoded in the navigation database.

On this particular procedure, lateral and vertical course guidance from the DA to the Runway Waypoint (Landing Threshold Point or LTP) is provided by the aircraft's FMS and onboard navigation database; however, any continued flight beyond and below the DA to the landing threshold is to be conducted under visual meteorological conditions (VMC).

RNP values for each individual leg of the procedure, defined by the procedure design criteria for containment purposes, are encoded into the aircraft's navigation database. Applicable landing minimums are shown in a normal manner along with the associated RNP value in the landing minimums section. When more than one set of RNP landing minimums is available and an aircrew is able to achieve lower RNP through approved means, the available (multiple) sets of RNP minimums are listed with the lowest set shown first; remaining sets shown in ascending order, based on the RNP value.

RNP SAAAR requirements are highlighted in large, bold print.

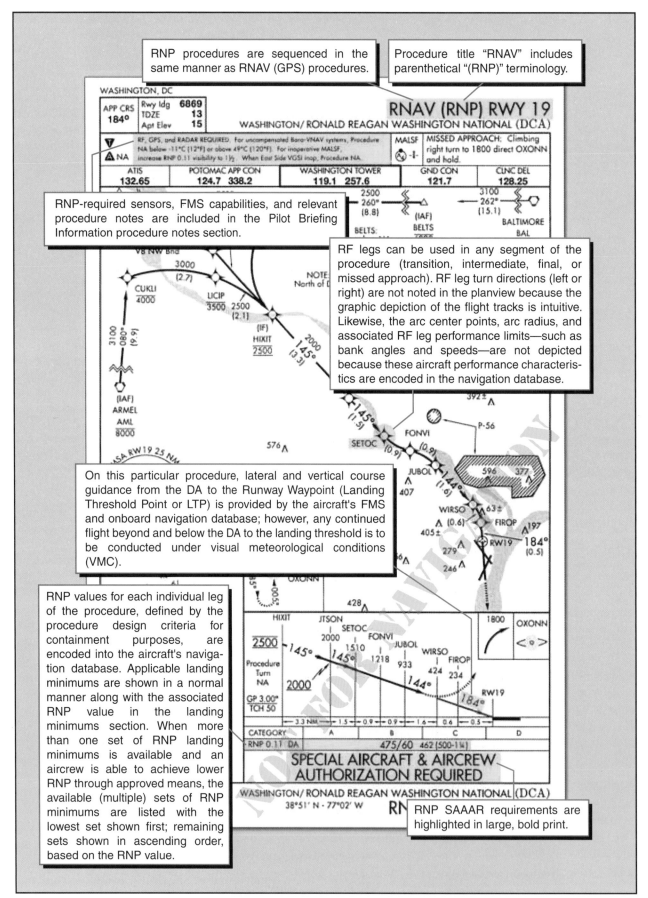

Figure 5-17. RNAV (RNP) Approach Procedure with Curved Flight Tracks.

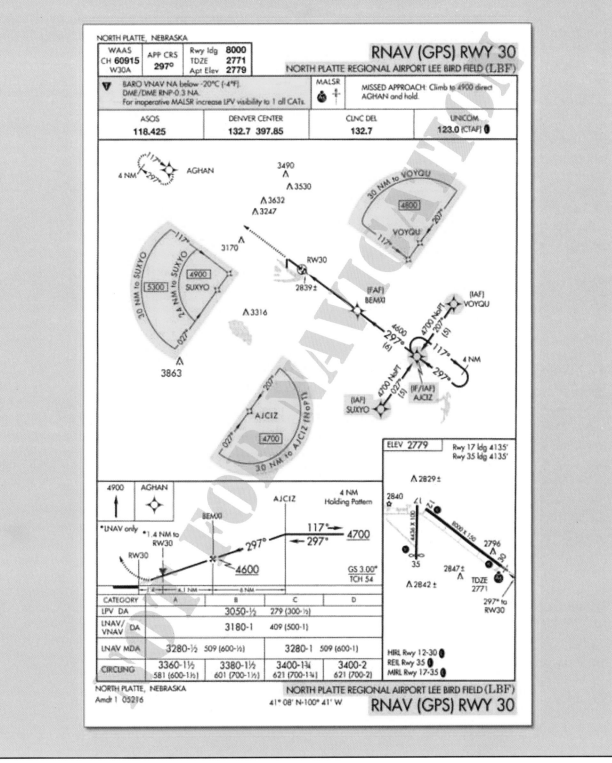

Figure 5-18. North Platte Regional (KLBF), North Platte, Nebraska, RNAV (GPS) RWY 30.

Baro-VNAV avionics provide advisory VNAV path indications to the pilot referencing a procedure's vertical path angle (VPA). The computer calculated vertical guidance is based on barometric altitude, and is either computed as a geometric path between two waypoints or an angle from a single waypoint. If a flight crew is authorized to conduct VNAV approaches using an RNAV system that falls into this category, the Baro-VNAV temperature limitations listed in the notes section of the approach procedure apply. Also, since Baro-VNAV is advisory guidance, the pilot must continuously crosscheck the primary barometric altimeter to ensure compliance with all altitude restrictions on an instrument procedure.

Considering the pronounced effect of cold temperatures on Baro-VNAV operations, a minimum temperature limitation is published for each procedure for which Baro-VNAV minimums are published. This temperature represents the airport temperature below which the use of Baro-VNAV is not authorized to the LNAV/VNAV DA. The note "Baro-VNAV NA below -20°C (-4°F)" implies that the approach may not be flown at all using Baro-VNAV when the temperature is below -20° Celsius. However, Baro-VNAV may be used for approach guidance down to the published LNAV MDA. This information can be seen in the notes section of the previous example.

In the example for the RNAV (GPS) RWY 30 approach, the note "DME/DME RNP-0.3 NA" prohibits aircraft that use only DME/DME sensors for RNAV from conducting the approach.

Because these procedures can be flown with an approach approved RNP system and "RNP" is not sensor specific, it was necessary to add this note to make it clear that those aircraft deriving RNP 0.3 using DME/DME only are not authorized to conduct the procedure.

The lowest performing sensor authorized for RNP navigation is DME/DME. The necessary DME NAVAID ground infrastructure may or may not be available at the airport of intended landing. The procedure designer has a computer program for determining the usability of DME based on geometry and coverage. Where FAA Flight Inspection successfully determines that the coverage and accuracy of DME facilities support RNP, and that the DME signal meets inspection tolerances, although there are none currently published, the note "DME/DME RNP 0.3 Authorized" would be charted. Where DME facility availability is a factor, the note would read, "DME/DME RNP 0.3 Authorized; ABC and XYZ required," meaning that ABC and XYZ DME facilities are required to assure RNP 0.3.

AIRPORT/RUNWAY INFORMATION
Another important piece of a thorough approach briefing is the discussion of the airport and runway environment. A detailed examination of the runway length (this must include the Airport/Facility Directory for the landing distance available), the intended turnoff taxiway, and the route of taxi to the parking area, are all important briefing items. In addition, runway conditions should be discussed. The effect on the aircraft's performance must be considered if the runway is wet or contaminated.

NACO approach charts include a runway sketch on each approach chart to make important airport information easily accessible to pilots. In addition, at airports that have complex runway/taxiway configurations, a separate full-page airport diagram will be published.

The airport diagram also includes the latitude/longitude information required for initial programming of FMS equipment. The included latitude/longitude grid shows the specific location of each parking area on the airport surface for use in initializing FMSs. Figure 5-19 shows the airport sketch and diagram for Chicago-O'Hare International Airport.

Pilots making approaches to airports that have this type of complex runway and taxiway configuration must ensure that they are familiar with the airport diagram prior to initiating an instrument approach. A combination of poor weather, high traffic volume, and high ground controller workload makes the pilot's job on the ground every bit as critical as the one just performed in the air.

INSTRUMENT APPROACH PROCEDURE BRIEFING
A thorough instrument approach briefing greatly increases the likelihood of a successful instrument approach. Most Part 121, 125, and 135 operators designate specific items to be included in an IAP briefing, as well as the order in which those items will be briefed.

Before an IAP briefing can begin, flight crews must decide which procedure is most likely to be flown from the information that is available to them. Most often, when the flight is being conducted into an airport that has ATIS information, the ATIS will provide the pilots with the approaches that are in use. If more than one approach is in use, the flight crew may have to make an educated guess as to which approach will be issued to them based on the weather, direction of their arrival into the area, any published airport NOTAMs, and previous experience at the specific airport. If the crew is in contact with the approach control facility, they can query ATC as to which approach is to be expected from the controller. Pilots may request specific approaches to meet the individual needs of their equipment or regulatory restrictions at any time and ATC will, in most cases, be able to accommodate those requests, providing that workload and traffic permit.

If the flight is operating into an airport without a control tower, the flight crew will occasionally be given the choice of any available instrument approach at the field. In these cases, the flight crew must choose an appropriate approach based on the expected weather, aircraft performance, direction of arrival, airport NOTAMs, and previous experience at the airport.

NAVIGATION AND COMMUNICATION RADIOS
Once the anticipated approach and runway have been selected, each crewmember sets up their "side" of the cockpit. The pilots use information gathered from ATIS, dispatch (if available), ATC, the specific approach chart for the approach selected, and any other

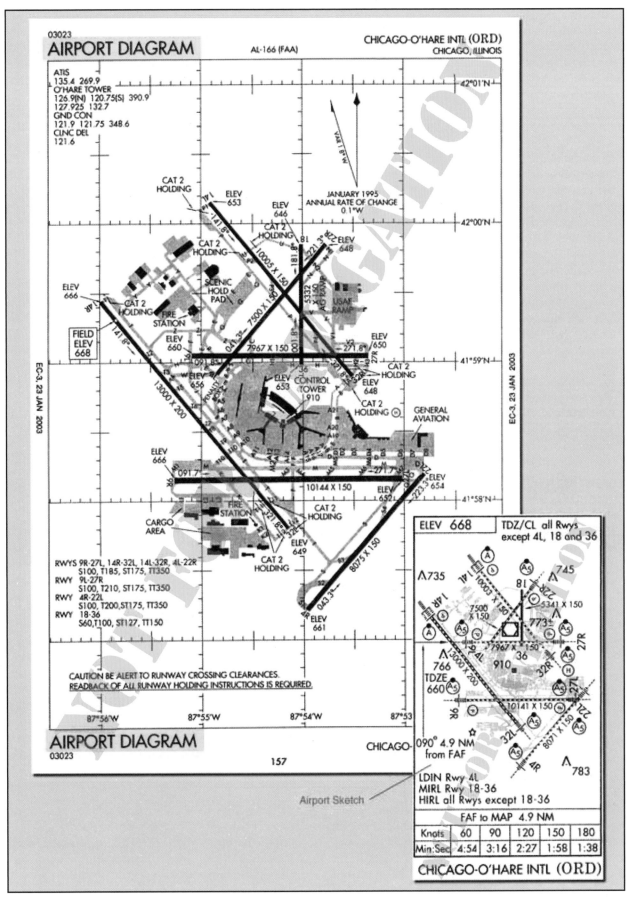

Figure 5-19. Airport Sketch and Diagram for Chicago-O'Hare International.

sources that are available. Company regulations dictate how certain things are set up and others are left up to pilot technique. In general, the techniques used at a specific company are similar. This section addresses two-pilot operations. During single-pilot IFR flights, the same items must be set up and the pilot should still do an approach briefing to verify that everything is set up correctly.

The number of items that can be set up ahead of time depends on the level of automation of the aircraft and the avionics available. In a conventional cockpit, the only things that can be set up, in general, are the airspeed bugs (based on performance calculations), altimeter bug (to DA, DH, or MDA), go around thrust/power setting, the radio altimeter bug (if installed and needed for the approach), and the navigation/communication radios (if a standby frequency selector is available). The standby side of the PF navigation radio should be set to the primary NAVAID for the approach and the PM navigation radio standby selector should be set to any other NAVAIDs that are required or available, and as dictated by company procedures, to add to the overall situational awareness of the crew. The automatic direction finder (ADF) should also be tuned to an appropriate frequency as required by the approach, or as selected by the crew.

FLIGHT MANAGEMENT SYSTEMS

In addition to the items that are available on a conventional cockpit aircraft, glass-cockpit aircraft, as well as aircraft with an approved RNAV (GPS) system, usually give the crew the ability to set the final approach course for the approach selected and many other options to increase situational awareness. Crews of FMS equipped aircraft have many options available as far as setting up the flight management computer (FMC), depending on the type of approach and company procedures. The PF usually programs the FMC for the approach and the PM verifies the information. A menu of available approaches is usually available to select from based on the destination airport programmed at the beginning of the flight or a new destination selected while en route.

The amount of information provided for the approach varies from aircraft to aircraft, but the crew can make modifications if something is not pre-programmed into the computer, such as adding a missed approach procedure or even building an entire approach for situational awareness purposes only. The PF can also program a VNAV profile for the descent and LNAV for segments that were not programmed during preflight, such as a standard terminal arrival route (STAR) or expected route to the planned approach. Any crossing restrictions for

the STAR might need to be programmed as well. The most common crossing restrictions, whether mandatory or "to be expected," are usually automatically programmed when the STAR is selected, but can be changed by ATC at any time. Other items that need to be set up are dictated by aircraft-specific procedures, such as autopilot, auto-throttles, auto-brakes, pressurization system, fuel system, seat belt signs, anti-icing/de-icing equipment, igniters, etc.

AUTOPILOT MODES

In general, an autopilot can be used to fly approaches even if the FMC is inoperative (refer to the specific airplane's minimum equipment list [MEL] to determine authorization for operating with the FMC inoperative). Whether or not the FMC is available, use of the autopilot should be discussed during the approach briefing, especially regarding the use of the altitude pre-selector and auto-throttles, if equipped. The AFM for the specific airplane outlines procedures and limitations required for the use of the autopilot during an instrument approach in that aircraft.

There are just as many different autopilot modes to climb or descend the airplane, as there are terms for these modes (ex. Level Change [LVL CHG], Vertical Speed [V/S], VNAV, Takeoff/Go Around [TO/GA], etc.). The pilot controls the airplane through the autopilot by selecting pitch modes and/or roll modes, as well as the associated auto-throttle modes. This panel, sometimes called a mode control panel, is normally accessible to both pilots. Most aircraft with sophisticated auto-flight systems and auto-throttles have the capability to select modes that climb the airplane with maximum climb thrust and descend the airplane with the throttles at idle (LVL CHG, Flight Level Change [FL CHG], Manage Level, etc.). They also have the capability to "capture," or level off at pre-selected altitudes, as well as track a LOC and glide slope (G/S) or a VOR course. If the airplane is RNAV equipped, the autopilot will also track the RNAV generated course. Most of these modes will be used at some point during an instrument approach using the autopilot. Additionally, these modes can be used to provide flight director (FD) guidance to the pilot while hand-flying the aircraft.

For the purposes of this precision approach example, the auto-throttles are engaged when the autopilot is engaged and specific airspeed and configuration changes will not be discussed. The PF controls airspeed with the speed selector on the mode control panel and calls for flaps and landing gear as needed, which the PM will select. The example in figure 5-20 begins with the airplane 5 NM northwest of BROWN at 4,500 feet with the autopilot engaged, and the flight has been cleared to track the Rwy 12 LOC inbound. The current roll mode is LOC with the PF's NAV radio tuned to the LOC frequency of 109.3; and the current pitch mode is altitude hold (ALT HOLD). Approach control clears the airplane

for the approach. The PF makes no immediate change to the autopilot mode to prevent the aircraft from capturing a false glide slope; but the PM resets the altitude selector to 2,200 feet. The aircraft will remain level because the pitch mode remains in ALT HOLD until another pitch mode is selected. Upon reaching BROWN, the PF selects LVL CHG as the pitch mode. The auto-throttles retard to idle as the

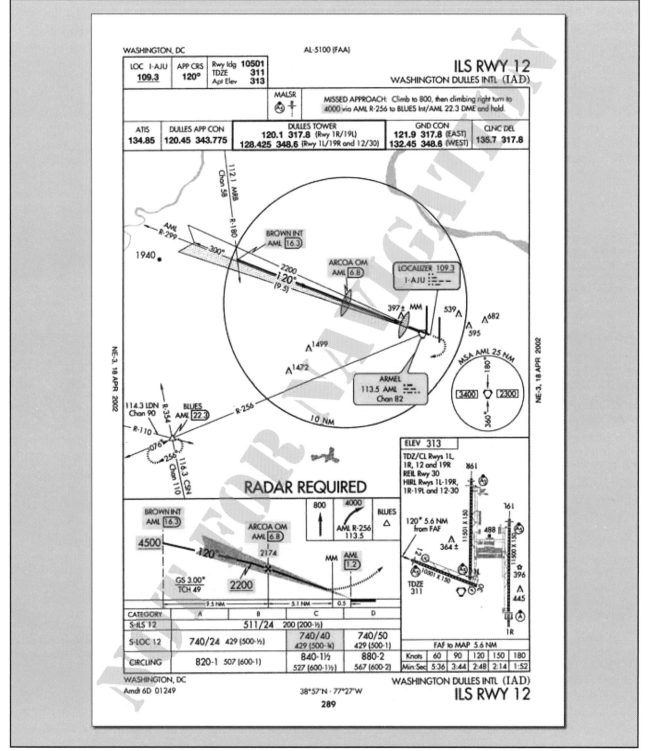

Figure 5-20. Example Approaches Using Autopilot.

airplane begins a descent. Approaching 2,200 feet, the pitch mode automatically changes to altitude acquire (ALT ACQ) then to ALT HOLD as the airplane levels at 2,200 feet. In addition to slowing the airplane and calling for configuration changes, the PF selects approach mode (APP). The roll mode continues to track the LOC and the pitch mode remains in ALT HOLD; however, the G/S mode arms. Selecting APP once the aircraft has leveled at the FAF altitude is a suggested technique to ensure that the airplane captures the glide slope from below, and that a false glide slope is not being tracked.

The PF should have the aircraft fully configured for landing before intercepting the glide slope to ensure a stabilized approach. As the airplane intercepts the glide slope, the pitch mode changes to G/S. Once the glide slope is "captured" by the autopilot, the PM can select the missed approach altitude in the altitude pre-selector, as requested by the PF. The airplane will continue to track the glide slope. The minimum altitude at which the PF is authorized to disconnect the autopilot is airplane specific (Example, 50 feet below DA, DH, or MDA but not less than 50 feet AGL). The PF can disconnect the autopilot at any time prior to reaching this altitude during a CAT I approach. The initial missed approach is normally hand flown with flight director guidance unless both autopilots are engaged for autoland during a CAT II or III approach.

The differences when flying the underlying nonprecision approach begin when the aircraft has leveled off at 2,200 feet. Once ALT HOLD is annunciated the MDA is selected by the PM as requested by the PF. **It is extremely important for both pilots to be absolutely sure that the correct altitude is selected for the MDA so that the airplane will not inadvertently descend below the MDA.** For aircraft that the altitude pre-selector can only select 100-foot increments, the MDA for this approach must be set at 800 feet instead of 740 feet.

Vertical speed mode is used from the FAF inbound to allow for more precise control of the descent. If the pilots had not selected the MDA in the altitude pre-selector window, the PF would not be able to input a V/S and the airplane would remain level. The autopilot mode will change from ALT ACQ to ALT HOLD as the airplane levels at 800 feet. Once ALT HOLD is annunciated, the PF calls for the missed approach altitude of 4,000 feet to be selected in the altitude pre-selector window. This step is very important because accurate FD guidance will not be available to the PF during a missed approach if the MDA is left in the window.

NOTE: See Maximum Acceptable Descent Rates under the heading Descent Rates and Glidepaths for Nonprecision Approaches.

STABILIZED APPROACH

In instrument meteorological conditions (IMC), you must continuously evaluate instrument information throughout an approach to properly maneuver the aircraft (or monitor autopilot performance) and to decide on the proper course of action at the decision point (DA, DH, or MAP). Significant speed and configuration changes during an approach can seriously degrade situational awareness and complicate the decision of the proper action to take at the decision point. The swept wing handling characteristics at low airspeeds and slow engine-response of many turbojets further complicate pilot tasks during approach and landing operations. You must begin to form a decision concerning the probable success of an approach before reaching the decision point. Your decision-making process requires you to be able to determine displacements from the course or glidepath centerline, to mentally project the aircraft's three-dimensional flight path by referring to flight instruments, and then apply control inputs as necessary to achieve and maintain the desired approach path. This process is simplified by maintaining a constant approach speed, descent rate, vertical flight path, and configuration during the final stages of an approach. This is referred to as the stabilized approach concept.

A stabilized approach is essential for safe turbojet operations and commercial turbojet operators must establish and use procedures that result in stabilized approaches. A stabilized approach is also strongly recommended for propeller-driven airplanes and helicopters. You should limit configuration changes at low altitudes to those changes that can be easily accommodated without adversely affecting your workload. For turbojets, the airplane must be in an approved configuration for landing or circling, if appropriate, with the engines spooled up, and on the correct speed and flight path with a descent rate of less than 1,000 FPM before descending below the following minimum stabilized approach heights:

- For all straight-in instrument approaches (this includes contact approaches) in IFR weather conditions, the approach must be stabilized before descending below 1,000 feet above the airport or TDZE.

- For visual approaches and straight-in instrument approaches in VFR weather conditions, the approach must be stabilized before descending below 500 feet above the airport elevation.

- For the final segment of a circling approach maneuver, the approach must be stabilized 500 feet above the airport elevation or at the MDA, whichever is lower.

These conditions must be maintained throughout the approach until touchdown for the approach to be considered a stabilized approach. This also helps you to recognize a windshear situation should abnormal indications exist during the approach.

DESCENT RATES AND GLIDEPATHS FOR NONPRECISION APPROACHES

Maximum Acceptable Descent Rates: Operational experience and research have shown that a descent rate of greater than approximately 1,000 FPM is unacceptable during the final stages of an approach (below 1,000 feet AGL). This is due to a human perceptual limitation that is independent of the type of airplane or helicopter. Therefore, the operational practices and techniques must ensure that descent rates greater than 1,000 FPM are not permitted in either the instrument or visual portions of an approach and landing operation.

For short runways, arriving at the MDA at the MAP when the MAP is located at the threshold may require a missed approach for some airplanes. For nonprecision approaches a descent rate should be used that will ensure that the airplane reaches the MDA at a distance from the threshold that will allow landing in the touchdown zone. On many IAPs this distance will be annotated by a VDP. To determine the required rate of descent, subtract the TDZE from the FAF altitude and divide this by the time inbound. For example if the FAF altitude is 2,000 feet MSL, the TDZE is 400 feet MSL and the time inbound is two minutes, an 800 FPM rate of descent should be used.

To verify the airplane is on an approximate 3° glidepath, use a calculation of "300-foot-to 1 NM." The glidepath height above TDZE is calculated by multiplying the NM distance from the threshold by 300. For example, at 10 NM the aircraft should be 3,000 feet above the TDZE, at 5 NM 1,500 feet, at 2 NM 600 feet, at 1.5 NM 450 feet, etc., until a safe landing can be made. In the above example the aircraft should arrive at the MDA (800 feet MSL) approximately 1.3 NM from the threshold and in a position to land in the touchdown zone. Techniques for deriving a "300-to-1" glidepath include using distance measuring equipment (DME), distance advisories provided by radar-equipped control towers, RNAV (exclusive of Omega navigation systems), GPS, dead reckoning, and pilotage when familiar features on the approach course are visible. The runway threshold should be crossed at a nominal height of 50 feet above the TDZE.

TRANSITION TO VISUAL

The transition from instrument flight to visual flight during an instrument approach can be very challenging, especially during low visibility operations. Additionally, single-pilot operations make the transition even more challenging. Approaches with vertical guidance add to the safety of the transition to visual because the approach is already stabilized upon visually acquiring the required references for the runway. One hundred to 200 feet prior to reaching the DA, DH, or MDA, most of the PM's attention should be outside of the aircraft in order to visually acquire at least one visual reference for the runway, as required by the regulations. The PF should stay focused on the instruments until the PM calls out any visual aids that can be seen, or states *"runway in sight."* The PF should then begin the transition to visual flight. It is common practice for the PM to call out the V/S during the transition to confirm to the PF that the instruments are being monitored, thus allowing more of the PF's attention to be focused on the visual portion of the approach and landing. Any deviations from the stabilized approach criteria should also be announced by the PM.

Single-pilot operations can be much more challenging because the pilot must continue to fly by the instruments while attempting to acquire a visual reference for the runway. While it is important for both pilots of a two-pilot aircraft to divide their attention between the instruments and visual references, it is even more critical for the single-pilot operation. The flight visibility must also be at least the visibility minimum stated on the instrument approach chart, or as required by regulations. CAT II and III approaches have specific requirements that may differ from CAT I precision or nonprecision approach requirements regarding transition to visual and landing. This information can be found in the operator's OpsSpecs or Flight Operations Manual.

The visibility published on an approach chart is dependent on many variables, including the height above touchdown for straight-in approaches, or height above airport elevation for circling approaches. Other factors include the approach light system coverage, and type of approach procedure, such as precision, nonprecision, circling or straight-in. Another factor determining the minimum visibility is the penetration of the 34:1 and 20:1 surfaces. These surfaces are inclined planes that begin 200 feet out from the runway and extend outward to 10,000 feet. If there is a penetration of the 34:1 surface, the published visibility can be no lower than 3/4 SM. If there is penetration of the 20:1 surface, the published visibility can be no lower than 1 SM with a note prohibiting approaches to the affected runway at night (both straight-in and circling). [Figure 5-21] Circling may be permitted at night if penetrating obstacles are marked and lighted. If the penetrating obstacles are not marked and lighted, a note is published that night circling is "Not Authorized." Pilots should be aware of these penetrating obstacles when entering the visual and/or circling segments of an approach and take adequate precautions to avoid them.

For RNAV approaches only, the presence of a grey shaded line from the MDA to the runway symbol in the profile view, is an indication that the visual segment below the MDA is clear of obstructions on the 34:1 slope. Absence of the gray shaded area indicates the 34:1 OCS is not free of obstructions.

MISSED APPROACH
Many reasons exist for executing a missed approach. The primary reason, of course, is that the required flight visibility prescribed in the IAP being used does not exist or the required visual references for the runway cannot be seen upon arrival at the DA, DH or MAP. In addition, according to Part 91, the aircraft must continuously be in a position from which a descent to a landing on the intended runway can be made at a normal rate of descent using normal maneuvers, and for operations conducted under Part 121 or 135, unless that descent rate will allow touchdown to occur within the touchdown zone of the runway of intended landing. [Figure 5-22] CAT II and III approaches call for different visibility requirements as prescribed by the Administrator.

Once descent below the DA, DH, or MDA is begun, a missed approach must be executed if the required visibility is lost or the runway environment is no longer visible, unless the loss of sight of the runway is a result of normal banking of the aircraft during a circling approach. A missed approach procedure is also required upon the execution of a rejected landing for any reason, such as men and equipment or animals on the runway, or if the approach becomes unstabilized and a normal landing cannot be performed. After the MAP in the visual segment of a nonprecision approach there may be hazards when executing a missed approach below the MDA. Any missed approach after a DA, DH, or MAP below the DA, DH, or MDA involves additional risk until established on the published missed approach procedure course and altitude.

At airports with control towers it is common for ATC to assign alternate missed approach instructions; even so, pilots should always be prepared to fly the published

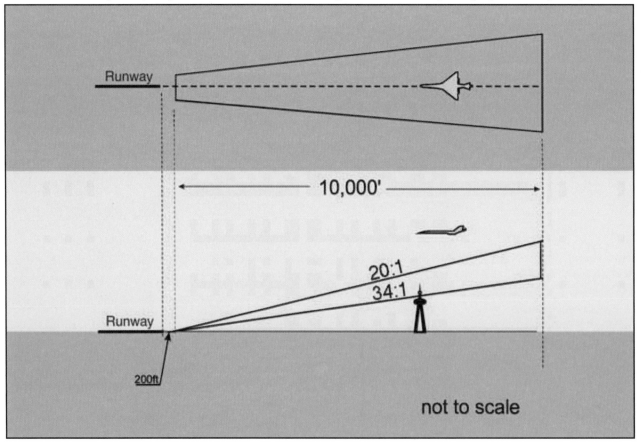

Figure 5-21. Determination of Visibility Minimums.

91.175 TAKEOFF AND LANDING UNDER IFR

(c)   Operation below DH or MDA. Where a DH or MDA is applicable, no pilot may operate an aircraft, except a military aircraft of the United States, at any airport below the authorized MDA or continue an approach below the authorized DH unless —

(1)   The aircraft is continuously in a position from which a descent to a landing on the intended runway can be made at a normal rate of descent using normal maneuvers, and for operations conducted under Part 121 or Part 135 unless that descent rate will allow touchdown to occur within the touchdown zone of the runway of intended landing.

(2)   The flight visibility is not less than the visibility prescribed in the standard instrument approach procedure being used; and

(3)   Except for a Category II or Category III approach where any necessary visual reference requirements are specified by the Administrator, at least one of the following visual references for the intended runway is distinctly visible and identifiable to the pilot:

(i)   The approach light system, except that the pilot may not descend below 100 feet above the touchdown zone elevation using the approach lights as a reference unless the red terminating bars or the red side row bars are also distinctly visible and identifiable.

(ii)   The threshold.

(iii)   The threshold markings.

(iv)   The threshold lights.

(v)   The runway end identifier lights.

(vi)   The visual approach slope indicator.

(vii)   The touchdown zone or touchdown zone markings.

(viii)   The touchdown zone lights.

(ix)   The runway or runway markings.

(x)   The runway lights.

Figure 5-22. Operation Below DA, DH, or MDA.

missed approach. When a missed approach is executed prior to reaching the MAP, the pilot is required to continue along the final approach course, at an altitude above the DA, DH, or MDA, until reaching the MAP before making any turns. If a turn is initiated prior to the MAP, obstacle clearance is not guaranteed. It is appropriate after passing the FAF, and recommended, where there aren't any climb restrictions, to begin a climb to the missed approach altitude without waiting to arrive at the MAP. Figure 5-23 gives an example of an altitude restriction that would prevent a climb between the FAF and MAP. In this situation, the Orlando Executive ILS or LOC RWY 7 approach altitude is restricted at the BUVAY 3 DME fix to prevent aircraft from penetrating the overlying protected airspace for approach routes into Orlando International Airport. If a missed approach is initiated before reaching BUVAY, a pilot may be required to continue descent to 1,200 feet before proceeding to the MAP and executing the missed approach climb instructions. In addition to the missed approach notes on the chart, the Pilot Briefing Information icons in the profile view indicate the initial vertical and lateral missed approach guidance.

The missed approach course begins at the MAP and continues until the aircraft has reached the designated fix and a holding pattern has been entered, unless there is no holding pattern published for the missed approach. It is common at large airports with high traf-

fic volume to not have a holding pattern depicted at the designated fix. [Figure 5-24 on page 5-35] In these circumstances, the departure controller will issue further instructions before the aircraft reaches the final fix of the missed approach course. It is also common for the designated fix to be an IAF so that another approach attempt can be made without having to fly from the holding fix to an IAF.

As shown in Figure 5-25 on page 5-36, there are many different ways that the MAP can be depicted, depending on the type of approach. On all approach charts it is depicted in the profile and planviews by the end of the solid course line and the beginning of the dotted missed approach course line for the "top-line"/lowest published minima. For a precision approach, the MAP is the point at which the aircraft reaches the DA or DH while on the glide slope. MAPs on nonprecision approaches can be determined in many different ways. If the primary NAVAID is on the airport, the MAP is normally the point at which the aircraft passes the NAVAID.

On some nonprecision approaches, the MAP is given as a fixed distance with an associated time from the FAF to the MAP based on the groundspeed of the aircraft. A table on the lower right hand side of the approach chart shows the distance in NM from the FAF to the MAP and the time it takes at specific groundspeeds, given in 30-knot increments. Pilots must determine the approximate groundspeed and time based on the approach speed and true airspeed of their aircraft and the current winds along the final approach course. A clock or stopwatch should be started at the FAF of an approach requiring this method. Many nonprecision approaches designate a specific fix as the MAP. These can be identified by a course (LOC or VOR) and DME, a cross radial from a VOR, or an RNAV (GPS) waypoint.

Obstacles or terrain in the missed approach segment may require a steeper climb gradient than the standard 200 feet per NM. If a steeper climb gradient is required, a note will be published on the approach chart plan view with the penetration description and examples of the required FPM rate of climb for a given groundspeed (future charting will use climb gradient). An alternative will normally be charted that allows using the standard climb gradient. [Figure 5-25 on page 5-36] In this example, if the missed approach climb requirements cannot be met for the Burbank ILS RWY 8 chart, the alternative is to use the LOC RWY 8 that is charted separately. The LOC RWY 8, S-8 procedure has a MDA that is 400 foot higher than the ILS RWY 8, S-LOC 8 MDA, and meets the standard climb gradient requirement over the terrain.

EXAMPLE APPROACH BRIEFING

During an instrument approach briefing, the name of the airport and the specific approach

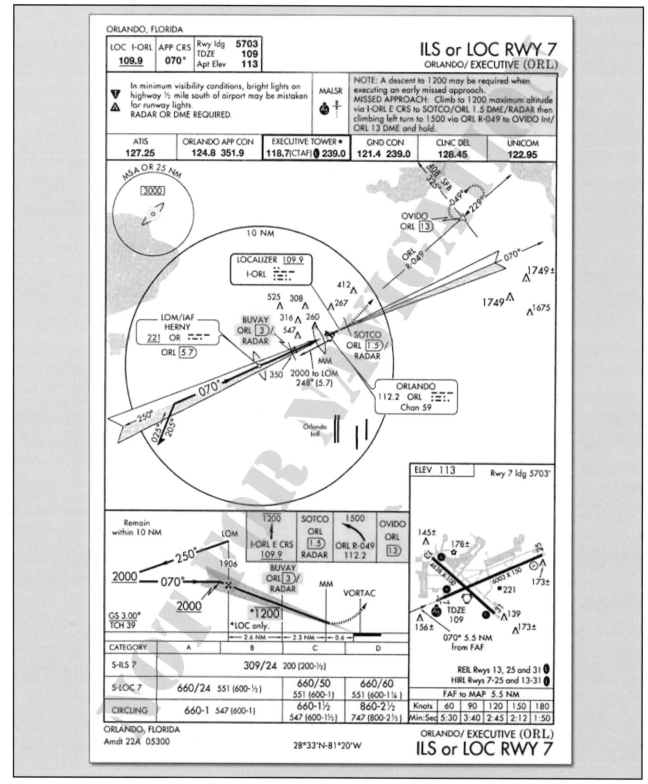

Figure 5-23. Orlando Executive Airport, Orlando, Florida, ILS RWY 7.

procedure should be identified to allow other crewmembers the opportunity to cross-reference the chart being used for the brief. This ensures that pilots intending to conduct an instrument approach have collectively reviewed and verified the information pertinent to the approach. Figure 5-26 on page 5-37 gives an example of the items to be briefed and their sequence. Although the following example is based on multi-crew aircraft, the process is also applicable to single-pilot operations. A complete instrument approach and operational briefing example follows.

The approach briefing begins with a general discussion of the ATIS information, weather, terrain, NOTAMs, approaches in use, runway conditions,

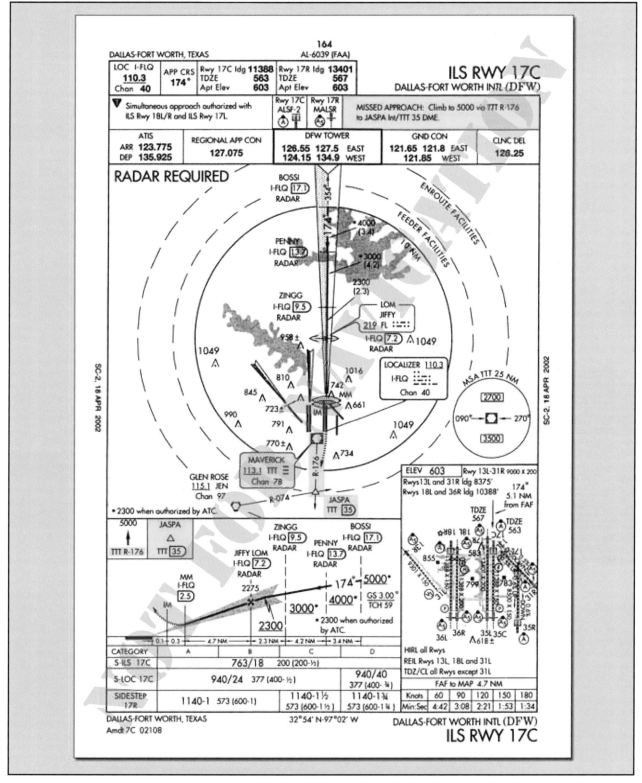

Figure 5-24. Missed Approach Procedure without Holding Pattern.

performance considerations, expected route to the final approach course, and the traffic situation. As the discussion progresses, the items and format of the briefing become more specific. The briefing can also be used as a checklist to ensure that all items have been set up correctly. Most pilots will verbally brief the specific missed approach procedure so that it is fresh in their minds and there is no confusion as to who is doing what during a missed approach. Also, it is a very good idea to brief the published missed approach even if the tower will most likely give you alternate instructions in the event of a missed approach. A typical approach briefing might sound like the following example for a flight inbound to the Monroe Regional Airport (KMLU):

5-35

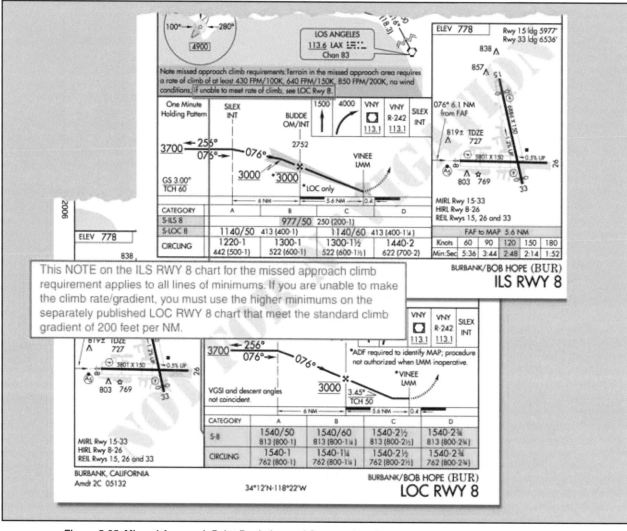

Figure 5-25. Missed Approach Point Depiction and Steeper than Standard Climb Gradient Requirements.

ATIS: *"Monroe Regional Airport Information Bravo, time 2253 Zulu, wind 360 at 10, visibility 1 mile, mist, ceiling 300 overcast, temperature 4, dew point 3, altimeter 29.73, ILS Runway 4 approach in use, landing and departing Runway 4, advise on initial contact that you have information Bravo."*

PF (F/O): *"We're planning an ILS approach to Runway 4 at Monroe Regional Airport, page 216, Amdt 21 Alpha. Localizer frequency is 109.5, SABAR Locator Outer Marker is 219, Monroe VOR is 117.2, final approach course is 042°, we'll cross SABAR at 1,483 feet barometric, decision altitude is 278 feet barometric, touchdown zone elevation is 78 feet with an airport elevation of 79 feet. Missed approach procedure is climb to 2,000 feet, then climbing right turn to 3,000 feet direct SABAR locator outer marker and hold. The MSA is 2,200 feet to the north and along our missed approach course, and 3,100 feet to the south along the final approach course. ADF is required for the approach and the airport has pilot controlled lighting when the tower is closed, which does not apply to this approach. The runway has a medium intensity approach lighting system with runway alignment indicator lights and no VGSI. We need a half-mile visi-*

bility so with one mile we should be fine. Runway length is 7,507 feet. I'm planning a flaps 30 approach, autobrakes 2, left turn on Alpha or Charlie 1 then Alpha, Golf to the ramp. With a left crosswind, the runway should be slightly to the right. I'll use the autopilot until we break out and, after landing, I'll slow the aircraft straight ahead until you say you have control and I'll contact ground once we are clear of the runway. In the case of a missed approach, I'll press TOGA (Take-off/Go-Around button used on some turbojets), call 'go-around thrust, flaps 15, positive climb, gear up, set me up,' climb straight ahead to 2,000 feet then climbing right turn to 3,000 feet toward SABAR or we'll follow the tower's instructions. Any questions?"

PM (CAP): *"I'll back up the auto-speedbrakes. Other than that, I don't have any questions."*

INSTRUMENT APPROACH PROCEDURE SEGMENTS

An instrument approach may be divided into as many as four approach segments: initial, intermediate, final, and missed approach. Additionally, feeder routes provide a transition from the en route structure to the IAF.

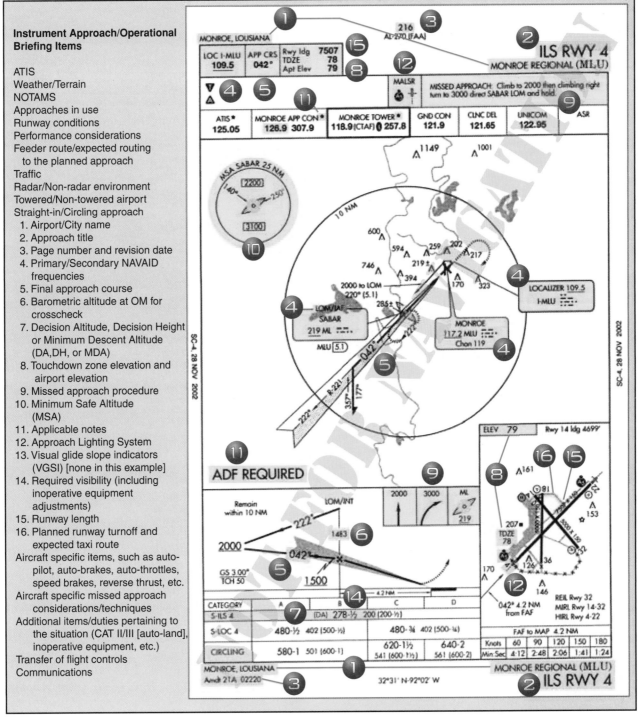

Figure 5-26. Approach Chart Briefing Sequence.

The U.S. Standard for Terminal Instrument Procedures (TERPS) criteria provides obstacle clearance for each segment of an approach procedure as shown in Figure 5-27 on page 5-38.

FEEDER ROUTES

By definition, a **feeder route** is a route depicted on IAP charts to designate courses for aircraft to proceed from the en route structure to the IAF. Feeder routes, also referred to as approach transitions, technically are not considered approach segments but are an integral part of

many IAPs. Although an approach procedure may have several feeder routes, pilots normally choose the one closest to the en route arrival point. When the IAF is part of the en route structure, there may be no need to designate additional routes for aircraft to proceed to the IAF.

When a feeder route is designated, the chart provides the course or bearing to be flown, the distance, and the minimum altitude. En route airway obstacle clearance criteria apply to feeder routes, providing 1,000 feet of obstacle clearance (2,000 feet in mountainous areas).

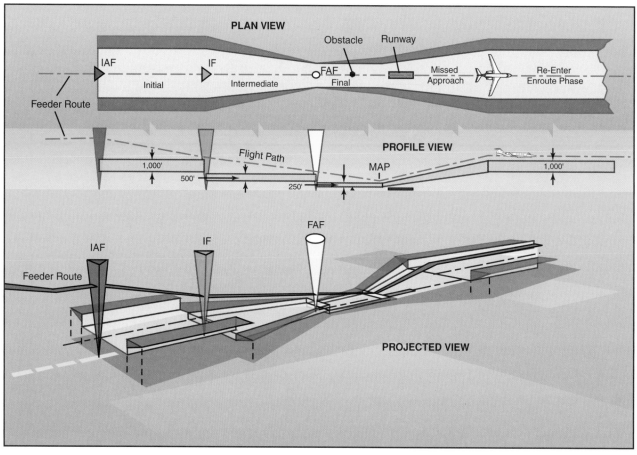

Figure 5-27. Approach Segments and Obstacle Clearance.

TERMINAL ROUTES

In cases where the IAF is part of the en route structure and feeder routes are not required, a transition or terminal route is still needed for aircraft to proceed from the IAF to the intermediate fix (IF). These routes are initial approach segments because they begin at the IAF. Like feeder routes, they are depicted with course, minimum altitude, and distance to the IF. Essentially, these routes accomplish the same thing as feeder routes but they originate at an IAF, whereas feeder routes terminate at an IAF.

DME ARCS

DME arcs also provide transitions to the approach course, but DME arcs are actually approach segments while feeder routes, by definition, are not. When established on a DME arc, the aircraft has departed the en route phase and has begun the approach and is maneuvering to enter an intermediate or final segment of the approach. DME arcs may also be used as an intermediate or a final segment, although they are extremely rare as final approach segments.

An arc may join a course at or before the IF. When joining a course at or before the IF, the angle of intersection of the arc and the course is designed so it does not exceed 120°. When the angle exceeds 90°, a radial that provides at least 2 NM of lead shall be identified to assist in leading the turn on to the intermediate course.

DME arcs are predicated on DME collocated with a facility providing omnidirectional course information, such as a VOR. A DME arc cannot be based on an ILS or LOC DME source because omnidirectional course information is not provided.

Required obstruction clearance (ROC) along the arc depends on the approach segment. For an initial approach segment, a ROC of 1,000 feet is required in the primary area, which extends to 4 NM on either side of the arc. For an intermediate segment primary area the ROC is 500 feet. The initial and intermediate segment secondary areas extend 2 NM from the primary boundary area edge. The ROC starts at the primary area boundary edge at 500 feet and tapers to zero feet at the secondary area outer edge. [Figure 5-28]

COURSE REVERSAL

Some approach procedures do not permit straight-in approaches unless pilots are being radar vectored. In these situations, pilots will be required to complete a procedure turn (PT) or other course reversal, generally within 10 NM of the PT fix, to establish the aircraft inbound on the intermediate or final approach segment.

If Category E airplanes are using the PT or there is a descent gradient problem, the PT distance available can be as much as 15 NM. During a procedure turn, a maximum speed of 200 knots indicated airspeed

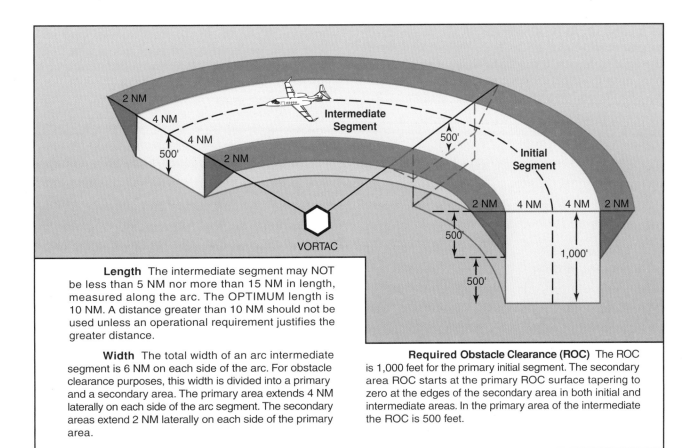

Length The intermediate segment may NOT be less than 5 NM nor more than 15 NM in length, measured along the arc. The OPTIMUM length is 10 NM. A distance greater than 10 NM should not be used unless an operational requirement justifies the greater distance.

Width The total width of an arc intermediate segment is 6 NM on each side of the arc. For obstacle clearance purposes, this width is divided into a primary and a secondary area. The primary area extends 4 NM laterally on each side of the arc segment. The secondary areas extend 2 NM laterally on each side of the primary area.

Required Obstacle Clearance (ROC) The ROC is 1,000 feet for the primary initial segment. The secondary area ROC starts at the primary ROC surface tapering to zero at the edges of the secondary area in both initial and intermediate areas. In the primary area of the intermediate the ROC is 500 feet.

Figure 5-28. DME Arc Obstruction Clearance.

(KIAS) should be observed from first crossing the course reversal IAF through the procedure turn maneuver to ensure containment within the obstruction clearance area. Unless a holding pattern or teardrop procedure is published, the point where pilots begin the turn and the type and rate of turn are optional. If above the procedure turn minimum altitude, pilots may begin descent as soon as they cross the IAF outbound.

The 45° procedure turn, the racetrack pattern (holding pattern), the teardrop procedure turn, or the 80°/260° course reversal are mentioned in the AIM as acceptable variations for course reversal. When a holding pattern is published in place of a procedure turn, pilots must make the standard entry and follow the depicted pattern to establish the aircraft on the inbound course. Additional circuits in the holding pattern are not necessary or expected by ATC if pilots are cleared for the approach prior to returning to the fix. In the event additional time is needed to lose altitude or become better established on course, pilots should advise ATC and obtain approval for any additional turns. When a teardrop is depicted and a course reversal is required, pilots also must fly the procedural track as published.

A procedure turn is the maneuver prescribed to perform a course reversal to establish the aircraft inbound on an intermediate or final approach course. The procedure turn or hold- in lieu- of- procedure

turn (PT) is a required maneuver when it is depicted on the approach chart. However, the procedure turn or the hold-in-lieu-of-PT is not permitted when the symbol "No PT" is depicted on the initial segment being flown, when a RADAR VECTOR to the final approach course is provided, or when conducting a timed approach from a holding fix. The altitude prescribed for the procedure turn is a minimum altitude until the aircraft is established on the inbound course. The maneuver must be completed within the distance specified in the profile view. The pilot may elect to use the procedure turn or hold-in-lieu-of-PT when it is not required by the procedure, but must first receive an amended clearance from ATC. When ATC is Radar vectoring to the final approach course, or to the Intermediate Fix as may occur with RNAV standard instrument approach procedures, ATC may specify in the approach clearance "CLEARED STRAIGHT-IN (type) APPROACH" to ensure that the pilot understands that the procedure turn or hold-in-lieu-of-PT is not to be flown. If the pilot is uncertain whether ATC intends for a procedure turn or a straight-in approach to be flown, the pilot shall immediately request clarification from ATC (14 CFR Part 91.123).

Approach charts provide headings, altitudes, and distances for a course reversal. Published altitudes are "minimum" altitudes, and pilots must complete the maneuver within the distance specified on the profile view (typically within 10 NM). Pilots also are required to maneuver the aircraft on the procedure turn side of the final approach course. These requirements are necessary to stay within the protected airspace and maintain adequate obstacle clearance. [Figure 5-29]

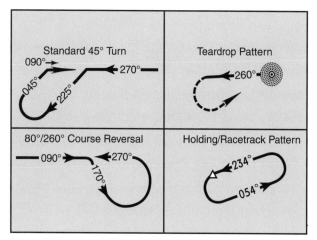

Figure 5-29. Course Reversal Methods.

A minimum of 1,000 feet of obstacle clearance is provided in the procedure turn primary area. [Figure 5-30] In the secondary area, 500 feet of obstacle clearance is provided at the inner edge, tapering uniformly to zero feet at the outer edge. The primary and secondary areas determine obstacle clearance in both the entry and maneuvering zones. The use of entry and maneuvering zones provides further relief from obstacles. The entry zone is established to control the obstacle clearance prior to proceeding outbound from the procedure turn fix. The maneuvering zone is established to control obstacle clearance after proceeding outbound from the procedure turn fix.

INITIAL APPROACH SEGMENT

The purpose of the initial approach segment is to provide a method for aligning the aircraft with the intermediate or final approach segment. This is accomplished by using a DME arc, a course reversal, such as a procedure turn or holding pattern, or by following a terminal route that intersects the final approach course. The initial approach segment

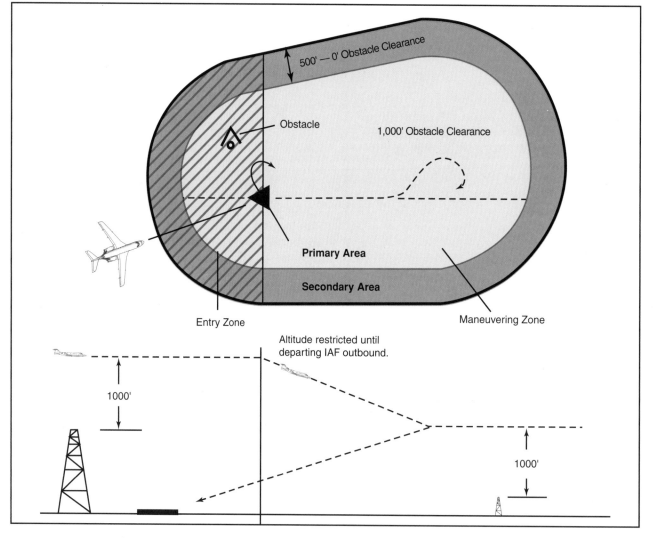

Figure 5-30. Procedure Turn Obstacle Clearance.

begins at an IAF and usually ends where it joins the intermediate approach segment or at an IF. The letters IAF on an approach chart indicate the location of an IAF and more than one may be available. Course, distance, and minimum altitudes are also provided for initial approach segments. A given procedure may have several initial approach segments. When more than one exists, each joins a common intermediate segment, although not necessarily at the same location.

Occasionally, a chart may depict an IAF, although there is no initial approach segment for the procedure. This usually occurs at a point located within the en route structure where the intermediate segment begins. In this situation, the IAF signals the beginning of the intermediate segment.

INTERMEDIATE APPROACH SEGMENT
The intermediate segment is designed primarily to position the aircraft for the final descent to the airport. Like the feeder route and initial approach segment, the chart depiction of the intermediate segment provides course, distance, and minimum altitude information.

The intermediate segment, normally aligned within 30° of the final approach course, begins at the IF, or intermediate point, and ends at the beginning of the final approach segment. In some cases, an IF is not shown on an approach chart. In this situation, the intermediate segment begins at a point where you are proceeding inbound to the FAF, are properly aligned with the final approach course, and are located within the prescribed distance prior to the FAF. An instrument approach that incorporates a procedure turn is the most common example of an approach that may not have a charted IF. The intermediate segment in this example begins when you intercept the inbound course after completing the procedure turn. [Figure 5-31]

FINAL APPROACH SEGMENT
The final approach segment for an approach with vertical guidance or a precision approach begins where the glide slope intercepts the minimum glide slope intercept altitude shown on the approach chart. If ATC authorizes a lower intercept altitude, the final approach segment begins upon glide slope interception at that altitude. For a nonprecision approach, the final approach segment begins either at a designated FAF, depicted as a cross on the profile view, or at the point where the aircraft is established inbound on the final approach course. When a FAF is not designated, such as on an approach that incorporates an on-airport VOR or NDB, this point is typically where the procedure turn intersects the final approach course inbound. This point is referred to as the final approach point (FAP). The final approach segment ends at either the designated MAP or upon landing.

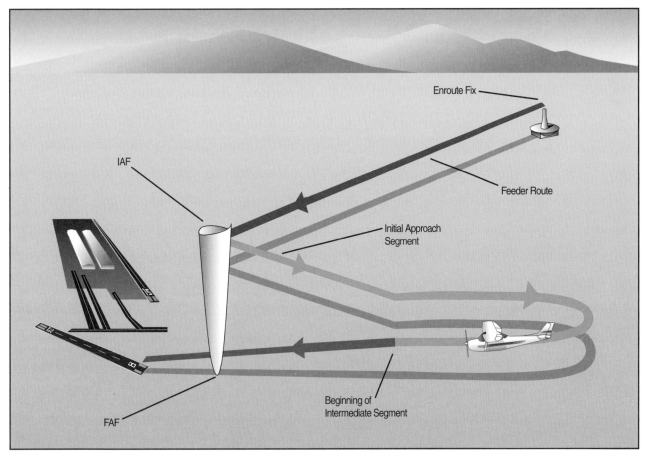

Figure 5-31. Approach without a Designated IF.

There are three types of procedures based on the final approach course guidance:

- Precision Approach (PA) — an instrument approach based on a navigation system that provides course and glidepath deviation information meeting precision standards. Precision Approach Radar (PAR), ILS, and Microwave Landing System (MLS) procedures are examples of PA procedures.

- Approach with Vertical Guidance (APV) — an instrument approach based on a navigation system that is not required to meet the precision approach standards but provides course and glidepath deviation information. Baro-VNAV, LDA with glidepath, and LPV are examples of APV approaches.

- Nonprecision Approach (NPA) — an instrument approach based on a navigation system that provides course deviation information but no glidepath deviation information is considered a NPA procedure. VOR, TACAN, LNAV, NDB, LOC and ASR approaches are examples of NPA procedures.

MISSED APPROACH SEGMENT

The missed approach segment begins at the MAP and ends at a point or fix where an initial or en route segment begins. The actual location of the MAP depends upon the type of approach you are flying. For example, during a precision or an APV approach, the MAP occurs at the DA or DH on the glide slope. For nonprecision approaches, the MAP is either a fix, NAVAID, or after a specified period of time has elapsed after crossing the FAF.

APPROACH CLEARANCE

According to *FAA Order 7110.65, Air Traffic Control*, clearances authorizing instrument approaches are issued on the basis that, if visual contact with the ground is made before the approach is completed, the entire approach procedure will be followed unless the pilot receives approval for a contact approach, is cleared for a visual approach, or cancels the IFR flight plan.

Approach clearances are issued based on known traffic. The receipt of an approach clearance does not relieve the pilot of his/her responsibility to comply with applicable Parts of the CFRs and notations on instrument approach charts, which impose on the pilot the responsibility to comply with or act on an instruction, such as "procedure not authorized at night." The name of the approach, as published, is used to identify the approach. Approach name items within parentheses are not included in approach clearance phraseology.

VECTORS TO FINAL APPROACH COURSE

The approach gate is an imaginary point used within ATC as a basis for vectoring aircraft to the final approach course. The gate will be established along the final approach course one mile from the FAF on the side away from the airport and will be no closer than 5 NM from the landing threshold. Controllers are also required to ensure the assigned altitude conforms to the following:

- For a precision approach, at an altitude not above the glide slope/glidepath or below the minimum glide slope intercept altitude specified on the approach procedure chart.

- For a nonprecision approach, at an altitude that will allow descent in accordance with the published procedure.

Further, controllers must assign headings that will permit final approach course interception without exceeding the following:

Distance from Interception Point to Approach Gate	Maximum Interception Angle
• Less than 2 NM or with triple simultaneous ILS/MLS approaches in use.	20°
• 2 NM or more	30° (45° for helicopters)

A typical vector to the final approach course and associated approach clearance is as follows:

"…four miles from LIMA, turn right heading three four zero, maintain two thousand until established on the localizer, cleared ILS runway three six approach."

Other clearance formats may be used to fit individual circumstances but the controller should always assign an altitude to maintain until the aircraft is established on a segment of a published route or IAP. The altitude assigned must guarantee IFR obstruction clearance from the point at which the approach clearance is issued until the aircraft is established on a published route. Part 91.175 (j) prohibits a pilot from making a procedure turn when vectored to a FAF or course, when conducting a timed approach, or when the procedure specifies "NO PT."

When vectoring aircraft to the final approach course, controllers are required to ensure the intercept is at least 2 NM outside the approach gate. Exceptions include the following situations, but do not apply to

RNAV aircraft being vectored for a GPS or RNAV approach:

- When the reported ceiling is at least 500 feet above the MVA/MIA and the visibility is at least 3 SM (may be a pilot report [PIREP] if no weather is reported for the airport), aircraft may be vectored to intercept the final approach course closer than 2 NM outside the approach gate but no closer than the approach gate.

- If specifically requested by the pilot, aircraft may be vectored to intercept the final approach course inside the approach gate but no closer than the FAF.

NONRADAR ENVIRONMENT

In the absence of radar vectors, an instrument approach begins at an IAF. An aircraft that has been cleared to a holding fix that, prior to reaching that fix, is issued a clearance for an approach, but not issued a revised routing, such as, *"proceed direct to…"* is expected to proceed via the last assigned route, a feeder route if one is published on the approach chart, and then to commence the approach as published. If, by following the route of flight to the holding fix, the aircraft would overfly an IAF or the fix associated with the beginning of a feeder route to be used, the aircraft is expected to commence the approach using the published feeder route to the IAF or from the IAF as appropriate. The aircraft would not be expected to overfly and return to the IAF or feeder route.

For aircraft operating on unpublished routes, an altitude is assigned to maintain until the aircraft is established on a segment of a published route or IAP. (Example: *"maintain 2,000 until established on the final approach course outbound, cleared VOR/DME runway 12."*) The International Civil Aviation Organization (ICAO) definition of established on course requires the aircraft to be within half scale deflection for the ILS and VOR, or within ±5° of the required bearing for the NDB. Generally, the controller assigns an altitude compatible with glide slope intercept prior to being cleared for the approach.

TYPES OF APPROACHES

In the NAS, there are approximately 1,033 VOR stations, 1,200 NDB stations, and 1,370 ILS installations, including 25 LOC-Type Directional Aids (LDAs), 23 Simplified Directional Facilities (SDFs), and 242 LOC only facilities. As time progresses, it is the intent of the FAA to reduce navigational dependence on VOR, NDB, and other ground-based NAVAIDs and, instead, to increase the use of satellite-based navigation.

To expedite the use of RNAV procedures for all instrument pilots, the FAA has begun an aggressive schedule to develop RNAV procedures. During 2002, the number of RNAV/GPS approaches published in the NAS exceeded 3,300, with additional procedures published every revision cycle. While it had originally been the plan of the FAA to begin decommissioning VORs, NDBs, and other ground-based NAVAIDs, the overall strategy has been changed to incorporate a majority dependence on augmented satellite navigation while maintaining a satisfactory backup system. This backup system will include retaining all CAT II and III ILS facilities and close to one-half of the existing VOR network.

Each approach is provided obstacle clearance based on the Order 8260.3 TERPS design criteria as appropriate for the surrounding terrain, obstacles, and NAVAID availability. Final approach obstacle clearance is different for every type of approach but is guaranteed from the start of the final approach segment to the runway (not below the MDA for nonprecision approaches) or MAP, whichever occurs last within the final approach area. Both pilots and ATC assume obstacle clearance responsibility, but it is dependent upon the pilot to maintain an appropriate flight path within the boundaries of the final approach area.

There are numerous types of instrument approaches available for use in the NAS including RNAV (GPS), ILS, MLS, LOC, VOR, NDB, SDF, and radar approaches. Each approach has separate and individual design criteria, equipment requirements, and system capabilities.

VISUAL AND CONTACT APPROACHES

To expedite traffic, ATC may clear pilots for a **visual approach** in lieu of the published approach procedure if flight conditions permit. Requesting a **contact approach** may be advantageous since it requires less time than the published IAP and provides separation from IFR and special visual flight rules (SVFR) traffic. A contact or visual approach may be used in lieu of conducting a SIAP, and both allow the flight to continue as an IFR flight to landing while increasing the efficiency of the arrival.

VISUAL APPROACHES

When it is operationally beneficial, ATC may authorize pilots to conduct a visual approach to the airport in lieu of the published IAP. A pilot or the controller can initiate a visual approach. Before issuing a visual approach clearance, the controller must verify that pilots have the airport, or a preceding aircraft that they are to follow, in sight. In the event pilots have the airport in sight but do not see the aircraft they are to follow, ATC may issue the visual approach clearance but will maintain responsibility for aircraft and wake turbulence separation. Once pilots report the aircraft in sight, they assume the responsibilities for their own separation and wake turbulence avoidance.

A visual approach is an ATC authorization for an aircraft on an IFR flight plan to proceed visually to the airport of intended landing; it is not an IAP. Also, there is no missed approach segment. An aircraft unable to complete a visual approach must be handled as any other go-around and appropriate separation must be provided. A vector for a visual approach may be initiated by ATC if the reported ceiling at the airport of intended landing is at least 500 feet above the MVA/MIA and the visibility is 3 SM or greater. At airports without weather reporting service there must be reasonable assurance (e.g. area weather reports, PIREPs, etc.) that descent and approach to the airport can be made visually, and the pilot must be informed that weather information is not available.

The visual approach clearance is issued to expedite the flow of traffic to an airport. It is authorized when the ceiling is reported or expected to be at least 1,000 feet AGL and the visibility is at least 3 SM. Pilots must remain clear of the clouds at all times while conducting a visual approach. At an airport with a control tower, pilots may be cleared to fly a visual approach to one runway while others are conducting VFR or IFR approaches to another parallel, intersecting, or converging runway. Also, when radar service is provided, it is automatically terminated when the controller advises pilots to change to the tower or advisory frequency.

CONTACT APPROACHES

If conditions permit, pilots can request a contact approach, which is then authorized by the controller. A contact approach cannot be initiated by ATC. This procedure may be used instead of the published procedure to expedite arrival, as long as the airport has a SIAP or **special instrument approach procedure** (special IAPs are approved by the FAA for individual operators, but are not published in Part 97 for public use), the reported ground visibility is at least 1 SM, and pilots are able to remain clear of clouds with at least one statute mile flight visibility throughout the approach. Some advantages of a contact approach are that it usually requires less time than the published instrument procedure, it allows pilots to retain the IFR clearance, and provides separation from IFR and SVFR traffic. On the other hand, obstruction clearances and VFR traffic avoidance becomes the pilot's responsibility. Unless otherwise restricted, the pilot may find it necessary to descend, climb, or fly a circuitous route to the airport to maintain cloud clearance or terrain/obstruction clearance.

The main differences between a visual approach and a contact approach are: a pilot must request a contact approach, while a visual approach may be assigned by ATC or requested by the pilot; and, a contact approach may be approved with 1 mile visibility if the flight can remain clear of clouds, while a visual approach requires the pilot to have the airport in sight, or a preceding aircraft to be followed, and the ceiling must be at least 1,000 feet AGL with at least 3 SM visibility.

CHARTED VISUAL FLIGHT PROCEDURES

A **charted visual flight procedure (CVFP)** may be established at some airports with control towers for environmental or noise considerations, as well as when necessary for the safety and efficiency of air traffic operations. Designed primarily for turbojet aircraft, CVFPs depict prominent landmarks, courses, and recommended altitudes to specific runways. When pilots are flying the Roaring Fork Visual RWY 15 shown in figure 5-32, mountains, rivers, and towns provide guidance to Aspen, Colorado's Sardy Field instead of VORs, NDBs, and DME fixes.

Pilots must have a charted visual landmark or a preceding aircraft in sight, and weather must be at or above the published minimums before ATC will issue a CVFP clearance. ATC will clear pilots for a CVFP if the reported ceiling at the airport of intended landing is at least 500 feet above the MVA/MIA, and the visibility is 3 SM or more, unless higher minimums are published for the particular CVFP. When accepting a clearance to follow a preceding aircraft, pilots are responsible for maintaining a safe approach interval and wake turbulence separation. Pilots must advise ATC if unable at any point to continue a charted visual approach or if the pilot loses sight of the preceding aircraft.

RNAV APPROACHES

Because of the complications with database coding, naming conventions were changed in January 2001 to accommodate all approaches using RNAV equipment into one classification — RNAV. This classification includes both ground-based and satellite dependent systems. Eventually all approaches that use some type of RNAV will reflect RNAV in the approach title. This changeover is being made to reflect two shifts in instrument approach technology. The first shift is the use of the RNP concept outlined in Chapter 2 — Departure Procedures, in which a single performance standard concept is being implemented for approach procedure design. Through the use of RNP, the underlying system of navigation may not be required, provided the aircraft can maintain the appropriate RNP standard. The second shift is that advanced avionics systems such as FMSs, used by most airlines, needed a new navigation standard by which RNAV could be fully integrated into the instrument approach system. An FMS uses multi-sensor navigation inputs to produce a composite position. Essentially, the FMS navigation function automatically blends or selects position

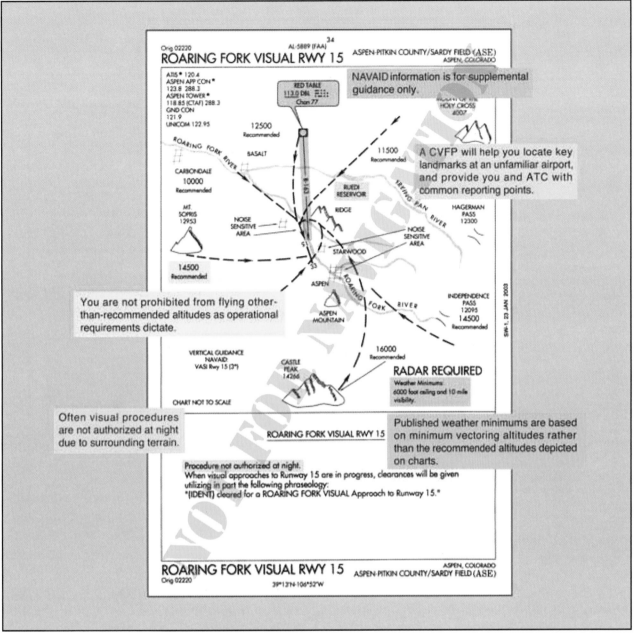

Figure 5-32. Charted Visual Flight Procedures.

sensors to compute aircraft position. Instrument approach charts and RNAV databases needed to change to reflect these issues. A complete discussion of airborne navigation databases is included in Appendix A — Airborne Navigation Databases.

Due to the multi-faceted nature of RNAV, new approach criteria have been developed to accommodate the design of RNAV instrument approaches. This includes criteria for TAAs, RNAV basic approach criteria, and specific final approach criteria for different types of RNAV approaches.

TERMINAL ARRIVAL AREAS

TAAs are the method by which aircraft are transitioned from the RNAV en route structure to the terminal area with minimal ATC interaction. Terminal arrival areas

are depicted in the planview of the approach chart, and each waypoint associated with them is also provided with a unique five character, pronounceable name. The TAA consists of a designated volume of airspace designed to allow aircraft to enter a protected area, offering guaranteed obstacle clearance where the initial approach course is intercepted based on the location of the aircraft relative to the airport. Where possible, TAAs are developed as a basic "T" shape that is divided into three separate arrival areas around the head of the "T": left base, right base, and straight-in. Typically, the TAA offers an IAF at each of these three arrival areas that are 3-6 NM from an IF, which often doubles as the IAF for straight-in approaches, a FAF located approximately 5 NM from the runway threshold, and a MAP. [Figure 5-33 on page 5-46]

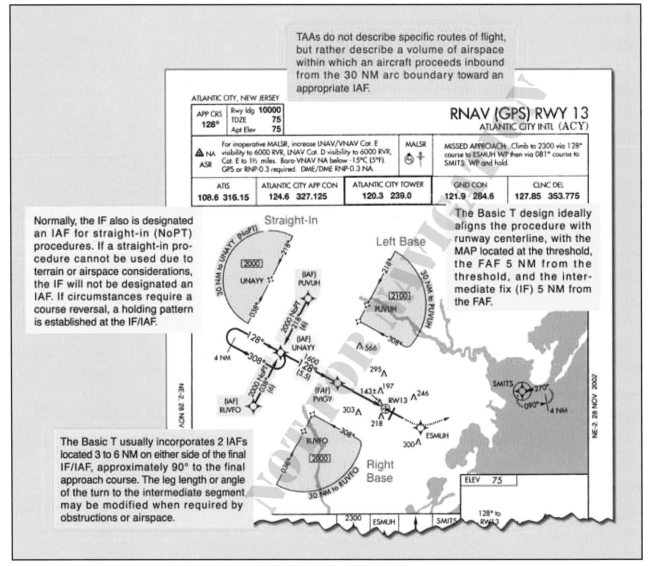

Figure 5-33. Terminal Arrival Area Design (Basic "T").

Procedurally, pilots may be cleared to an IAF associated with the TAA. ATC expects the flight to proceed to the IAF and maintain the altitude depicted for that area of the TAA, unless cleared otherwise. An obstacle clearance of at least 1,000 feet is guaranteed within the boundaries of the TAA.

TAAs are modified or even eliminated if necessary to meet the requirements of a specific airport and surrounding terrain, or airspace considerations negating the use of the "T" approach design concept. Alternative designs are addressed in *FAA Order 8260.45A, Terminal Arrival Area (TAA) Design Criteria*. Variations may eliminate one or both base areas, and/or limit or modify the angular size of the straight-in area. When both base areas are eliminated, TAAs are not depicted in the planview. Normally, a portion of the TAA underlies an airway. If this is not the case, at least one feeder route is provided from an airway fix or NAVAID to the TAA boundary. The feeder route provides a direct course from the en route fix/NAVAID to

the appropriate IF/IAF. Multiple feeder routes may also be established. In some cases, TAAs may not be depicted because of airspace congestion or other operational requirements. [Figure 5-34]

RNAV FINAL APPROACH DESIGN CRITERIA
RNAV encompasses a variety of underlying navigation systems and, therefore, approach criteria. This results in different sets of criteria for the final approach segment of various RNAV approaches. RNAV instrument approach criteria address the following procedures:

- GPS overlay of pre-existing nonprecision approaches.

- VOR/DME based RNAV approaches.

- Stand-alone RNAV (GPS) approaches.

- RNAV (GPS) approaches with vertical guidance (APV).

- RNAV (GPS) precision approaches (WAAS and LAAS).

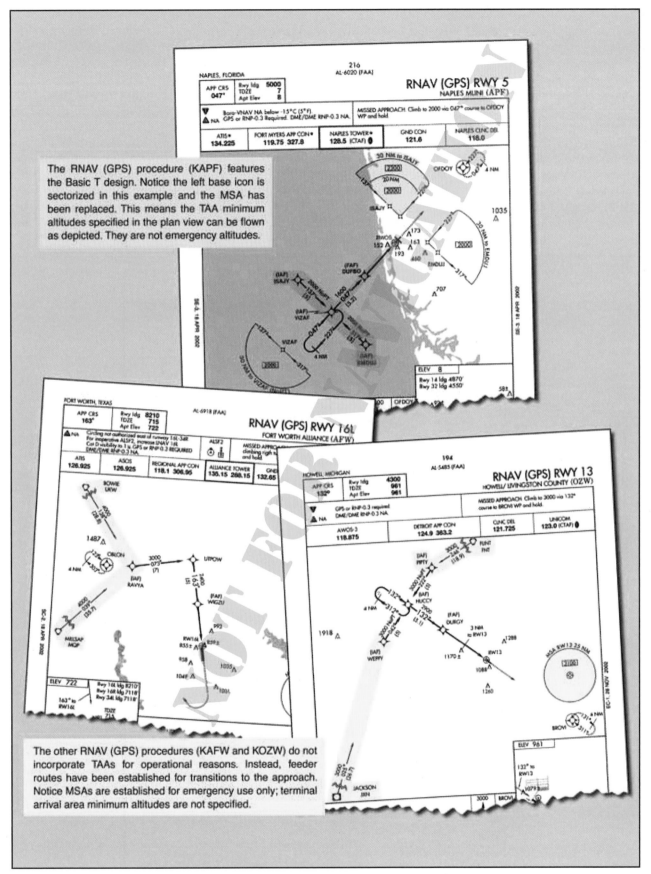

The RNAV (GPS) procedure (KAPF) features the Basic T design. Notice the left base icon is sectorized in this example and the MSA has been replaced. This means the TAA minimum altitudes specified in the plan view can be flown as depicted. They are not emergency altitudes.

The other RNAV (GPS) procedures (KAFW and KOZW) do not incorporate TAAs for operational reasons. Instead, feeder routes have been established for transitions to the approach. Notice MSAs are established for emergency use only; terminal arrival area minimum altitudes are not specified.

Figure 5-34. RNAV Approaches with and without TAAs.

GPS OVERLAY OF NONPRECISION APPROACH

The original GPS approach procedures provided authorization to fly nonprecision approaches based on conventional, ground-based NAVAIDs. Many of these approaches have been converted to stand-alone approaches, and the few that remain are identified by the name of the procedure and "or GPS." These GPS nonprecision approaches are predicated upon the design criteria of the ground-based NAVAID used as the basis of the approach. As such, they do not adhere to the RNAV design criteria for stand-alone GPS approaches, and are not considered part of the RNAV (GPS) approach classification for determining design criteria. [Figure 5-35]

GPS STAND-ALONE/RNAV (GPS) APPROACH

RNAV (GPS) approaches are named so that airborne navigation databases can use either GPS or RNAV as the title of the approach. This is required for non-GPS approach systems such as VOR/DME based RNAV systems. In the past, these approaches were often referred to as stand-alone GPSs. They are considered nonprecision approaches, offering only LNAV and circling minimums. Precision minimums are not authorized, although LNAV/VNAV minimums may be published and used as long as the on-board system is capable of providing approach approved VNAV. The RNAV (GPS) Runway 18 approach for Alexandria, Louisiana incorporates only LNAV and circling minimums. [Figure 5-36]

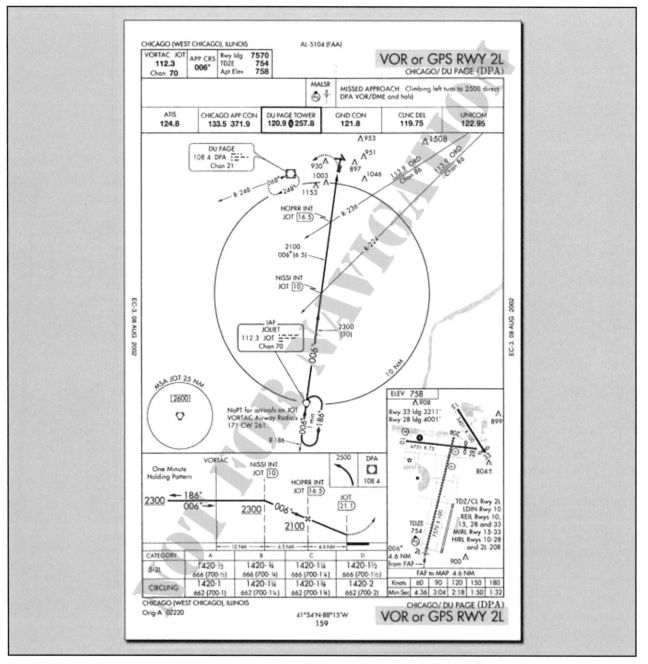

Figure 5-35. Traditional GPS Overlay Approach.

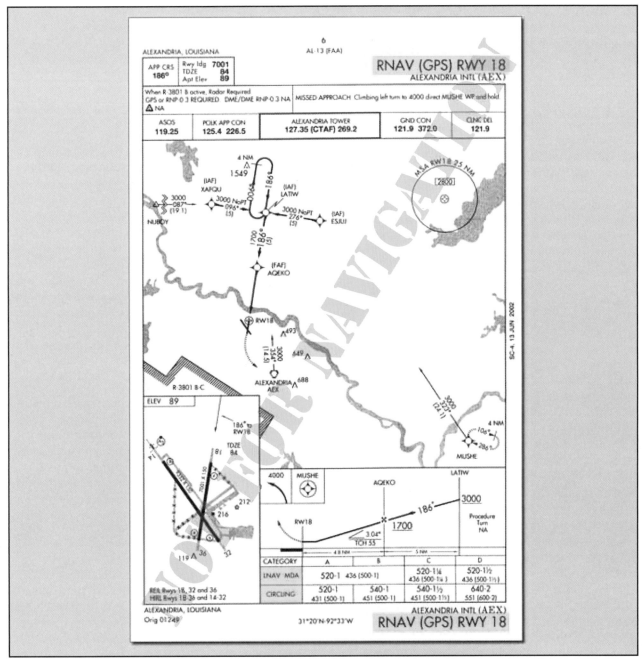

Figure 5-36. Alexandria International (KAEX), Alexandria, Louisiana, RNAV (GPS) RWY 18.

For a non-vertically guided straight-in RNAV (GPS) approach, the final approach course must be aligned within 15° of the extended runway centerline. The final approach segment should not exceed 10 NM, and when it exceeds 6 NM, a stepdown fix is typically incorporated. A minimum of 250 feet obstacle clearance is also incorporated into the final approach segment for straight-in approaches, and a maximum 400-foot per NM descent gradient is permitted.

The approach design criteria are different for approaches that use vertical guidance provided by a Baro-VNAV system. Because the Baro-VNAV guidance is advisory and not primary, Baro-VNAV approaches are not authorized in areas of hazardous ter-

rain, nor are they authorized when a remote altimeter setting is required. Due to the inherent problems associated with barometric readings and cold temperatures, these procedures are also temperature limited. Additional approach design criteria for RNAV Approach Construction Criteria can be found in the appropriate Order 8260 series directives.

RNAV (GPS) APPROACH USING WAAS

WAAS was commissioned in July, 2003, with initial operational capability (IOC). Although precision approach capability is still in the future, initial WAAS currently provides a new type of approach with vertical guidance (APV) known as LPV. Approach minimums as low as 200 feet HAT and 1/2 SM visibility are possible, even though LPV is semi-precision, and not considered a precision approach. WAAS covers 95 percent of the country 95 percent of the time.

NOTE: WAAS avionics receive an airworthiness approval in accordance with Technical Standard Order (TSO) C-145A, *Airborne Navigation Sensors Using the (GPS) Augmented by the Wide Area Augmentation System (WAAS),* or TSO-146A, *Stand-Alone Airborne Navigation Equipment Using the Global Positioning System (GPS) Augmented by the Wide Area Augmentation System (WAAS)*, and installed in accordance with AC 20-130A, *Airworthiness Approval of Navigation or Flight Management Systems Integrating Multiple Navigation Sensors,* or AC 20-138A, *Airworthiness Approval of Global Positioning System (GPS) Navigation Equipment for Use as a VFR and IFR Navigation System.*

Precision approach capability will become available when LAAS becomes operational. LAAS further increases the accuracy of GPS and improves signal integrity warnings. Precision approach capability requires obstruction planes and approach lighting systems to meet Part 77 standards for ILS approaches. This will delay the implementation of RNAV (GPS) precision approach capability due to the cost of certifying each runway.

ILS APPROACHES

Notwithstanding emerging RNAV technology, the ILS is the most precise and accurate approach NAVAID currently in use throughout the NAS. An ILS CAT I precision approach allows approaches to be made to 200 feet above the TDZE and with visibilities as low as 1,800 RVR; with CAT II and CAT III approaches allowing descents and visibility minimums that are even lower. Nonprecision approach alternatives cannot begin to offer the precision or flexibility offered by an ILS. In order to further increase the approach capacity of busy airports and exploit the maximum potential of ILS technology, many different applications are in use.

A single ILS system can accommodate 29 arrivals per hour on a single runway. Two or three parallel runways operating consecutively can double or triple the capacity of the airport. For air commerce this means greater flexibility in scheduling passenger and cargo service. Capacity is increased through the use of parallel (dependent) ILS, simultaneous parallel (independent) ILS, simultaneous close parallel (independent) ILS, **precision runway monitor (PRM)**, and converging ILS approaches. A parallel (dependent) approach differs from a simultaneous (independent) approach in that the minimum distance between parallel runway centerlines is reduced; there is no requirement for radar monitoring or advisories; and a staggered separation of aircraft on the adjacent localizer/azimuth course is required.

In order to successfully accomplish parallel, simultaneous parallel, and converging ILS approaches, flight crews and air traffic controllers have additional responsibilities. When multiple instrument approaches are in use, ATC will advise flight crews either directly or through ATIS. It is the pilot's responsibility to inform ATC if unable or unwilling to execute a simultaneous approach. Pilots must comply with all ATC requests in a timely manner, and maintain strict radio discipline, including using complete aircraft call signs. It is also incumbent upon the flight crew to notify ATC immediately of any problems relating to aircraft communications or navigation systems. At the very least, the approach procedure briefing should cover the entire approach procedure including the approach name, runway number, frequencies, final approach course, glide slope intercept altitude, DA or DH, and the missed approach instructions. The review of autopilot procedures is also appropriate when making coupled ILS or MLS approaches.

As with all approaches, the primary navigation responsibility falls upon the pilot in command. ATC instructions will be limited to ensuring aircraft separation. Additionally, missed approach procedures are normally designed to diverge in order to protect all involved aircraft. ILS approaches of all types are afforded the same obstacle clearance protection and design criteria, no matter how capacity is affected by multiple ILS approaches. [Figure 5-37]

ILS APPROACH CATEGORIES

There are three general classifications of ILS approaches — CAT I, CAT II, and CAT III (autoland). The basic ILS approach is a CAT I approach and requires only that pilots be instrument rated and current, and that the aircraft be equipped appropriately. CAT II and CAT III ILS approaches typically have lower minimums and require special certification for operators, pilots, aircraft, and airborne/ground equipment. Because of the complexity and high cost of the equipment, CAT III ILS approaches are used primarily in air carrier and military operations. [Figure 5-38]

CAT II AND III APPROACHES

The primary authorization and minimum RVRs allowed for an air carrier to conduct CAT II and III approaches can be found in OpsSpecs – Part C. CAT II and III operations allow authorized pilots to make instrument approaches in weather that would otherwise be prohibitive.

While CAT I ILS operations permit substitution of midfield RVR for TDZ RVR (when TDZ RVR is not

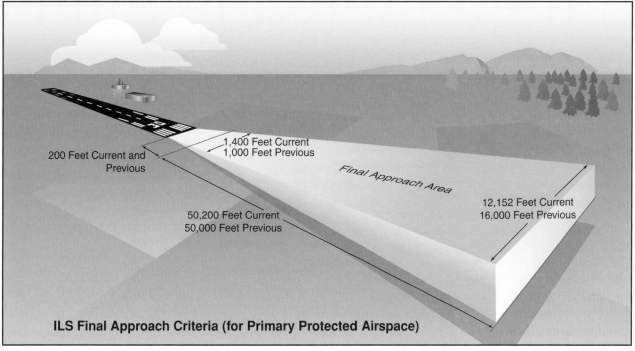

Figure 5-37. ILS Final Approach Segment Design Criteria.

available), CAT II ILS operations do not permit any substitutions for TDZ RVR. The touchdown zone RVR system is required and must be used. Touchdown zone RVR is controlling for all CAT II ILS operations.

The lowest authorized ILS minimums, with all required ground and airborne systems components operative, are

- CAT I — Decision Height (DH) 200 feet and Runway Visual Range (RVR) 2,400 feet (with touchdown zone and centerline lighting, RVR 1800 feet),
- CAT II — DH 100 feet and RVR 1,200 feet,
- CAT IIIa — No DH or DH below 100 feet and RVR not less than 700 feet,
- CAT IIIb — No DH or DH below 50 feet and RVR less than 700 feet but not less than 150 feet, and
- CAT IIIc — No DH and no RVR limitation.

NOTE: Special authorization and equipment are required for CAT II and III.

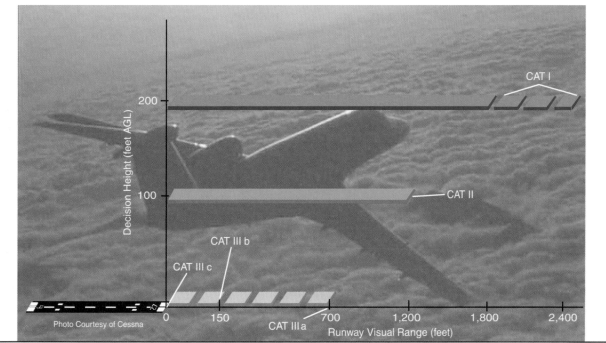

Figure 5-38. ILS Approach Categories.

The weather conditions encountered in CAT III operations range from an area where visual references are adequate for manual rollout in CAT IIIa, to an area where visual references are inadequate even for taxi operations in CAT IIIc. To date, no U.S. operator has received approval for CAT IIIc in OpsSpecs. Depending on the auto-flight systems, some airplanes require a DH to ensure that the airplane is going to land in the touchdown zone and some require an Alert Height as a final crosscheck of the performance of the auto-flight systems. These heights are based on radio altitude (RA) and can be found in the specific aircraft's AFM. [Figure 5-39]

Both CAT II and III approaches require special ground and airborne equipment to be installed and operational, as well as special aircrew training and authorization. The OpsSpecs of individual air carriers detail the requirements of these types of approaches as well as their performance criteria. Lists of locations where each operator is approved to conduct CAT II and III approaches can also be found in the OpsSpecs.

ILS APPROACHES TO PARALLEL RUNWAYS
Airports that have two or three parallel runways may be authorized to use parallel approaches to maximize the capacity of the airport. There are three classifications of parallel ILS approaches, depending on the runway centerline separation and ATC procedures.

PARALLEL
Parallel (dependent) ILS approaches are allowed at airports with parallel runways that have centerlines

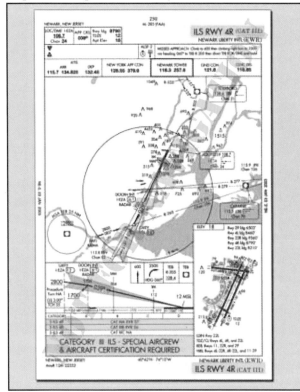

Figure 5-39. Category III Approach Procedure.

separated by at least 2,500 feet. Aircraft are allowed to fly ILS approaches to parallel runways; however, the aircraft must be staggered by a minimum of 1 1/2 NM diagonally. Aircraft are staggered by 2 NM diagonally for runway centerlines that are separated by more than 4,300 feet and up to but not including 9,000 feet, and that do not have final monitor air traffic controllers. Separation for this type of approach is provided by radar. [Figure 5-40]

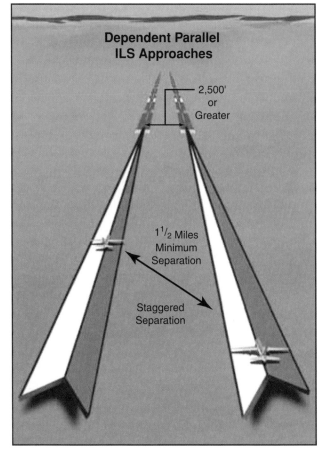

Figure 5-40. Parallel (Dependent) ILS Approach Separation Criteria.

Though this type of approach procedure is approved for several airports, it is not required that the approach chart contain information notifying flight crews of the use of parallel approaches. Therefore, a pilot may not know that parallel approaches are approved or used at a specific airport based on the information contained on the chart. ATC normally communicates an advisory over ATIS that parallel approach procedures are in effect. For example, pilots flying into Sacramento, California may encounter parallel approach procedures. [Figure 5-41]

SIMULTANEOUS
Simultaneous parallel ILS approaches are used at authorized airports that have between 4,300 feet and 9,000 feet separation between runway centerlines. A dedicated final monitor controller is required to monitor separation for this type of approach, which

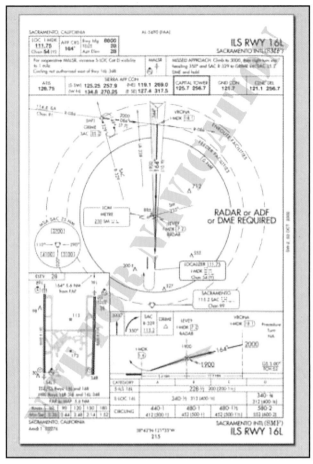

Figure 5-41. Sacramento International (KSMF), Sacramento, California, ILS RWY 16L.

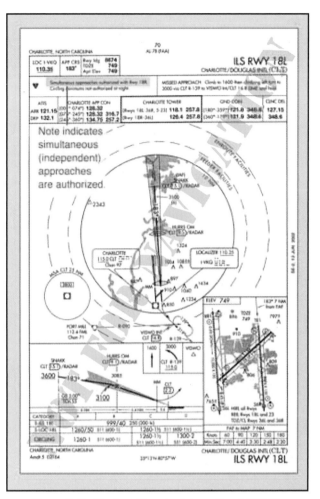

Figure 5-42. Charlotte/Douglas International (KCLT), Charlotte, North Carolina, ILS RWY 18.

eliminates the need for staggered approaches. Final monitor controllers track aircraft positions and issue instructions to pilots of aircraft observed deviating from the LOC course. [Figure 5-42]

Triple simultaneous approaches are authorized provided the runway centerlines are separated by at least 5,000 feet and are below 1,000 feet MSL airport elevation. Additionally, for triple parallel approaches above airport elevations of 1,000 feet MSL, ASR with high-resolution final monitor aids or high update RADAR with associated final monitor aids is required.

As a part of the simultaneous parallel approach approval, normal operating zones and non-transgression zones must be established to ensure proper flight track boundaries for all aircraft. The normal operating zone (NOZ) is the operating zone within which aircraft remain during normal approach operations. The NOZ is typically no less than 1,400 feet wide, with 700 feet of space on either side of the runway centerline. A no transgression zone (NTZ) is a 2,000-foot wide area located between the parallel runway final approach courses. It is equidistant between the runways and indicates an area within which flight is not authorized. [Figure 5-43 on page 5-54] Any time an aircraft breaches the NTZ, ATC issues instructions for all aircraft to break off the approach to avoid potential conflict.

PRECISION RUNWAY MONITOR
Simultaneous close parallel (independent) ILS PRM approaches are authorized for use at airports that have parallel runways separated by at least 3,400 feet and no more than 4,300 feet. [Figure 5-44 on page 5-54] They are also approved for airports with parallel runways separated by at least 3,000 feet with an offset LOC where the offset angle is at least 2.5 degrees but no more than 3 degrees. The offset LOC approaches are referred to as Simultaneous Offset Instrument Approaches (SOIA) and are discussed in depth later in this chapter.

The PRM system provides the ability to accomplish simultaneous close parallel (independent) ILS approaches and enables reduced delays and fuel savings during reduced visibility operations. It is also the safest method of increasing ILS capacity through the use of parallel approaches. The PRM system incorporates high-update radar with one second or better update time and a high resolution ATC radar

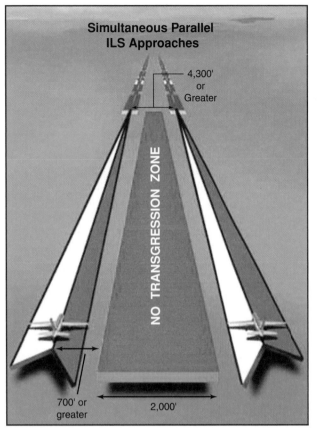

Figure 5-43. Simultaneous Parallel ILS Approach Criteria.

Figure 5-44. Simultaneous Close Parallel ILS Approach (ILS PRM) Criteria.

display that contains automated tracking software that can track aircraft in real time. Position and velocity is updated each second and a ten second projected position is displayed. The system also incorporates visual and aural alerts for the controllers.

Approval for ILS PRM approaches requires the airport to have a precision runway monitoring system and a final monitor controller who can only communicate with aircraft on the final approach course. Additionally, two tower frequencies are required to be used and the controller broadcasts over both frequencies to reduce the chance of instructions being missed. Pilot training is also required for pilots using the PRM system. Part 121 and 135 operators are required to complete training that includes the viewing of one of two videos available from the FAA through the Flight Standards District Office (FSDO) or current employer:

- "RDU Precision Runway Monitor: A Pilot's Approach."

- "ILS PRM Approaches, Information for Pilots."

When pilots or flight crews wish to decline a PRM approach, ATC must be notified immediately and the flight will be transitioned into the area at the convenience of ATC. Flight crews should advise ATC within 200 NM of the landing airport if they are not qualified or not equipped to fly a PRM approach.

The approach chart for the PRM approach typically requires two pages and outlines pilot, aircraft, and procedure requirements necessary to participate in PRM operations. [Figure 5-45] Pilots need to be aware of the differences associated with this type of ILS approach:

- Immediately follow break out instructions as soon as safety permits.

- Listen to both tower frequencies to avoid missed instructions from stuck mikes or blocked transmissions. The final ATC controller can override the radio frequency if necessary.

- Broadcast only over the main tower frequency.

- Disengage the autopilot for breakouts because hand-flown breakouts are quicker.

- Set the Traffic Alert and Collision Avoidance System (TCAS) to the appropriate TA (traffic advisory) or RA (resolution advisory) mode in compliance with current operational guidance on the attention all users page (AAUP), or other authorized guidance, i.e., approved flight manual, flight operations manual.

It is important to note that descending breakouts may be issued. Additionally, flight crews will never be issued breakout instructions that clear them below the MVA, and they will not be required to descend at more than 1,000 FPM.

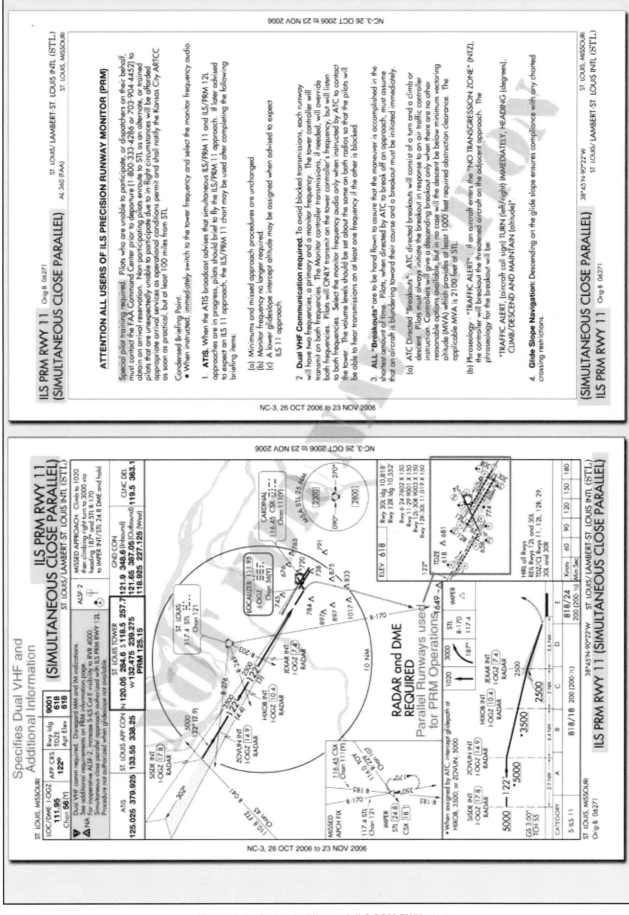

Figure 5-45. St. Louis, Missouri, ILS PRM RWY 11.

5-55

SIMULTANEOUS OFFSET INSTRUMENT APPROACHES

SOIAs allow simultaneous approaches to two parallel runways spaced at least 750 feet apart, but less than 3,000 feet. The SOIA procedure utilizes an ILS/PRM approach to one runway and an offset Localizer-Type Directional Aid (LDA)/PRM approach with glide slope to the adjacent runway. The use of PRM technology is also required with these operations; therefore, the approach charts will include procedural notes such as "Simultaneous approach authorized with LDA PRM RWY XXX." San Francisco has the first published SOIA approach. [Figure 5-46]

The training, procedures, and system requirements for SOIA ILS/PRM and LDA/PRM approaches are identical with those used for simultaneous close parallel ILS/PRM approaches until near the LDA/PRM approach MAP, except where visual acquisition of the ILS aircraft by the LDA aircraft must be accomplished. If visual acquisition is not accomplished a missed approach must be executed. A visual segment for the LDA/PRM approach is established between the LDA MAP and the runway threshold. Aircraft transition in visual conditions from the LDA course, beginning at the LDA MAP, to align with the runway and can be stabilized by 500 feet above ground level (AGL) on the extended runway centerline.

The FAA website has additional information about PRM and SOIA, including instructional videos at: http://www.faa.gov/education_research/training/prm/

CONVERGING

Another method by which ILS approach capacity can be increased is through the use of converging approaches. Converging approaches may be established at airports that have runways with an angle between 15 and 100 degrees and each runway must have an ILS. Additionally, separate procedures must be established for each approach and each approach must have a MAP at least 3 NM apart with no overlapping of the protected missed approach airspace. Only straight-in approaches are approved for converging ILS procedures. If the runways intersect, the controller must be able to visually separate intersecting runway traffic. Approaches to intersecting runways also have higher minimums with a 700-foot minimum and no less than 2 SM visibility. Pilots are informed of the use of converging ILS approaches by the controller upon initial contact or through ATIS. [Figure 5-47 on page 5-58]

Dallas/Fort Worth International airport is one of the few airports that makes use of converging ILS approaches because its runway configuration has multiple parallel runways and two offset runways. [Figure 5-48 on page 5-58] The approach chart title indicates the use of con-

verging approaches and the notes section highlights other runways that are authorized for converging approach procedures.

MICROWAVE LANDING SYSTEM

The MLS is a precision instrument approach alternative to the ILS. It provides azimuth, elevation, and distance information, as well as a back azimuth capable of providing guidance for missed approach procedures and departures. In addition to straight-in approaches, the MLS system can also provide three-dimensional RNAV type approaches in both computed straight and curved paths. It was initially designed to replace the ILS system and it provided inherent flexibility and broader reception range with the greatest limitation being the capabilities of the airborne equipment installed in individual aircraft.

The MLS has multiple advantages including an increased number of frequencies, compact ground equipment, and complex approach paths. For a variety of reasons, particularly the advent of civil use GPS, MLS installation was deferred, and by 1994 it was officially cancelled by the FAA. Today there are few MLS installations in the U.S. and currently there are no plans for further installations. Futhermore, the MLS equipment required for an MLS approach was not widely installed in aircraft, whereas most new aircraft produced today come with GPS systems. With the limited number of MLS installations around the country, it is highly unlikely that most pilots will ever encounter the MLS approach, and if they do, it is even less likely that the proper equipment would be installed in the aircraft.

Like the ILS, the basic MLS approach requires the final approach course alignment to be within 3 degrees of the extended runway centerline. This type of approach uses a glide slope between 3 and 6.40 degrees and provides precision landing minimums to 200 feet HAT. Obstacle clearance is based on the glide slope angle used in the approach design. The design criteria differ for each type of MLS approach and incorporate numerous formulas for the derivation of specific course criteria. This information is contained in *FAA Order 8260.3 Volume 3, Chapters 2 and 3.*

In the front of the TPP, there is a page containing additional information pertaining to the use of an MLS system. The MLS Channeling and Frequency Pairing Table cross references the appropriate MLS channel with its paired VHF and TACAN frequencies. Ground equipment associated with the MLS operates on the MLS channels, while the MLS angle/data and DME is required to operate using one of the paired VHF or TACAN frequencies.

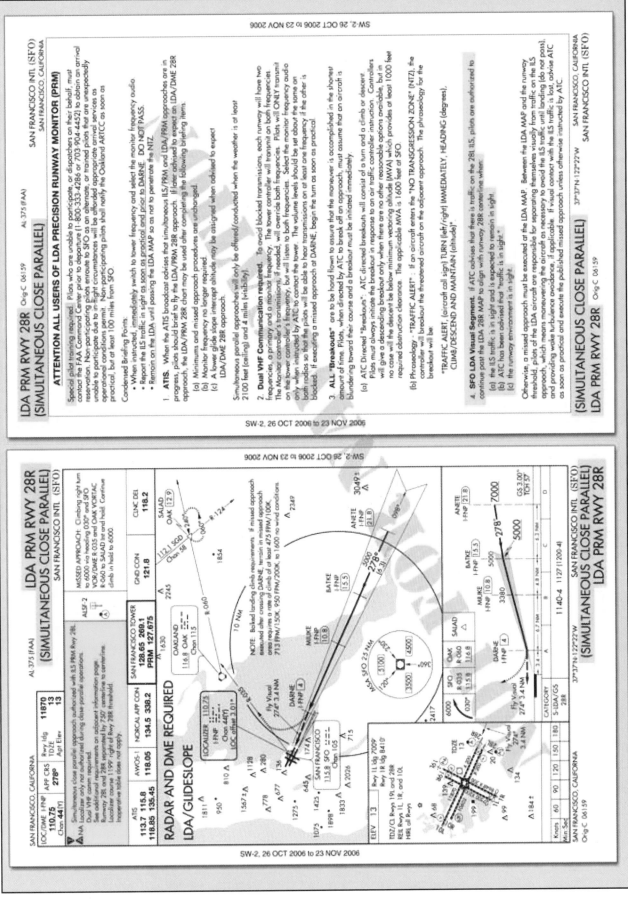

Figure 5-46. Simultaneous Offset Instrument Approach Procedure.

5-57

Figure 5-47. Converging Approach Criteria.

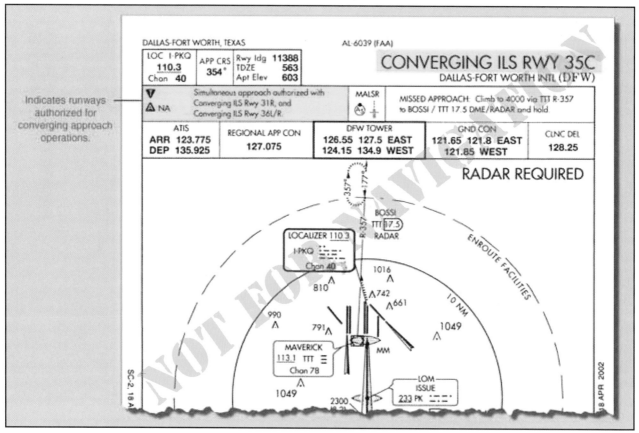

Figure 5-48. Dallas/Fort Worth (KDFW), Dallas/Fort Worth, Texas, CONVERGING ILS RWY 35C.

VOR APPROACH

The VOR is one of the most widely used nonprecision approach types in the NAS. VOR approaches use VOR facilities both on and off the airport to establish approaches and include the use of a wide variety of equipment such as DME and TACAN. Due to the wide variety of options included in a VOR approach, TERPS outlines design criteria for both on and off airport VOR facilities as well as VOR approaches with and without a FAF. Despite the various configurations, all VOR approaches are nonprecision approaches, require the presence of properly operating VOR equipment, and can provide MDAs as low as 250 feet above the runway. VOR also offers a flexible advantage in that an approach can be made toward or away from the navigational facility.

The VOR approach into Missoula International in Missoula, Montana, is an example of a VOR approach where the VOR facility is on the airport and there is no specified FAF. [Figure 5-49] For a straight-in approach, the final approach course is typically aligned to intersect the extended runway centerline 3,000 feet from the runway threshold, and the angle of convergence between the two does not exceed 30 degrees. This type of VOR

approach also includes a minimum of 300 feet of obstacle clearance in the final approach area. The final approach area criteria include a 2 NM wide primary area at the facility that expands to 6 NM wide at a distance of 10 NM from the facility. Additional approach criteria are established for courses that require a high altitude teardrop approach penetration.

When DME is included in the title of the VOR approach, operable DME must be installed in the aircraft in order to fly the approach from the FAF. The use of DME allows for an accurate determination of position without timing, which greatly increases situational awareness throughout the approach. Alexandria, Louisiana, is an excellent example of a VOR/DME approach in which the VOR is off the airport and a FAF is depicted. [Figure 5-50 on page 5-60] In this case, the final approach course is a radial or straight-in final approach and is designed to intersect the runway centerline at the runway threshold with the angle of convergence not exceeding 30 degrees.

The criteria for an arc final approach segment associated with a VOR/DME approach is based on the arc being beyond 7 NM and no farther than 30 NM from the VOR,

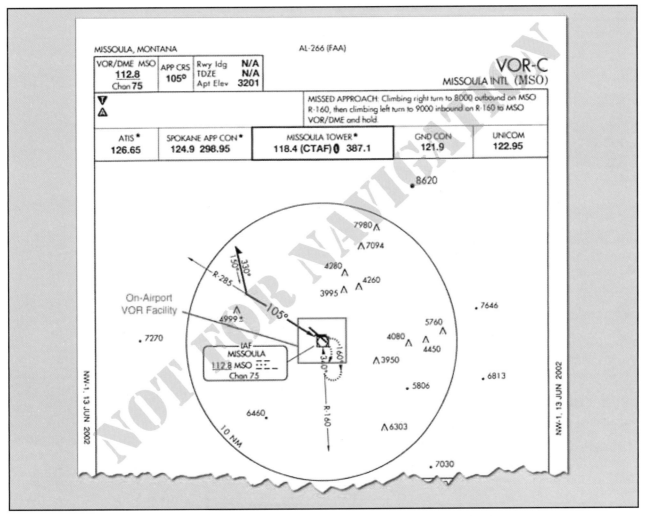

Figure 5-49. Missoula International, Missoula, Montana (KMSO), VOR–C.

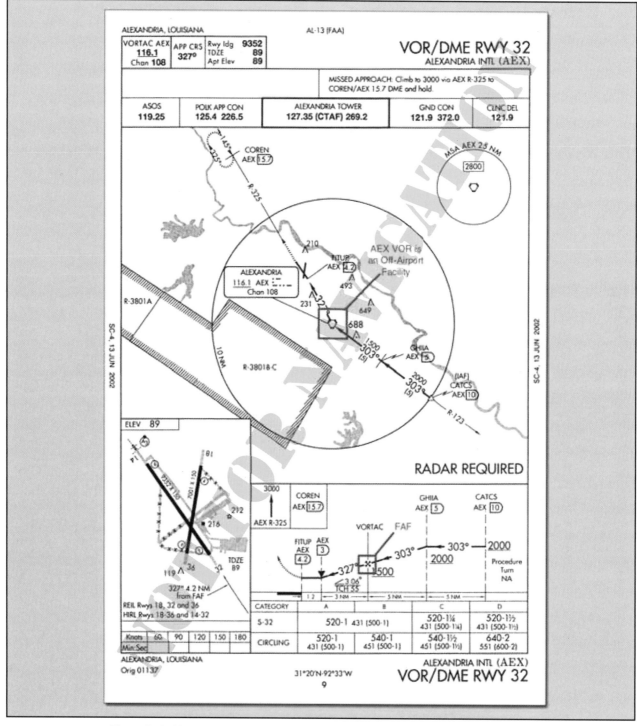

Figure 5-50. Alexandria International, Alexandria, Louisiana (KAEX), VOR/DME RWY 32.

and depends on the angle of convergence between the runway centerline and the tangent of the arc. Obstacle clearance in the primary area, which is considered the area 4 NM on either side of the arc centerline, is guaranteed by at least 500 feet.

NDB APPROACH

Like the VOR approach, an NDB approach can be designed using facilities both on and off the airport, with or without a FAF, and with or without DME availability. At one time it was commonplace for an instrument student to learn how to fly an NDB approach, but

with the growing use of GPS, many pilots no longer use the NDB for instrument approaches. New RNAV approaches are also rapidly being constructed into airports that are served only by NDB. The long-term plan includes the gradual phase out of NDB facilities, and eventually, the NDB approach will become nonexistent. Until that time, the NDB provides additional availability for instrument pilots into many smaller, remotely located airports.

The NDB Runway 9 approach at Charleston Executive Airport, is an example of an NDB approach established with an on-airport NDB that does not incorporate a FAF. [Figure 5-51] In this case, a procedure turn or penetration turn is required to be a part of the approach design. For the NDB to be considered an on-airport facility, the facility must be located within one mile of any portion of the landing runway for straight-in approaches and within one mile of any portion of usable landing surface for circling approaches. The final approach segment of the approach is designed with a final approach area that is 2.5 NM wide at the facility, and increases to 8 NM wide at 10 NM from the facility. Additionally, the final approach course and the extended runway centerline angle of convergence cannot exceed 30 degrees for straight-in approaches. This type of NDB approach is afforded a minimum of 350 feet obstacle clearance.

When a FAF is established for an NDB approach, the approach design criteria changes. It also takes into account whether or not the NDB is located on or off the airport. Additionally, this type of approach can be made both moving toward or away from the NDB facility. The St. Mary's, Alaska, NDB DME RWY 16 [Figure 5-52 on page 5-62] is an approach with a FAF using an on-airport NDB facility that also incorporates the use of DME. In this case, the NDB has DME capabilities from the LOC approach system installed on the airport. While the alignment criteria and obstacle clearance remain the same as an NDB approach without a FAF, the final approach segment area criteria changes to an area that is 2.5 NM wide at the facility and increases to 5 NM wide, 15 NM from the NDB.

RADAR APPROACHES

The two types of radar approaches available to pilots when operating in the NAS are PAR and ASR. Radar approaches may be given to any aircraft at the pilot's request. ATC may also offer radar approach options to aircraft in distress regardless of the weather conditions, or as necessary to expedite traffic. Despite the control exercised by ATC in a radar approach environment, it remains the pilot's responsibility to ensure the approach and landing minimums listed for the approach are appropriate for the existing weather conditions considering personal approach criteria certification and company OpsSpecs.

Perhaps the greatest benefit of either type of radar approach is the ability to use radar to execute a "no-gyro" approach. Assuming standard rate turns, an air traffic controller can indicate when to begin and end turns. If available, pilots should make use of this approach when the heading indicator has failed and partial panel instrument flying is required.

Information about radar approaches is published in tabular form in the front of the TPP booklet. PAR, ASR, and circling approach information including runway, DA, DH, or MDA, height above airport (HAA), HAT, ceiling, and visibility criteria are outlined and listed by specific airport.

Regardless of the type of radar approach in use, ATC monitors aircraft position and issues specific heading and altitude information throughout the entire

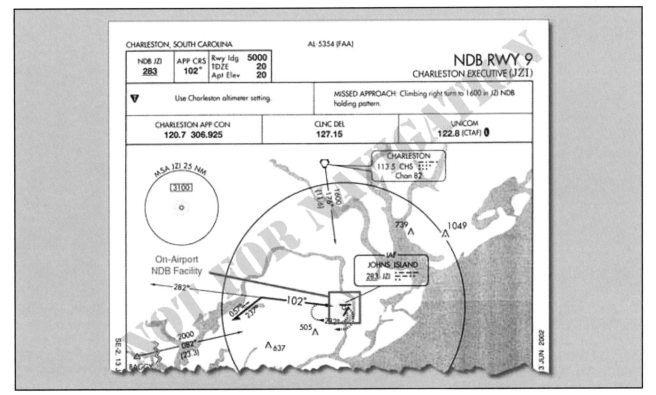

Figure 5-51. Charleston Executive (KJZI), Charleston, South Carolina, NDB RWY 9.

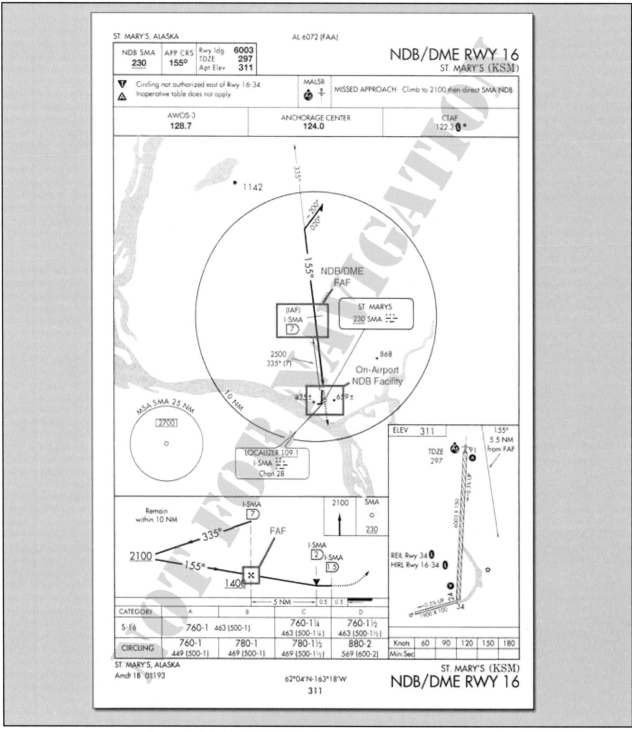

Figure 5-52. St. Mary's (PASM), St. Mary's, Alaska, NDB DME RWY 16.

approach. Particularly, lost communications procedures should be briefed prior to execution to ensure pilots have a comprehensive understanding of ATC expectations if radio communication were lost. ATC also provides additional information concerning weather and missed approach instructions when beginning a radar approach. [Figure 5-53]

PRECISION APPROACH RADAR

PAR provides both vertical and lateral guidance, as well as range, much like an ILS, making it the most

precise radar approach available. The radar approach, however, is not able to provide visual approach indications in the cockpit. This requires the flight crew to listen and comply with controller instructions. PAR approaches are rare, with most of the approaches used in a military setting; any opportunity to practice this type of approach is beneficial to any flight crew.

The final approach course of a PAR approach is always directly aligned with the runway centerline, and the associated glide slope is typically no less than 2 degrees

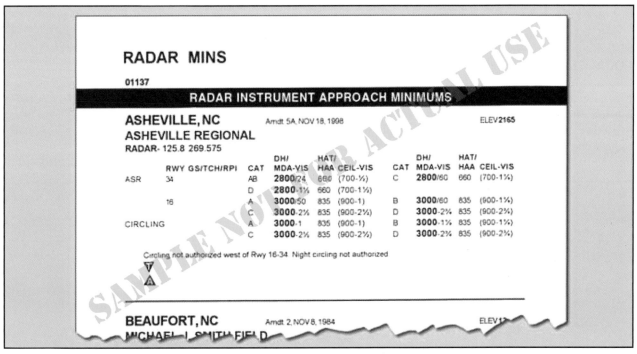

RADAR MINS

01137

RADAR INSTRUMENT APPROACH MINIMUMS

ASHEVILLE, NC Amdt 5A, NOV 18, 1998 ELEV 2165
ASHEVILLE REGIONAL
RADAR- 125.8 269.575

	RWY GS/TCH/RPI	CAT	DH/ MDA-VIS	HAT/ HAA	CEIL-VIS	CAT	DH/ MDA-VIS	HAT/ HAA	CEIL-VIS
ASR	34	AB	2800/24	660	(700-½)	C	2800/60	660	(700-1¼)
		D	2800-1¼	660	(700-1¼)				
	16	A	3000/50	835	(900-1)	B	3000/60	835	(900-1¼)
		C	3000-2¼	835	(900-2¼)	D	3000-2¼	835	(900-2¼)
CIRCLING		A	3000-1	835	(900-1)	B	3000-1¼	835	(900-1¼)
		C	3000-2¼	835	(900-2¼)	D	3000-2¼	835	(900-2¼)

Circling not authorized west of Rwy 16-34. Night circling not authorized

BEAUFORT, NC Amdt 2, NOV 8, 1984 ELEV 12
MICHAEL J. SMITH FIELD

Figure 5-53. Asheville Regional (KAVL), Asheville, NC, Radar Instrument Approach Minimums.

and no more than 3 degrees. Obstacle clearance for the final approach area is based on the particular established glide slope angle and the exact formula is outlined in TERPS Volume 1, Chapter 10. [Figure 5-54]

AIRPORT SURVEILLANCE RADAR

ASR approaches are typically only approved when necessitated for an ATC operational requirement, or in an unusual or emergency situation. This type of radar only provides heading and range information, although the controller can advise the pilot of the altitude where the aircraft should be based on the distance from the runway. An ASR approach procedure can be established at any radar facility that has an antenna within 20 NM of the airport and meets the equipment requirements outlined in Order 8200.1, *U.S. Standard Flight Inspection Manual* (latest version). ASR approaches are not authorized for use when **Center Radar ARTS processing (CENRAP)** procedures are in use due to diminished radar capability.

The final approach course for an ASR approach is aligned with the runway centerline for straight-in approaches and aligned with the center of the airport for circling approaches. Within the final approach area, the pilot is also guaranteed a minimum of 250 feet obstacle clearance. ASR descent gradients are designed to be relatively flat, with an optimal gradient of 150 feet per mile and never exceeding 300 feet per mile.

LOCALIZER APPROACHES

As an approach system, the localizer is an extremely flexible approach aid that, due to its inherent design, provides many applications for a variety of needs in instrument flying. An ILS glide slope installation may be impossible due to surrounding terrain. For whatever reason, the localizer is able to provide four separate applications from one approach system:

- Localizer Approach.

- Localizer/DME Approach.

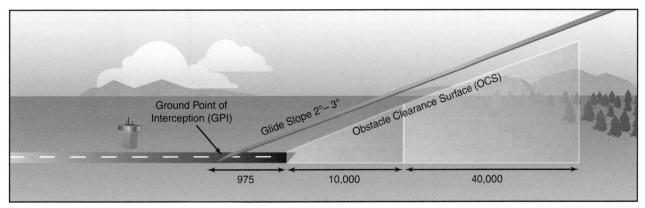

Figure 5-54. PAR Final Approach Area Criteria.

- Localizer Back Course Approach.
- Localizer-type Directional Aid (LDA).

LOCALIZER AND LOCALIZER DME

The localizer approach system can provide both precision and nonprecision approach capabilities to a pilot. As a part of the ILS system, the localizer provides horizontal guidance for a precision approach. Typically, when the localizer is discussed, it is thought of as a nonprecision approach due to the fact that either it is the only approach system installed, or the glide slope is out of service on the ILS. In either case, the localizer provides a nonprecision approach using a localizer transmitter installed at a specific airport. [Figure 5-55]

TERPS provide the same alignment criteria for a localizer approach as it does for the ILS since it is essentially the same approach without vertical guidance stemming from the glide slope. A localizer is always aligned within

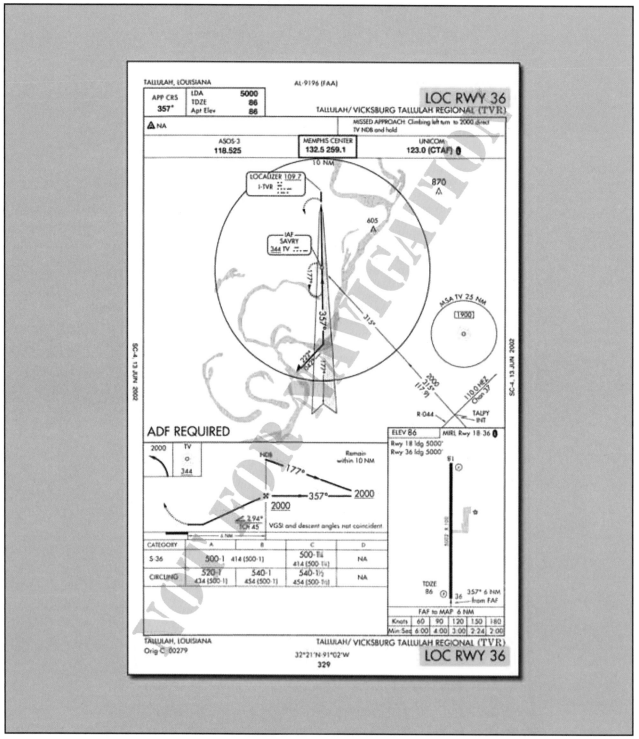

Figure 5-55. Vicksburg Tallulah Regional (KTVR), Tallulah/Vicksburg, Louisiana, LOC RWY 36.

3 degrees of the runway, and it is afforded a minimum of 250 feet obstacle clearance in the final approach area. In the case of a localizer DME (LOC DME) approach, the localizer installation has a collocated DME installation that provides distance information required for the approach. [Figure 5-56]

LOCALIZER BACK COURSE

In cases where an ILS is installed, a back course may be available in conjunction with the localizer. Like the localizer, the back course does not offer a glide slope, but remember that the back course can project

a false glide slope signal and the glide slope should be ignored. Reverse sensing will occur on the back course using standard VOR equipment. With an HSI (horizontal situation indicator) system, reverse sensing is eliminated if it is set appropriately to the front course. [Figure 5-57 on page 5-66]

LOCALIZER-TYPE DIRECTIONAL AID

An LDA is a NAVAID that provides nonprecision approach capabilities. The LDA is essentially a localizer. It is termed LDA because the course alignment with the runway exceeds 3 degrees. Typically, an LDA

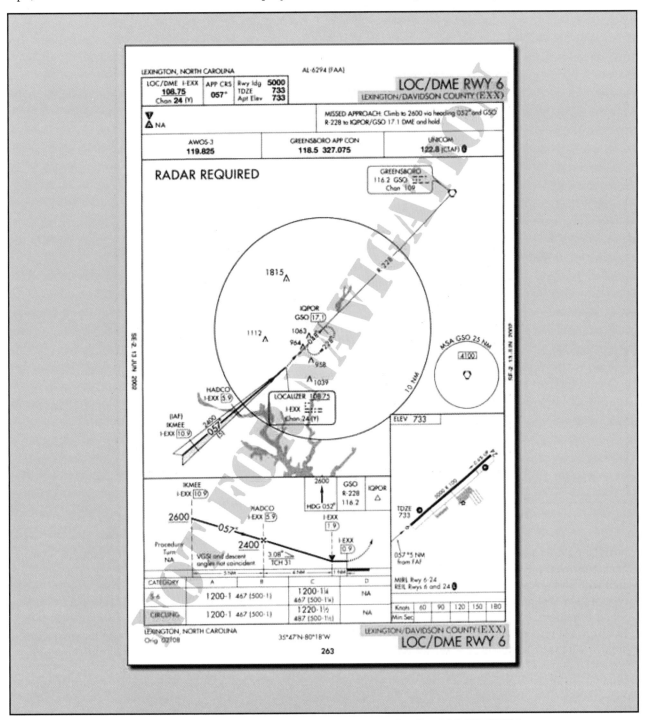

Figure 5-56. Davidson County (KEXX), Lexington, North Carolina, LOC DME RWY 6.

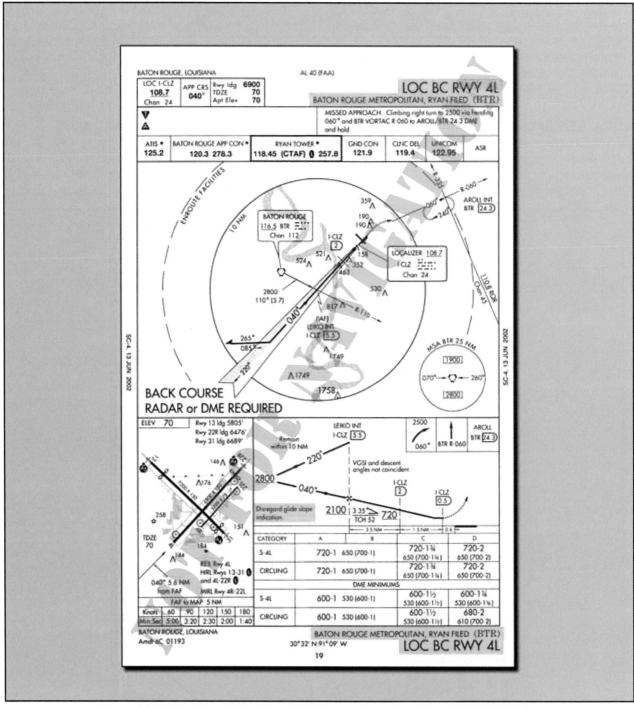

Figure 5-57. Baton Rouge Metro/Ryan (KBTR), Baton Rouge, Louisiana, LOC BC RWY 4L.

installation does not incorporate a glide slope component. However, the availability of a glide slope associated with an LDA is noted on the approach chart. This type of NAVAID provides an approach course between 3 and 6 degrees, making it similar in accuracy to a localizer, but remember that the LDA is not as closely aligned with the runway and it does not offer a navigable back course. Currently there are less than 30 LDA installations in the U.S., and as a result, most pilots are

not familiar with this type of instrument approach. [Figure 5-58]

SIMPLIFIED DIRECTIONAL FACILITY

The SDF is another instrument approach system that is not as accurate as the LOC approach facilities. Like the LOC type approaches, the SDF is an alternative approach that may be installed at an airport for a variety of reasons, including terrain. The final approach

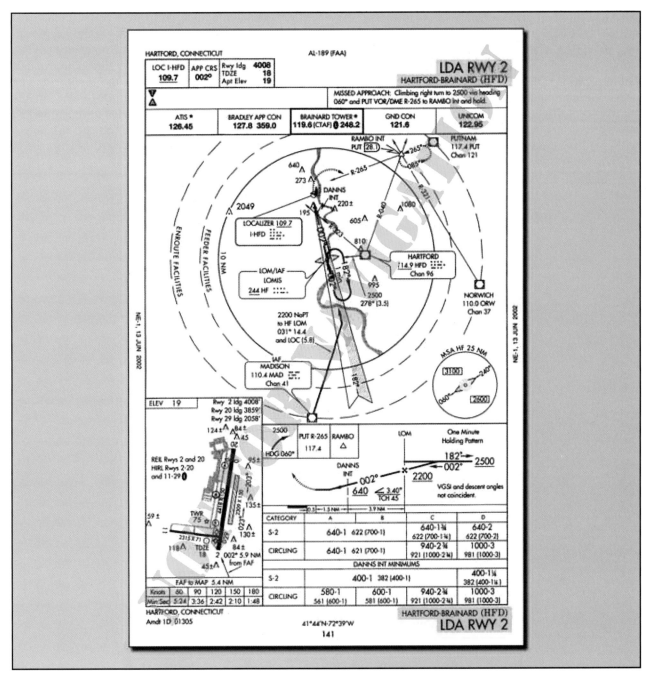

Figure 5-58. Hartford-Brainard (KHFD), Hartford, Connecticut, LDA RWY 2.

course width of an SDF system is set at either 6 or 12 degrees. The SDF is a nonprecision approach since it only provides lateral guidance to the runway.

For straight-in SDF approaches, the angle of convergence for the final approach course and the extended runway centerline is 30 degrees or less, and if the angle of convergence is beyond 30 degrees, the SDF will only have circling minimums. An SDF approach is provided a minimum of 250 feet obstacle clearance for straight-in approaches while in the final approach area, which is an area defined for a 6 degrees course: 1,000 feet at or abeam the runway threshold expanding to 19,228 feet 10 NM from the threshold. The same final approach area for a 12 degrees course is larger. This type of approach is also designed with a maximum descent gradient of 400 feet per NM, unless circling only minimums are authorized. [Figure 5-59]

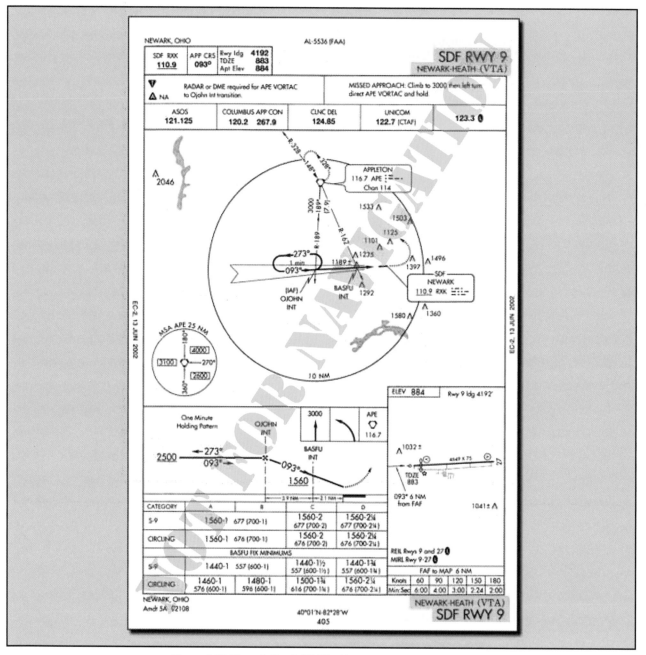

Figure 5-59. Newark-Heath (KVTA), Newark, Ohio, SDF RWY 9.

CHAPTER 6
SYSTEM IMPROVEMENT PLANS

In the next ten years, exciting new technologies will be implemented to help ease air traffic congestion, add to system capacity, and enhance safety. Some of these changes will be invisible to pilots and will be made seamlessly. Others will entail changing some old habits and learning new procedures. New aircraft equipment will bring powerful new capabilities, but will require training and practice to master.

FLEET IMPROVEMENT

Airlines and other operators will continue trying to find more efficient ways to use the National Airspace System (NAS). More and more users are working with federal agencies to write new policies and develop exchanges of real-time flight information, all in the interest of improving their service as well as their bottom lines. As new business strategies emerge, there also will be changes in the aircraft fleet. For example, as regional jets continue to increase in popularity, they have significant potential to reduce traffic at major airports as well as on the most crowded airways. Providing service along underused area navigation (RNAV) routes directly between smaller city pairs, they can bypass congested hubs and avoid airborne choke points. The number of regional jets is forecast to increase by more than 80 percent in the next decade. Compared to the turboprop airplanes they will replace, RJs fly at similar speeds and altitudes as larger jets, so they mix into traffic streams more smoothly, making en route traffic management easier for controllers. [Figure 6-1]

At the other end of the spectrum, larger airplanes capable of carrying over 500 passengers are now flying. These "super-jumbos" have the potential to reduce airway and terminal congestion by transporting more people in fewer airplanes.

Figure 6-1. Regional Jets.

This ability is especially valuable at major hubs, where the number of flight operations exceeds capacity at certain times of day. On the other hand, some of these airplanes have a double-deck configuration that might require extensive changes to terminals so that large numbers of passengers can board and deplane quickly and safely. Their size may require increased separation of taxiways and hold lines from runways due to increased wingspans and tail heights. Their weight also may require stronger runways and taxiways, as well as increased separation requirements for wake turbulence. [Figure 6-2]

Other innovative airplanes include the turbofan-powered very light jets (VLJs), which are relatively small turbo-

Figure 6-2. Superjumbo Airplanes.

fan-powered aircraft with 6 to 8 seats, with cruising speeds between 300 and 500 knots, and with a range of around 1,000 nautical miles (NMs). [Figure 6-3] If initial orders are an accurate indicator of their popularity, they will soon form a significant segment of the general aviation fleet. The FAA predicts that the business jet fleet will nearly double over the next ten years, approaching 16,000 airplanes by 2016. At least eight manufacturers are planning VLJs, several prototypes are flying, and the first new airplanes are being delivered to customers. Most are intended for single-pilot operation, and most will be certified for flight up to FL410. All will be technically advanced aircraft, with advanced glass cockpit avionics, digital engine controls, and sophisticated autopilots. These new airplanes will be capable of RNAV, required navigation performance (RNP), and reduced vertical separation minimum (RVSM) operations, and will operate mostly point-to-point, either on Q-Routes or random off-airways routes. With prices well below other business jets and competitive with turboprop singles, VLJs will appeal to many customers who could not otherwise justify the cost of a jet aircraft. VLJs have the potential of providing air taxi/air limousine services at costs comparable to commercial airlines, but with greater schedule flexibility, relatively luxurious accommodations, faster travel times, and the ability to fly into thousands more airports.

Courtesy Eclipse Aviation.

Figure 6-3. Very Light Jets are expected to become a sizeable segment of the high-altitude fleet.

ELECTRONIC FLIGHT BAG

As part of an ongoing effort to use the best technology available, industry has improved the timeliness and accuracy of information available to the pilot by converting it from a paper to a digital medium. An electronic flight bag (EFB) is an electronic display system intended primarily for cockpit/flightdeck or cabin use. EFBs can display a variety of aviation data or perform basic calculations, such as determining performance data or computing fuel requirements. In the past, paper references or an airline's flight dispatch department provided these functions. The EFB system may also include various other hosted databases and applications. These devices are sometimes referred to as auxiliary performance computers or laptop auxiliary performance computers.

The EFB is designed to improve efficiency and safety by providing real-time and stored data to pilots electronically. Use of an EFB can reduce some of a pilot's time-consuming communications with ground controllers while eliminating considerable weight in paper. EFBs can electronically store and retrieve many required documents, such as the General Operations Manual (GOM), Operations Specifications (OpSpecs), company procedures, Airplane Flight Manual (AFM), maintenance manuals and records, and dozens of other documents. [Figure 6-4]

In addition, advanced EFBs can also provide interactive features and perform automatic calculations, including performance calculations, power settings, weight and balance computations, and flight plans. They can also display images from cabin-mounted video and aircraft exterior surveillance cameras.

An EFB may store airport maps that can help a pilot avoid making a wrong turn on a confusing path of runways and taxiways, particularly in poor visibility or at an unfamiliar airport. Many runway incursions are due to confusion about taxi routes or pilots not being quite sure where they are on the airport. [Figure 6-5]

The FAA neither accepts or approves Class 1 or 2 EFBs which contain Types A, B, or C application software. Those who operate under 14 CFR parts 91K, 121, 125, 129, or 135 must obtain authorization for use. Advisory Circular 120-76, Guidelines for the Certification, Airworthiness, and Operational Approval of Electronic Flight Bag Computing Devices, sets forth the acceptable means for obtaining both certification and approval for operational use of Class 3 EFBs. It also outlines the capabilities and limitations of each of the three classes of EFBs, which are grouped according to purpose and function. Depending on the features of the specific unit, these devices are able to display a wide range of flight-related information. The most capable EFBs are able to display checklists, flight operations manuals (FOMs), CFRs, minimum equipment lists, en route navigation and approach charts, airport diagrams, flight plans, logbooks, and operating procedures. Besides serving as a cockpit library, they can also make performance calculations and perform many of the tasks traditionally handled by a dispatch department. Some units can also accept satellite weather data or input from global positioning system (GPS) receivers, combining the aircraft position and graphic weather information on a moving map display.

Figure 6-4. Electronic Flight Bag. The EFB has the potential to replace many paper charts and manuals in the cockpit.

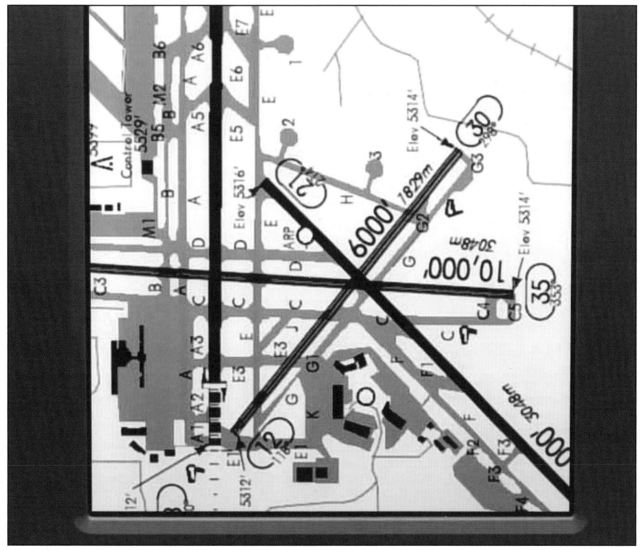

Figure 6-5. Moving Map Taxi Diagram on EFB.

Class 1 EFBs are portable. They can be used both on the ground and during flight, but must be stowed for takeoff and landing. They are limited to providing supplemental information only and cannot replace any required system or equipment. It may be connected to aircraft power through a certified power source, to operate the EFB and recharge its batteries. They are allowed to read data from other aircraft systems, and may receive and transmit data through a data link. Class 1 EFBs can display many different kinds of tabular data, such as performance tables, checklists, the FOM, AFM, and pilot's operating handbook (POH).

While a Class 2 EFB is also removable from the aircraft, it is installed in a structural-mounting bracket. This ensures that the EFB will not interfere with other aircraft systems. While Class 1 and 2 EFBs are both considered portable electronic devices, a logbook entry is required to remove the Class 2 EFB from the aircraft. It can be connected to aircraft power and to the aircraft's datalink port. The EFB can exchange data with aircraft systems, enabling it to make interactive performance calculations. It can be used to compute weight and balance information as well as takeoff and landing V-speeds, and to display flight critical pre-composed data, such as navigation charts. Since it is not necessarily stowed for takeoff and landing, pilots can use it to display departure, arrival, and approach charts.

The most capable EFBs are Class 3. These are built into the panel and require a Supplemental Type Certification (STC) or certification design approval with the aircraft as part of its equipment. Paper charts may not be required. Depending on the model, it may be connected to the GPS or Flight Management System (FMS), and it may be able to combine GPS position with the locations and speed vectors of other aircraft and graphic weather information into a single, detailed moving map display. Its detailed database can also provide obstacle and terrain warnings. It is important to remember that an EFB does not replace any system or equipment required by the regulations.

INCREASING CAPACITY AND SAFETY

Safety is, and will remain, the highest priority in all plans to increase capacity for the future. As demand for air travel continues to rise, it is clear that the NAS capacity must grow. Both the number of airport operations and en route capacity must increase simultaneously to accommodate the expanding needs. Neither can realistically be treated separately from the other, but for the sake of convenience, this chapter first discusses increasing the arrival/departure rate, then en route issues.

The number of aircraft operations is expected to increase by about 30 percent over the next decade. Although most parts of the NAS are able to handle current traffic, increasing operations will strain system capabilities

unless capacity grows to match demand. The FAA has identified and corrected several existing "choke points" in the NAS. While relatively few airports and airways experience large numbers of delays, the effects snowball into disruptions throughout the rest of the system, especially in adverse weather. Capacity must be increased to manage future growth. The FAA is implementing a number of programs to increase the capacity and efficiency of the NAS. Industry itself is also taking specific actions to address some of the problems.

INCREASING THE DEPARTURE/ARRIVAL RATE

Relatively few routes and airports experience the majority of congestion and delays. In the case of airports, peak demand occurs for only a few, isolated hours each day, so even the busiest hubs are able to handle their traffic load most of the time. Adjusting the number of arrivals and departures to get rid of those peak demand times would ease congestion throughout the system.

MORE RUNWAYS

At some major hubs, adding new runways or improving existing runways can increase capacity by as much as 50 percent, but the process is complex and time-consuming. During the planning phase, the appropriate FAA offices must review the new runway's impact on airspace, air traffic control (ATC) procedures, navigational aids (NAVAIDs), and obstructions. New instrument procedures must be developed, and economic feasibility and risk analysis may be required.

The next phase includes land acquisition and environmental assessment. Often, the airports that most need new runways are "landlocked" by surrounding developed areas, so obtaining land can be difficult. On top of that, residents and businesses in the area sometimes resist the idea of building a new runway. Concerns range from increased noise to safety and environmental impact. While environmental assessments and impact statements are essential, they take time. The FAA is working with other federal authorities to streamline the process of obtaining permits. Good community relations are extremely important, and working with airport neighbors can often address many of the questions and concerns.

The next phase of development involves obtaining the funding. A new runway typically costs between 100 million and one billion dollars. Money comes from airport cash flow, revenue and general obligation bonds, airport improvement program grants, passenger facility charges, and state and local funding programs.

The last phase includes the actual construction of the new runway, which may take as many as three years to complete. In all, over 350 activities are necessary to commission one new runway. The FAA has created the

Runway Template Action Plan to help airport authorities coordinate the process.

SURFACE TRAFFIC MANAGEMENT

In cooperation with the FAA, the National Aeronautics and Space Administration (NASA) is studying automation for aiding surface traffic management at major airport facilities. The surface management system is an enhanced decision support tool that will help controllers and airlines manage aircraft surface traffic at busy airports, thus improving safety, efficiency, and flexibility. The surface management system provides tower controllers and air carriers with accurate predictions of the future departure demand and how the situation on the airport surface, such as takeoff queues and delays at each runway, will evolve in response to that demand. To make these predictions, the surface management system will use real-time surface surveillance, air carrier predictions of when each flight will want to push back, and computer software that accurately predicts how aircraft will be directed to their departure runways.

In addition to predictions, the surface management system also provides advisories to help manage surface movements and departure operations. For example, the surface management system advises a departure sequence to the ground and local controllers that efficiently satisfies various departure restrictions such as miles-in-trail and expected departure clearance times (EDCTs). Information from the surface management system is displayed in ATC towers and airline ramp towers, using either dedicated surface management system displays or by adding information to the displays of other systems.

Parts of the system were tested in 2003 and 2004, and are now ready for deployment. Other capabilities are accepted in concept, but are still under development. Depending on the outcome of the research, the surface management system might also provide information to the terminal radar approach control (TRACON) and center traffic management units (TMUs), airline operations centers (AOCs), and ATC system command centers (ATCSCCs). In the future, additional developments may enable the surface management system to work with arrival and departure traffic management decision support tools.

The surface movement advisor (SMA) is another program now being tested in some locations. This project facilitates the sharing of information with airlines to augment decision-making regarding the surface movement of aircraft, but is concerned with arrivals rather than departures. The airlines are given automated radar terminal system (ARTS) data to help them predict an aircraft's estimated touchdown time. This enhances airline gate and ramp operations, resulting in more efficient movement of aircraft while they are on the ground. Airline customers reported reduced gate delays and diversions at the six locations where SMA is in use.

TERMINAL AIRSPACE REDESIGN

The FAA is implementing several changes to improve efficiency within terminal airspace. While some methods increase capacity without changing existing routes and procedures, others involve redesigning portions of the airspace system. One way of increasing capacity without major procedural changes is to fill the gaps in arrival and departure streams. Traffic management advisor (TMA) is ATC software that helps controllers by automatically sequencing arriving traffic. Based on flight plans, radar data, and other information, the software computes very accurate aircraft trajectories as much as an hour before the aircraft arrives at the TRACON. It can potentially increase operational capacity by up to ten percent, and has improved capacity by 3 to 5 percent for traffic into the Dallas/Ft. Worth, Los Angeles, Minneapolis, Denver, and Atlanta airports.

One limitation of TMA is that it uses information on incoming flights from a single Air Route Traffic Control Center (ARTCC). Another version is under development that will integrate information from more than one ARTCC. It is called multi-center traffic management advisor (McTMA). This system is being tested in the busy Northeastern area, and the results are promising.

Another software-based solution is the passive final approach spacing tool (pFAST). This software analyzes the arriving traffic at a TRACON and suggests appropriate runway assignment and landing sequence numbers to the controller. Controllers can accept or reject the advisories using their keyboards. The early version carries the "passive" designation because it provides only runway and sequence number advisories. A more advanced version, called active FAST (aFAST), is currently under development at NASA Ames Research Center. In addition to the information provided by pFAST, aFAST will display heading and speed, and it is expected to improve capacity by an additional 10 percent over pFAST.

Airlines can help ease congestion on shorter routes by filing for lower altitudes. Although the airplane uses more fuel at a lower cruising altitude, the flight may prove faster and more economical if weather or high traffic volume is delaying flights at higher levels. The tactical altitude assignment program consists of published routes from hubs to airports 200 to 400 NM away. Based on results of evaluation, it is not expected to be implemented nationally, although it may remain available in local areas.

Beyond using existing facilities and procedures more effectively, capacity can often be increased by making relatively minor changes in air traffic procedures. For example, in some instances, departure and arrival patterns have remained unchanged from when there was very little air traffic, and congestion results when today's traffic tries to use them. Likewise, arrival and departure procedures may overlap, either because they were based on lower volumes and staffing or because they are based on ground-based navigation. The interdependence of arrival and departure routes tends to limit throughput in both directions.

Separating departures from incoming traffic can simplify the work of controllers, reduce vectoring, and make more efficient use of terminal airspace. In the **four corner post configuration**, four NAVAIDs form the four corners of the TRACON area, roughly 60 NM from the primary airport. All arrivals to the area fly over one of these "corner posts" (also called arrival meters or feeder fixes). The outbound departure streams are spaced between the arrival streams. [Figure 6-6]

As more and more aircraft are equipped for RNAV, new arrival and departure routes are being created that do not depend on very high frequency omni-directional range (VOR) airways or ground-based NAVAIDs. Shifting traffic to new RNAV routes eases congestion on existing airways. There are already several new RNAV routes in use and many more are being developed.

SEPARATION STANDARDS

Current regulations permit a 3 NM separation within 40 NM of a single radar sensor. The FAA is looking at ways to increase the use of the 3 NM separation standard to improve efficiency and maximize the volume of traffic that can be safely moved into busy terminal areas. The methods involve increasing the size of terminal areas to include more en route airspace, redesigning airspace to encompass multiple airports within a single ATC facility, and consolidating certain TRACON facilities. This will involve major changes on the ground for ATC facilities, and changes in charts and procedures for pilots.

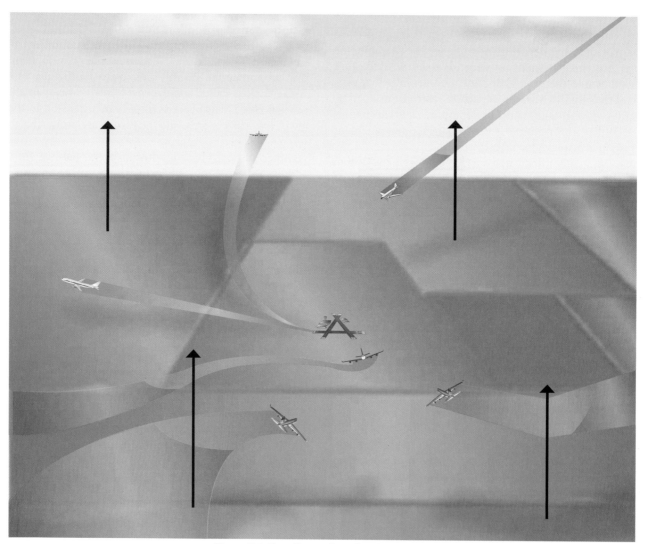

Figure 6-6. Four Corner Post Configuration.

As gaps are filled in arrival and departure streams and the 3 NM separation standard is applied more extensively, traffic advisories from the traffic alert and collision avoidance system (TCAS) are bound to increase. While newer software enhances functionality, provides more timely resolution advisories, and eliminates many nuisance alerts, data link technology based on GPS position information may offer even better results.

MAINTAINING RUNWAY USE IN REDUCED VISIBILITY

Although traffic in congested airspace typically operates under instrument flight rules (IFR), adverse weather and actual instrument meteorological conditions (IMC) can drastically reduce system capacity. Many parallel runways cannot be used simultaneously in IMC because of the time delay and limited accuracy of terminal area radar, and the runways are spaced closer than the minimum allowable distance for wake vortex separation.

LAAS AND WAAS IMPLEMENTATION

The wide area augmentation system (WAAS) became available at most locations in 2003. Additional ground reference stations are expected to become operational in Canada, Mexico, and Alaska by 2008, providing more complete WAAS coverage for the continental United States. The local area augmentation system (LAAS) provides even greater accuracy and may be certified for use in precision approaches at some locations beginning in 2007.

Another benefit of LAAS and WAAS is that better position information can be sent to controllers and other aircraft. Automatic dependant surveillance-broadcast (ADS-B) uses GPS to provide much more accurate location information than radar and transponder systems. This position information is broadcast to other ADS-equipped aircraft (as well as ground facilities), providing pilots and controllers with a more accurate real-time picture of traffic.

For full safety and effectiveness, every aircraft under the control of ATC will need ADS-B. Until that occurs, controllers must deal with a mix of ADS-B and transponder-equipped aircraft. Equipment is already available that can fuse the information from both sources and show it on the same display. **Traffic information service-broadcast (TIS-B)** does just that. Although TIS-B is primarily intended for use on the ground by controllers, the information can be transmitted to suitably equipped aircraft and displayed to pilots in the cockpit. The **cockpit display of traffic information (CDTI)** provides information for both ADS-B and non-ADS-B aircraft on a single cockpit display. [Figure 6-7] Since this information is shown even while the aircraft is on the ground, it also improves situational awareness during surface movement, and can help prevent or resolve taxiing conflicts.

Figure 6-7. Cockpit Display of Traffic Information. This display shows both ADS-B and other aircraft radar targets.

REDUCING EN ROUTE CONGESTION

In addition to the congestion experienced at major hubs and terminal areas, certain parts of the en route structure have reached capacity. Easing the burden on high-volume airways and eliminating airborne choke points are some of the challenges addressed by new airspace plans.

MATCHING AIRSPACE DESIGN TO DEMANDS

More new RNAV routes are being created, which are essentially airways that use RNAV for guidance instead of VORs. They are straighter than the old VOR airways, so they save flight time and fuel costs. By creating additional routes, they reduce traffic on existing airways, adding en route capacity. As new routes are created near existing airways, chart clutter will become more of an issue. Electronic chart presentations are being developed that will allow pilots to suppress information that is irrelevant to their flight, while ensuring that all information necessary for safety is displayed. The high degree of accuracy and reliability of RNP procedures offers another means of increasing capacity along popular RNAV routes. Instead of having all the aircraft that are using the route fly along the same ground track, RNP allows several closely spaced parallel tracks to be created for the same route. In essence, this changes a one-lane road into a multi-lane highway. [Figure 6-8]

REDUCING VOICE COMMUNICATION

Many runway incursions and airborne clearance mistakes are due to misunderstood voice communications. During busy periods, the necessity of exchanging dozens of detailed instructions and reports leads pilots and controllers to shorten and abbreviate standard phraseology, often leading to errors. It stands to reason that better ways to transfer information could reduce voice communications, and thus reduce the incidence of communication errors. One such innovation is similar to the display screen at fast-food drive-up windows. As the cashier punches in the order, it is displayed on the monitor so the customer can verify the order. This kind of feedback reduces the common problem of hearing what is expected to be heard, which is particularly problematic in ATC clearances and read backs. Not only does reducing voice communications reduce frequency congestion, it also eliminates certain opportunities for misunderstanding.

Controller pilot data link communication (CPDLC) augments voice communications by providing a second communication channel for use by the pilot and controller, using data messages that are displayed in the cockpit. This reduces delays resulting from congestion on voice channels. The initial version of CPDLC will display a limited number of air traffic messages, but future versions will have expanded message capabilities and permit pilot-initiated requests.

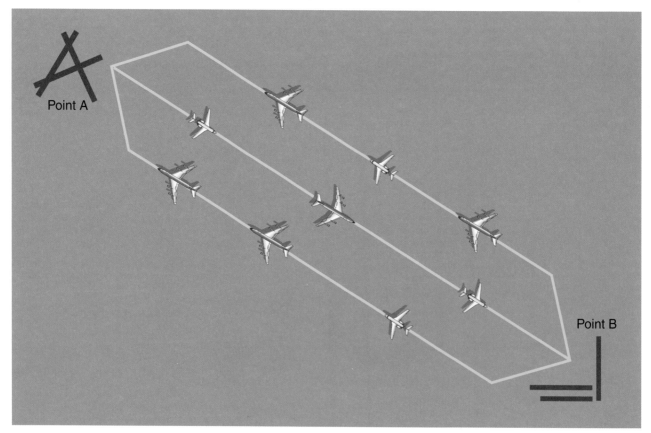

Figure 6-8. RNP allows parallel tracks along the same route, multiplying capacity along that route.

AIRCRAFT COMMUNICATIONS ADDRESSING AND REPORTING SYSTEM

Of course, pilot-controller communication is compromised when the crew is listening to other frequencies or engaged in other communications, such as talking to their company. If these communications could be accomplished silently and digitally, voice communications with ATC would improve. The Aircraft Communications Addressing and Reporting System (ACARS) is a commercial system that enables the crew to communicate with company personnel on the ground. It is often used to exchange routine flight status messages, weather information, and can serve as a non-voice communication channel in the event of an emergency. Many of the messages are sent and received automatically, such as the time the flight leaves the gate (triggered by the release of the parking brake), takeoff and touchdown times (triggered by landing gear switches), and arrival time (triggered when a cabin door is opened). Other information may include flight plans, significant meteorological information (SIGMETs), crew lists, cargo manifests, automatic terminal information service (ATIS) reports, en route and destination weather, clearances, and fuel reports. Some ACARS units can interface with onboard engine and performance-monitoring systems to inform company ground personnel of maintenance or operations related issues. [Figure 6-9]

Significant valuable meteorological data can be obtained by collecting data from aircraft fitted with appropriate software packages. To date, the predominant sources of automated aviation data have been from aircraft equipped with aircraft to satellite data relay (ASDAR) and ACARS, which routes data back via general purpose information processing and transmitting systems now fitted to many commercial aircraft. These systems offer the potential for a vast increase in the provision of aircraft observations of wind and temperature. Making an increasingly important contribution to the observational database, it is envisioned that ACARS data will inevitably supersede manual pilot reports (PIREPS).

Another use of ACARS is in conjunction with Digital ATIS (D-ATIS), which provides an automated process for the assembly and transmission of ATIS messages. ACARS enables audio messages to be displayed in text form in the flight decks of aircraft equipped with ACARS. A printout is also provided if the aircraft is equipped with an on-board printer. D-ATIS is operational at over 57 airports that now have pre-departure clearance (PDC) capability.

AUTOMATIC DEPENDENT SURVEILLANCE-BROADCAST

Unlike TCAS and terrain awareness and warning systems (TAWS), which have been used in airline and military air-

Figure 6-9. ACARS Communications Display.

craft for at least a decade, ADS-B is a relatively new air traffic technology. It is an onboard system that uses Mode S transponder technology to periodically broadcast an aircraft's position, along with some supporting information like aircraft identification and short-term intent. By picking up broadcast position information on the ground instead of using ground radar stations, ADS-B represents a significant advancement over the existing ATC system by providing increased accuracy and safety. This is possible because ADS-B addresses the major deficiency of TCAS - accuracy. In the TCAS system, aircraft positions are only accurate to a few degrees; thus, the accuracy of TCAS decreases with distance. Moreover, the reliance on transmission timing for range data in TCAS is error-prone. The method used by ADS-B avoids this problem.

In addition to the broadcast of position to the ground, ADS-B can be used to enable a new collection of aircraft-based applications. Unlike conventional radar, ADS-B works at low altitudes and on the ground. It is effective in remote areas or in mountainous terrain where there is no radar coverage, or where radar coverage is limited. One of the greatest benefits of ADS-B is its ability to provide the same real-time information to pilots in the aircraft cockpit and to ground controllers, so that for the first time, both can view the same data.

ADS-B will also enable aircraft to send messages to each other to provide surveillance and collision avoidance through data link. Other aircraft in the immediate vicinity can pick up position information broadcasts from equipped aircraft. This enables equipped aircraft to formulate a display of nearby aircraft for the pilot; the pilot's awareness of the current situation is enhanced. Combined with databases of current maps and charts, the onboard displays can show terrain as well as proximate aircraft. This is a powerful inducement for change. The heightened situational awareness offered by satellite navigation in conjunction with modern database applications and map displays, combined with the position of proximate aircraft, builds a picture in the cockpit equivalent to that on the ground used by the controller. This is particularly important in places like Alaska where aviation is vital, NAS infrastructure is minimal (because of the harsh conditions), and weather changes quickly and in unpredictable fashions.

Eventually, as the fleets equip, it may be possible to save money by retiring expensive long-range radars. Identified by the FAA as the future model for ATC, ADS-B is a major step in the direction of free flight. While ADS-B shows great promise for both air-to-air and air-to-ground surveillance, current aircraft transponders will continue to support surveillance operations in the NAS for the foreseeable future. If enough users equip with ADS-B avionics, the FAA will install a compatible ADS ground system to provide more accurate surveillance information to ATC compared to radar-based surveillance.

In the United States, two different data links have been adopted for use with ADS-B: 1090 MHz Extended Squitter (1090 ES) and the Universal Access Transceiver (UAT). The 1090 ES link is intended for aircraft that primarily operate at FL180 and above, whereas the UAT link is intended for use by aircraft that primarily operate at 18,000 feet and below. From a pilot's standpoint, the two links operate similarly and both support ADS-B and TIS-B. The UAT link additionally supports Flight Information Service-Broadcast (FIS-B) at any altitude when within ground based transmitter (GBT) coverage. FIS-B is the weather information component, and provides displays of graphical and textual weather information. Areas of approved use for the UAT include the United States (including oceanic airspace where air traffic services are provided), Guam, Puerto Rico, American Samoa, and the U.S. Virgin Islands. The UAT is approved for both air and airport surface use. ADS-B broadcast over the 1090 MHz data link has been approved for global use.

REDUCING VERTICAL SEPARATION

Current vertical separation minima (2,000 feet) were created more than 40 years ago when altimeters were not very accurate above FL 290. With better flight and navigation instruments, vertical separation has been safely reduced to 1,000 feet in most parts of the world, except Africa and China.

RVSM airspace has already been implemented over the Atlantic and Pacific Oceans, South China Sea, Australia, Europe, the Middle East and Asia south of the Himalayas. Domestic RVSM (DRVSM) in the United States was implemented in January 2005 when FL 300, 320, 340, 360, 380, and 400 were added to the existing structure. To fly at any of the flight levels from FL 290 to FL 410, aircraft and operator must be RVSM-approved. [Figure 6-10]

REDUCING HORIZONTAL SEPARATION

The current oceanic air traffic control system uses filed flight plans and position reports to track an aircraft's progress and ensure separation. Pilots send position reports by high frequency (HF) radio through a private radio service that then relays the messages to the air traffic control system. Position reports are made at intervals of approximately one hour. HF radio communication is subject to interference and disruption. Further delay is added as radio operators relay messages between pilots and controllers. These deficiencies in communications and surveillance have necessitated larger horizontal separation minimums when flying over the ocean out of radar range.

As a result of improved navigational capabilities made possible by technologies such as GPS and CPDLC, both lateral and longitudinal oceanic horizontal separation standards are being reduced. Oceanic lateral separation standards were reduced from 100 to 50 NM in the Northern and Central Pacific regions in 1998 and in the Central East Pacific in 2000. The FAA plans to extend the 50 NM separation standard to the South Pacific. Because flight times along the South Pacific routes often exceed 15 hours, the fuel and time savings resulting from more airplanes flying closer to the ideal wind route in this region are expected to be substantial. Separation standards of 30 NM are already undergoing operational trials in parts of South Pacific airspace for properly authorized airplanes and operators.

DIRECT ROUTING

Based on preliminary evaluations, FAA research has evidenced tremendous potential for the airlines to benefit from expected routing initiatives. Specifically, direct routing or "Free Flight" is the most promising for reducing total flight time and distance as well as minimizing congestion on heavily traveled airways. Traditionally,

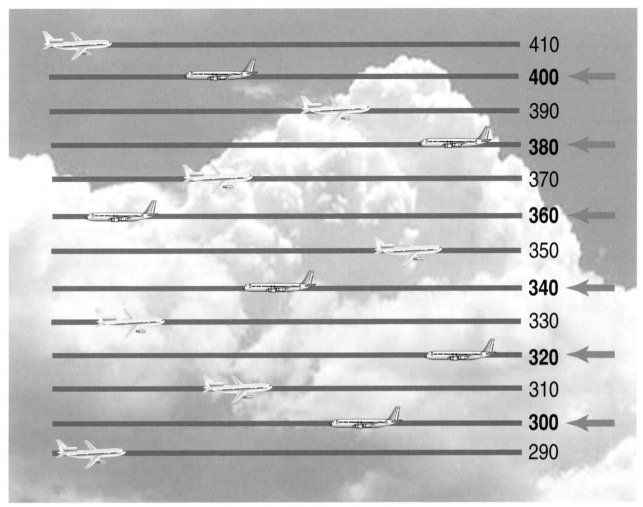

410
400 ←
390
380 ←
370
360 ←
350
340 ←
330
320 ←
310
300 ←
290

Figure 6-10. DRVSM High Altitude Routes.

pilots fly fixed routes that often are less direct due to their dependence on ground-based NAVAIDs. Through Free Flight, the FAA hopes to increase the capacity, efficiency, and safety of the NAS to meet growing demand as well as enhance the controller's productivity. The aviation industry, particularly the airlines, is seeking to shorten flight times and reduce fuel consumption. According to the FAA's preliminary estimates, the benefits to the flying public and the aviation industry could reach into the billions of dollars once the program is fully operational.

Free Flight Phase 1 began in October 1998 and launched five software tools over the next four years. These were Collaborative Decision Making (CDM), the User Request Evaluation Tool (URET), and the previously discussed SMA, TMA, and pFAST.

CDM allows airspace users and the FAA to share information, enabling the best use of available resources. It provides detailed, real-time information about weather, delays, cancellations, and equipment to airlines and major FAA air traffic control facilities. This shared data helps to manage the airspace system more efficiently, thereby reducing delays.

CDM consists of three components. The first component allows airlines and the FAA's System Command Center in Herndon, Virginia, to share the latest information on schedules, airport demand, and capacity at times (usually during bad weather) when airport capacity is reduced. This shared information is critical to getting the maximum number of takeoffs and landings at airports. The second component creates and assesses possible rerouting around bad weather. This tool enables the Command Center and busy major ATC facilities to share real-time information on high-altitude traffic flows with airline operations centers, thus developing the most efficient ways to avoid bad weather. The third component provides data on the operational status of the national airspace system. Examples include runway visibility at major airports and the current availability of Special Use Airspace.

URET allows controllers to plot changes in the projected flight paths of specific airplanes to see if they will get too close to other aircraft within the next 20 minutes. URET means that controllers can safely and quickly respond to pilots' requests for changes in altitude or direction, which leads to smoother, safer flights and more direct routings. During trials in the Memphis and

Indianapolis en route centers, the use of more direct routes made possible by URET was found to save airlines about $1.5 million per month.

ACCOMMODATING USER PREFERRED ROUTING

Free Flight Phase 2 builds on the successes of Free Flight Phase 1 to improve safety and efficiency within the NAS. Implementation of Phase 2 will include the expansion of Phase 1 elements to additional FAA facilities. This program will deploy a number of additional capabilities, such as CDM with collaborative routing coordination tool (CRCT) enhancements and CPDLC.

The National Airspace System status information (NASSI) tool is the most recent CDM element to be introduced. NASSI enables the real-time sharing of a wide variety of information about the operational status of the NAS. Much of this information has previously been unavailable to most airspace users. NASSI currently includes information on maintenance status and runway visual range at over 30 airports.

The CRCT is a set of automation capabilities that can evaluate the impact of traffic flow management rerouting strategies. The major focus of this tool is management of en route congestion.

IMPROVING ACCESS TO SPECIAL USE AIRSPACE

Special use airspace (SUA) includes prohibited, restricted, warning, and alert areas, as well as military operations areas (MOAs), controlled firing areas, and national security areas. The FAA and the Department of Defense are working together to make maximum use of SUA by opening these areas to civilian traffic when they are not being used by the military. The **military airspace management system (MAMS)** keeps an extensive database of information on the historical use of SUA, as well as schedules describing when each area is expected to be active. MAMS transmits this data to the **special use airspace management system (SAMS)**, an FAA program that provides current and scheduled status information on SUA to civilian users. This information is available at the following link **http://sua.faa.gov/**. The two systems work together to ensure that the FAA and system users have current information on a daily basis.

A prototype system called SUA in-flight service enhancement (SUA/ISE) provides graphic, near-real-time depictions of SUA to automated flight service station (AFSS) specialists who can use the information to help pilots during flight planning as well as during flight. Pronounced "Suzy," this tool can display individual aircraft on visual flight rule (VFR) flight plans (with data blocks), plot routes of flight,

identify active SUA and display weather radar echoes. Using information from the enhanced traffic management system, AFSS specialists will see this information on a combined graphic display. This data may also be transmitted and shown on cockpit displays in general and commercial aviation aircraft.

The central altitude reservation function (CARF) coordinates military, war plans, and national security use of the NAS. While SAMS handles the schedule information regarding fixed or charted SUA, CARF handles unscheduled time and altitude reservations. Both subsystems deal with planning and tracking the military's use of the NAS.

The FAA and the U.S. Navy have been working together to allow civilian use of offshore warning areas. When adverse weather prevents the use of normal air routes along the eastern seaboard, congestion and delays can result as flights are diverted to the remaining airways. When offshore warning areas are not in use by the Navy, the airspace could be used to ease the demand on inland airways. To facilitate the use of this airspace, the FAA established waypoints in offshore airspace along four routes for conducting point-to-point navigation when the Navy has released that airspace to the FAA. The waypoints take advantage of RNAV capabilities and provide better demarcation of airspace boundaries, resulting in more flexible release of airspace in response to changing weather. These new offshore routes, which stretch from northern Florida to Maine, are an excellent example of how close coordination between military and civil authorities can maximize the utility of limited airspace.

HANDLING EN ROUTE SEVERE WEATHER

Interpreting written or spoken weather information is not difficult, nor is visualizing the relationship of the weather to the aircraft's route, although verbal or textual descriptions of weather have inevitable limitations. Color graphics can show more detail and convey more information, but obtaining them in flight has been impractical, until recently. The graphical weather service (GWS) provides a nationwide precipitation mosaic, updated frequently, and transmitted to the aircraft and displayed in the cockpit. Whether the display is used to strategize navigation, to avoid weather en route, or for departures and approaches, consideration must always be given to the timeliness of the graphic update. Pilots can select any portion of the nationwide mosaic with range options of 25, 50, 100, and 200 NM. In addition to providing information on precipitation, this service can be expanded to include other graphical data. Some systems will place the detailed weather graphics directly on a moving map display, removing another step of interpretation and enabling pilots to see the weather in relation to their flight path. [Figure 6-11]

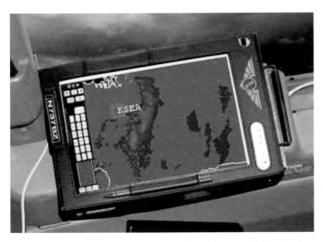

Figure 6-11. Prototype Data Link Equipment. This display shows a radar image of weather within 50 NM of the Seattle-Tacoma International Airport (KSEA).

NATIONAL ROUTE PROGRAM

In the U.S., the national route program (NRP), also known as "Free Flight," is an example of applying RNAV techniques. The NRP is a set of rules and procedures that are designed to increase the flexibility of user flight planning within published guidelines. The Free Flight program allows dispatchers and pilots to choose the most efficient and economical route for flights operating at or above FL 290 between city pairs, without being constrained to airways and preferred routes.

Free Flight is a concept that allows you the same type of freedom you have during a VFR flight. Instead of a NAS that is rigid in design, pilots are allowed to choose their own routes, or even change routes and altitudes at will to avoid icing, turbulence, or to take advantage of winds aloft. Complicated clearances become unnecessary, although flight plans are required for traffic planning purposes and as a fallback in the event of lost communication.

Free Flight is made possible with the use of advanced avionics, such as GPS navigation and datalinks between your aircraft, other aircraft, and controllers. Separation is maintained by establishing two airspace zones around each aircraft, as shown in Figure 6-12. The protected zone, which is the one closest to the aircraft, never meets the protected zone of another aircraft. The alert zone extends well beyond the protected zone, and aircraft can maneuver freely until alert zones touch. If alert zones do touch, a controller may provide the pilots with course suggestions, or onboard traffic displays may be

used to resolve the conflict. The size of the zones is based on the aircraft's speed, performance, and equipment. Free Flight is operational in Alaska, Hawaii, and part of the Pacific Ocean, using about 2,000 aircraft. Full implementation is projected to take about 20 years.

As the FAA and industry work together, the technology to help Free Flight become a reality is being placed into position, especially through the use of the GPS satellite system. Equipment such as ADS-B allows pilots in their cockpits and air traffic controllers on the ground to "see" aircraft traffic with more precision than has previously been possible. The FAA has identified more than 20 ways that ADS-B can make flying safer. It can provide a more efficient use of the airspace and improve your situational awareness.

DEVELOPING TECHNOLOGY

Head-up displays (HUDs) grew out of the reflector gun sights used in fighter airplanes before World War II. The early devices functioned by projecting light onto a slanted piece of glass above the instrument panel, between the pilot and the windscreen. At first, the display was simply a dot showing where bullets would go, surrounded by circles or dots to help the pilot determine the range to the target. By the 1970s, the gun sight had become a complete display of flight information. By showing airspeed, altitude, heading, and aircraft attitude on the HUD glass, pilots were able to keep their eyes outside the cockpit more of the time. Collimators make the image on the glass appear to be far out in front of the aircraft, so that the pilot need not change eye focus to view the relatively nearby HUD. Today's head-up guidance systems (HGS) use holographic displays. Everything from weapons status to approach information can be shown on current military HGS displays. This technology has

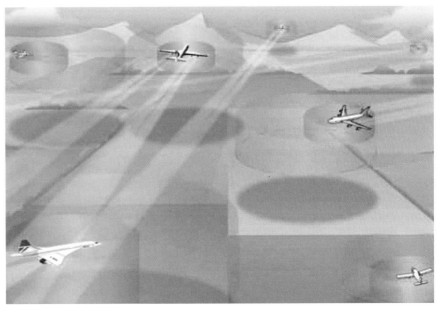

Figure 6-12. Free Flight.

obvious value for civilian aviation, but until 1993 no civilian HGS systems were available. This is changing, and application of HGS technology in airline and corporate aircraft is becoming widespread. [Figure 6-13]

Figure 6-13. Head-up Guidance System.

A large fraction of aircraft accidents are due to poor visibility. While conventional flight and navigation instruments generally provide pilots with accurate flight attitude and geographic position information, their use and interpretation requires skill, experience, and constant training. NASA is working with other members of the aerospace community to make flight in low visibility conditions more like flight in visual meteorological conditions (VMC). **Synthetic vision** is the name for systems that create a visual picture similar to what the pilot would see out the window in good weather, essentially allowing a flight crew to see through atmospheric obscurations like haze, clouds, fog, rain, snow, dust, or smoke.

The principle is relatively simple. GPS position information gives an accurate three-dimensional location, onboard databases provide detailed information on terrain, obstructions, runways, and other surface features, and virtual reality software combines the information to generate a visual representation of what would be visible

from that particular position in space. The dynamic image can be displayed on a head-down display (HDD) on the instrument panel, or projected onto a HGS in such a way that it exactly matches what the pilot would see in clear weather. Even items that are normally invisible, such as the boundaries of special use airspace or airport traffic patterns, could be incorporated into such a display. While the main elements of such a system already exist, work is continuing to combine them into a reliable, safe, and practical system. Some of the challenges include choosing the most effective graphics and symbology, as well as making the synthetic vision visible enough to be useful, but not so bright that it overwhelms the real view as actual terrain becomes visible. Integrating ADS-B information may make it possible for synthetic vision systems to show other aircraft. [Figure 6-14]

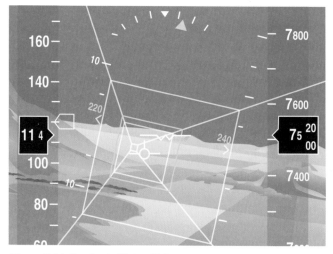

Figure 6-14. Synthetic Vision. This system uses projected images to provide a virtual view of terrain and other data in reduced visibility.

A natural extension of the synthetic vision concept is the **highway in the sky (HITS)** program. This technology adds an easy-to-interpret flight path depiction to an electronic flight instrument system (EFIS) type of cockpit display, which may be located on the instrument panel or projected on a HUD. The intended flight path is shown as a series of virtual rectangles that appear to stand like a series of window frames in front of the aircraft. The pilot maneuvers the aircraft so that it flies "through" each rectangle, essentially following a visible path through the sky. When installed as part of a general aviation "glass cockpit," this simple graphic computer display replaces many of the conventional cockpit instruments, including the attitude indicator, horizontal situation indicator, turn coordinator, airspeed indicator, altimeter, vertical speed indicator, and navigation indicators. Engine and aircraft systems information may also be incorporated. [Figure 6-15]

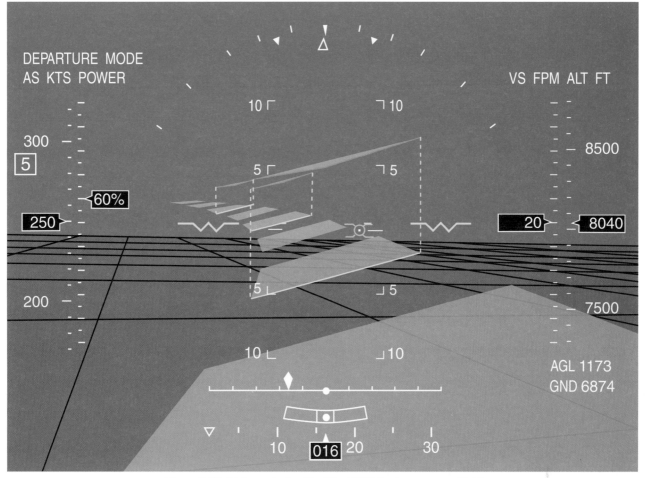

Figure 6-15. Highway in the Sky. The HITS display conveys flight
path and attitude information using an intuitive graphic interface.

CHAPTER 7
HELICOPTER INSTRUMENT PROCEDURES

This chapter presents information on instrument flight rule (IFR) helicopter operations in the National Airspace System (NAS). Although helicopter instrument flight is relatively new when compared to airplane instrument flight, the global positioning system (GPS) and the developing Wide Area Augmentation System (WAAS) are bringing approach procedures to heliports around the country. As of February 2006 there were approximately 45 public "Copter" instrument flight procedures, including 23 instrument landing system (ILS), 5 RNAV (GPS) point-in-space (PinS), 6 non-directional beacon (NDB), 8 VHF Omni-directional Range (VOR), and 227 private RNAV (GPS) "Specials" either to runways or PinS approaches to heliports. This does not include approach procedures that are located five miles or more from shore in the Gulf of Mexico and other locations.

The ability to operate helicopters under IFR increases their utility and safety. Helicopter IFR operators have an excellent safety record due to the investment in IFR equipped helicopters, development of instrument approach procedures, and IFR trained flight crews. The safety record of IFR operations in the Gulf of Mexico is equivalent to the safety record of the best-rated airlines. Manufacturers are working to increase IFR all-weather capabilities of helicopters by providing slower minimum instrument airspeeds (V_{MINI}), faster cruising speeds, and better autopilots and flight management systems (FMS). As a result, in October 2005, the first civil helicopter in the United States was certified for flight into known icing conditions. [Figure 7-1]

HELICOPTER IFR CERTIFICATION
It is very important that pilots be familiar with the IFR requirements for their particular helicopter. Within the same make, model and series of helicopter, variations in the installed avionics may change the required equipment or the level of augmentation for a particular operation. The Automatic Flight Control System/Autopilot/Flight Director (AFCS/AP/FD) equipment installed in IFR helicopters can be very complex. For some helicopters, the AFCS/AP/FD complexity will require formal training in order for the pilot(s) to obtain and maintain a high level of knowledge of system operation, limitations, failure indications and reversionary modes. For a helicopter to be certified to conduct operations in instrument meteorological

Figure 7-1. Icing Tests. To safely provide an all-weather capability and flight into known icing conditions that would otherwise delay or cancel winter flight operations, the digital control of the S-92 rotor ice protection system (RIPS) determines the temperature and moisture content of the air and removes any ice buildup by heating the main and tail rotor blades. The system is shown here during testing.

conditions (IMC), it must meet the design and installation requirements of Title 14 Code of Federal Regulations (14 CFR) Part 27, Appendix B (Normal Category) and Part 29, Appendix B (Transport Category), which are in addition to the visual flight rule (VFR) requirements.

These requirements are broken down into the following categories: flight and navigation equipment, miscellaneous requirements, stability, helicopter flight manual limitations, operations specifications, and minimum equipment list (MEL).

FLIGHT AND NAVIGATION EQUIPMENT
The basic installed flight and navigation equipment for helicopter IFR operations is listed under Part 29.1303, with amendments and additions in Appendix B of Parts 27 and 29 under which they are certified. The list includes:

- Clock.
- Airspeed indicator.
- Sensitive altimeter adjustable for barometric pressure[1].
- Magnetic direction indicator.
- Free-air temperature indicator.
- Rate-of-climb (vertical speed) indicator.
- Magnetic gyroscopic direction indicator.
- Standby bank and pitch (attitude) indicator.
- Non-tumbling gyroscopic bank and pitch (attitude) indicator.
- Speed warning device (if required by Part 29).

MISCELLANEOUS REQUIREMENTS

- Overvoltage disconnect.
- Instrument power source indicator.
- Adequate ice protection of IFR systems.
- Alternate static source (single pilot configuration).
- Thunderstorm lights (transport category helicopters).

STABILIZATION AND AUTOMATIC FLIGHT CONTROL SYSTEM (AFCS)

Helicopter manufacturers normally use a combination of a stabilization and/or AFCS in order to meet the IFR stability requirements of Parts 27 and 29. These systems include:

- **Aerodynamic surfaces**, which impart some stability or control capability that generally is not found in the basic VFR configuration.

- **Trim systems**, which provide a cyclic centering effect. These systems typically involve a magnetic brake/spring device, and may be controlled by a four-way switch on the cyclic. This system supports "hands on" flying of the helicopter.

- **Stability Augmentation Systems (SASs)**, which provide short-term rate damping control inputs to increase helicopter stability. Like trim systems, SAS supports "hands on" flying.

- **Attitude Retention Systems (ATTs)**, which return the helicopter to a selected attitude after a disturbance. Changes in attitude can be accomplished usually through a four-way "beep" switch, or by actuating a "force trim" switch on the cyclic, which sets the desired attitude manually. Attitude retention may be a SAS function, or may be the basic "hands off" autopilot function.

- **Autopilot Systems (APs)** provide for "hands off" flight along specified lateral and vertical paths. The functional modes may include heading, altitude, vertical speed, navigation tracking, and approach. APs typically have a control panel for mode selection and indication of mode status. APs may or may not be installed with an associated flight director (FD). APs typically control the helicopter about the roll and pitch axes (cyclic control) but may also include yaw axis (pedal control) and collective control servos.

- **Flight Directors (FDs)**, which provide visual guidance to the pilot to fly selected lateral and vertical modes of operation. The visual guidance is typically provided by a "single cue," commonly known as a "vee bar," which provides the indicated attitude to fly and is superimposed on the attitude indicator. Other flight directors may use a "two cue" presentation known as a "cross pointer system." These two presentations only provide attitude information. A third system, known as a "three cue" system, provides information to position the collective as well as attitude (roll and pitch) cues. The collective control cue system identifies and cues the pilot which collective control inputs to use when path errors are produced, or when airspeed errors exceed preset values. The three-cue system pitch command provides the required cues to control airspeed when flying an approach with vertical guidance at speeds slower than the best-rate-of-climb (BROC) speed. The pilot manipulates the helicopter's controls to satisfy these commands, yielding the desired flight path, or may couple the autopilot to the flight director to fly along the desired flight path. Typically, flight director mode control and indication are shared with the autopilot.

Pilots must be aware of the mode of operation of the augmentation systems, and the control logic and functions in use. For example, on an ILS approach and using the three-cue mode (lateral, vertical and collective cues), the flight director collective cue responds to glideslope deviation, while the horizontal bar cue of the "cross-pointer" responds to airspeed deviations. However, the same system when operated in the two-cue mode on an ILS, the flight director horizontal bar cue responds to glideslope deviations. The need to be aware of the flight director mode of operation is particularly significant when operating using two pilots. Pilots should have an established set of procedures and responsibilities for the control of flight director/autopilot modes for the various phases of flight. Not only does a full understanding of the system modes provide for a higher degree of accuracy in control of the helicopter, it is the basis for crew identification of a faulty system.

---

[1] A "sensitive" altimeter relates to the instrument's displayed change in altitude over its range. For "Copter" Category II operations the scale must be in 20-foot intervals.

HELICOPTER FLIGHT MANUAL LIMITATIONS

Helicopters are certified for IFR operations with either one or two pilots. Certain equipment is required to be installed and functional for two-pilot operations and additional equipment is required for single pilot operation.

In addition, the Helicopter Flight Manual defines systems and functions that are required to be in operation or engaged for IFR flight in either the single or two-pilot configurations. Often, in a two-pilot operation, this level of augmentation is less than the full capability of the installed systems. Likewise, a single-pilot operation may require a higher level of augmentation.

The Helicopter Flight Manual also identifies other specific limitations associated with IFR flight. Typically, these limitations include, but are not limited to:

- Minimum equipment required for IFR flight (in some cases, for both single-pilot and two-pilot operations).

- V_{MINI} (minimum speed - IFR). [Figure 7-2]

- V_{NEI} (never exceed speed - IFR).

- Maximum approach angle.

- Weight and center of gravity limits.

- Helicopter configuration limitations (such as door positions and external loads).

- Helicopter system limitations (generators, inverters, etc.).

- System testing requirements (many avionics and AFCS, AP, and FD systems incorporate a self-test feature).

- Pilot action requirements (for example, the pilot must have hands and feet on the controls during certain operations, such as an instrument approach below certain altitudes).

Final approach angles/descent gradient for public approach procedures can be as high as 7.5 degrees/795

Manufacturer	V_{MINI} Limitations	MAX IFR Approach Angle	G/A Mode Speed
Agusta			
A-109	60 (80 coupled)		
A-109C	40	9.0	
Bell			
BH 212	40		
BH 214ST`	70		
BH 222	50		
BH 222B	50		
BH 412	60	5.0	
BH 430	50 (65 coupled)	4.0	
Eurocopter			
AS-355	55	4.5	
AS-365	75	4.5	
BK-117	45 (70 coupled)	6.0	
EC-135	60	4.6	
EC-155	70	4.0	
Sikorsky			
S-76A	60 (AFCS Phase II)	3.5	75 KIAS
S-76A	50 (AFCS Phase III)	7.5	75 KIAS
S-76B	60	7.5	75 KIAS
S-76C	60		
SK-76C++	50 (60 coupled)	6.5	

NOTE: The V_{MINI}, MAX IFR Approach Angle, and G/A Mode Speed for a specific helicopter may vary with avionics/autopilot installation. Pilots are therefore cautioned to refer only to the Rotorcraft Flight Manual limitations for their specific helicopter. The maximum rate of descent for many autopilots is 1,000 FPM.

Figure 7-2. V_{MINI} Limitations, Maximum IFR Approach Angles and G/A Mode Speeds for selected IFR-certified helicopters.

feet per NM. At 70 KIAS (no wind) this equates to a descent rate of 925 FPM. With a 10-knot tailwind the descent rate increases to 1,056 FPM. "Copter" PinS approach procedures are restricted to helicopters with a maximum V$_{MINI}$ of 70 KIAS and an IFR approach angle that will enable them to meet the final approach angle/descent gradient. Pilots of helicopters with a V$_{MINI}$ of 70 KIAS may have inadequate control margins to fly an approach that is designed with the maximum allowable angle/descent gradient or minimum allowable deceleration distance from the MAP to the heliport. The "Copter" PinS final approach segment is limited to 70 KIAS since turn containment and the deceleration distance from the MAP to the heliport may not be adequate at faster speeds. For some helicopters, (highlighted yellow in Figure 7-2) engaging the autopilot may increase the V$_{MINI}$ to a speed greater than 70 KIAS, or in the "go-around" mode require a speed faster than 70 KIAS. It may be possible for these helicopters to be flown manually on the approach, or on the missed approach in a mode other than the G/A mode.

Since slower IFR approach speeds enable the helicopter to fly steeper approaches and reduces the distance from the heliport that is required to decelerate the helicopter, you may want to operate your helicopter at speeds slower than its established V$_{MINI}$. The provision to apply for a determination of equivalent safety for instrument flight below V$_{MINI}$ and the minimum helicopter requirements are specified in Advisory Circulars (AC) 27-1, Certification of Normal Category Rotorcraft and AC 29-2C, Certification of Transport Category Rotorcraft. Application guidance is available from the Rotorcraft Directorate Standards Staff, ASW-110, 2601 Meacham Blvd. Fort Worth, Texas 76137-4298, (817) 222-5111.

Performance data may not be available in the Helicopter Flight Manual for speeds other than the best rate of climb speed. To meet missed approach climb gradients pilots may use observed performance for similar weight, altitude, temperature, and speed conditions to determine equivalent performance. When missed approaches utilizing a climbing turn are flown with an autopilot, set the heading bug on the missed approach heading, and then at the MAP, engage the indicated airspeed mode, followed immediately by applying climb power and selecting the heading mode. This is important since the autopilot roll rate and maximum bank angle in the Heading Select mode are significantly more robust than in the NAV mode. Figure 7-3 represents the bank angle and roll limits of the S-76 used by the FAA for flight testing. It has a roll rate in the Heading Select mode of 5 degrees per second with only 1 degree per second in the NAV mode. The bank angle in the Heading Select mode is 20 degrees with only 17 degrees in the NAV Change Over mode. Furthermore, if the Airspeed Hold mode is not selected on some autopilots when commencing the missed approach, the helicopter will accelerate in level flight until the best rate of climb is attained, and only then will a climb begin.

Wide area augmentation system (WAAS) localizer performance (LP) lateral-only PinS testing conducted in 2005 by the FAA at the William J. Hughes Technical Center in New Jersey for helicopter PinS also captured the flight tracks for turning missed approaches. [Figure 7-4] The large flight tracks that resulted during the turning missed approach were attributed in part to operating the autopilot in the NAV mode and exceeding the 70 KIAS limit.

OPERATIONS SPECIFICATIONS

A flight operated under Part 135 has minimums and procedures more restrictive than a flight operated under Part 91. These Part 135 requirements are detailed in their operations specifications (OpsSpecs). Helicopter Emergency Medical Service (HEMS) operators have even more restrictive OpsSpecs. Figure 7-5 on page 7-6 is an excerpt from an OpsSpecs detailing the minimums for precision approaches. The inlay in Figure 7-5 shows the minimums for the ILS Rwy 3R approach at Detroit Metro Airport. With all lighting operative, the minimums for helicopter Part 91 operations are a 200-foot ceiling, and 1200-feet runway visual range (RVR) (one-half airplane Category A visibility but no less than 1/4 SM/1200 RVR). However, as shown in the OpsSpecs, the minimum visibility this Part 135 operator must adhere to is 1600 RVR. Pilots operating under Part 91 are encouraged to develop their own personal OpsSpecs based on their own equipment, training, and experience.

Autopilot Mode	Bank Angle Limit (Degrees)	Roll Rate Limit (Degrees/ Sec)
Heading hold	< 6	None specified
VOR/RNAV (Capture)	+/- 22	5
VOR/RNAV (On Course)	+/- 13	1 5 VOR/RNAV Approach
Heading Select	+/-20	5
VOR/RNAV (Course Change Over Station/Fix)	+/- 17	1

Figure 7-3. Autopilot Bank Angle and Roll Rate Limits for the S-76 used by the William J. Hughes Technical Center for Flight Tests.

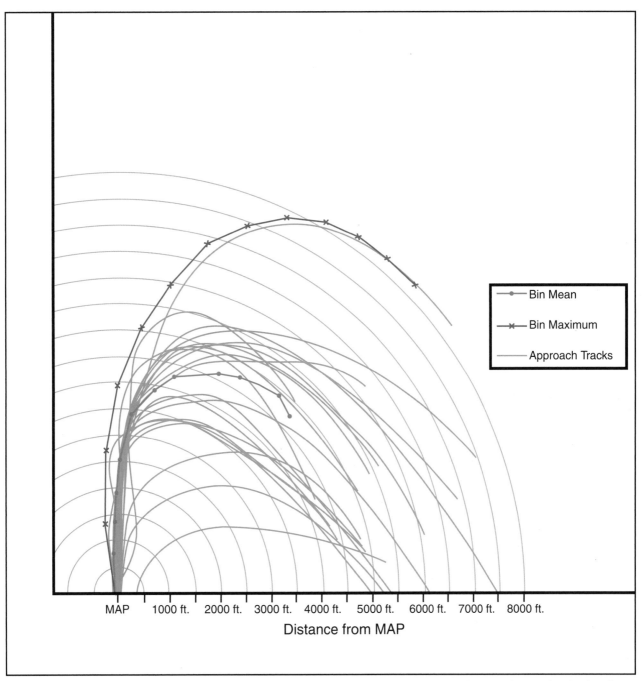

Figure 7-4. Flight tests at the William J. Hughes Technical Center point out the importance of airspeed control and using the correct technique to make a turning missed approach.

MINIMUM EQUIPMENT LIST

A helicopter operating under Part 135 with certain installed equipment inoperative is prohibited from taking off unless the operation is authorized in the approved MEL. The MEL provides for some equipment to be inoperative if certain conditions are met [Figure 7-6 on page 7-7]. In many cases, a helicopter configured for single-pilot IFR may depart IFR with certain equipment inoperative, provided a crew of two pilots is used. Under Part 91, a pilot may defer certain items without an MEL if those items are not required by the type certificate, CFRs, or airworthiness directives (ADs), and the flight can be performed safely

without them. If the item is disabled, or removed, or marked inoperative, a logbook entry is made.

PILOT PROFICIENCY

Helicopters of the same make and model may have variations in installed avionics that change the required equipment or the level of augmentation for a particular operation. The complexity of modern AFCS, AP, and FD systems requires a high degree of understanding to safely and efficiently control the helicopter in IFR operations. Formal training in the use of these systems is highly recommended for all pilots.

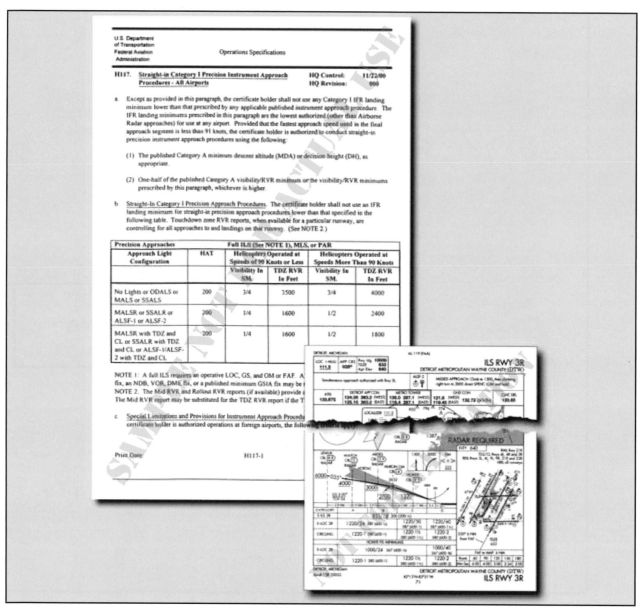

Figure 7-5. Operations Specifications.

During flight operations, you must be aware of the mode of operation of the augmentation system, and the control logic and functions employed. For example, during an ILS approach using a particular system in the three-cue mode (lateral, vertical, and collective cues), the flight director collective cue responds to glide slope deviation, while the horizontal bar of the "cross-pointer" responds to airspeed deviations. The same system, while flying an ILS in the two-cue mode, provides for the horizontal bar to respond to glide slope deviations. This concern is particularly significant when the crew consists of two pilots. Pilots should establish a set of procedures and division of responsibility for the control of flight director/autopilot and FMS modes for the various phases of flight. Not only is a full understanding of the system modes essential in order to provide for a high degree of accuracy in control of the helicopter, it is the basis for identification of system failures

HELICOPTER VFR MINIMUMS

Helicopters have the same VFR minimums as airplanes with two exceptions. In Class G airspace or under a special visual flight rule (SVFR) clearance, helicopters have no minimum visibility requirement but must remain clear of clouds and operate at a speed that is slow enough to give the pilot an adequate opportunity to see other aircraft or an obstruction in time to avoid a collision. Helicopters are also authorized (Part 91, appendix D, section 3) to obtain SVFR clearances at airports with the designation NO SVFR in the Airport Facility Directory (A/FD) or on the sectional chart. Figure 7-7 on page 7-8 shows the visibility and cloud clearance requirements for VFR and SVFR. However, lower minimums associated with Class G airspace and SVFR do not take the place of the VFR minimum requirements of either Part 135 regulations or respective OpsSpecs.

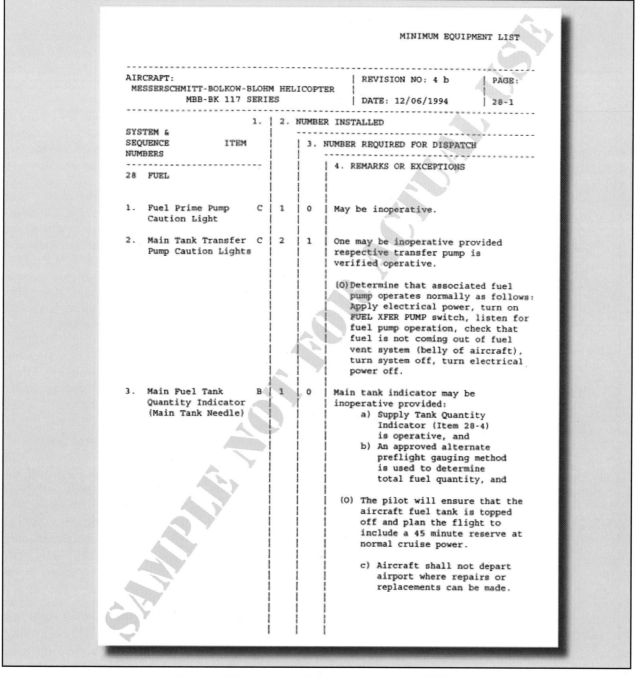

Figure 7-6. Example of a Minimum Equipment List (MEL).

Knowledge of all VFR minimums is required in order to determine if a **Point-in-Space (PinS) approach** can be conducted, or if a SVFR clearance is required to continue past the missed approach point (MAP). These approaches and procedures will be discussed in detail later.

HELICOPTER IFR TAKEOFF MINIMUMS

A pilot operating under Part 91 has no takeoff minimums with which to comply other than the requirement to attain VMINI before entering instrument meteorological conditions (IMC). For most helicopters, this requires a distance of approximately 1/2 mile and an altitude of 100 feet. If departing with a steeper climb gradient, some helicopters may require additional altitude to accelerate to V_{MINI}. To maximize safety, always consider using the Part 135 operator standard takeoff visibility minimum of 1/2 statute mile (SM) or the charted departure minima, whichever is higher. A charted departure that provides protection from obstacles will either have a higher visibility requirement, climb gradient, and/or departure path. Part 135 operators are required to adhere to the takeoff minimums prescribed in the instrument approach procedures (IAPs) for the airport.

Helicopter VFR Minimums

Airspace	Flight visibility	Distance from clouds
Class A	Not applicable	Not Applicable.
Class B	3 SM	Clear of Clouds.
Class C	3 SM	500 feet below. 1,000 feet above. 2,000 feet horizontal.
Class D	3 SM	500 feet below. 1,000 feet above. 2,000 feet horizontal.
Class E: Less than 10,000 feet MSL	3 SM	500 feet below. 1,000 feet above. 2,000 feet horizontal.
At or above 10,000 feet MSL	5 SM	1,000 feet below. 1,000 feet above. 1 statute mile horizontal.
Class G: 1,200 feet or less above the surface (regardless of MSL altitude).		
Day, except as provided in §91.155(b)	None	Clear of clouds.
Night, except as provided in §91.155(b)	None	Clear of clouds.
More than 1,200 feet above the surface but less than 10,000 feet MSL		
Day	1 SM	500 feet below. 1,000 feet above. 2,000 feet horizontal.
Night	3 SM	500 feet below. 1,000 feet above. 2,000 feet horizontal.
More than 1,200 feet above the surface and at or above 10,000 feet MSL	5 SM	1,000 feet below. 1,000 feet above. 1 statute mile horizontal.
B, C, D, E Surface Area Airspace SVFR Minimums		
Day	None	Clear of clouds.
Night	None	Clear of clouds.

Figure 7-7. Helicopter VFR Minimums.

PART 91 OPERATORS

Part 91 operators are not required to file an alternate if, at the estimated time of arrival (ETA) and for 1 hour after, the ceiling will be at least 1,000 feet above the airport elevation or 400 feet above the lowest applicable approach minima, whichever is higher, and the visibility is at least 2 SM. If an alternate is required, an airport can be used if the ceiling is at least 200 feet above the minimum for the approach to be flown and visibility is at least 1 SM, but never less than the minimum required for the approach to be flown. If no instrument approach procedure has been published for the alternate airport, the ceiling and visibility minima are those allowing descent from the MEA, approach, and landing under basic VFR.

PART 135 OPERATORS

Part 135 operators are not required to file an alternate if, for at least 1 hour before and 1 hour after the ETA, the ceiling will be at least 1,500 feet above the lowest circling approach minimum descent altitude (MDA). If a circling instrument approach is not authorized for the airport, the ceiling must be at least 1,500 feet above the lowest published minimum or 2,000 feet above the airport elevation, whichever is higher. For the instrument approach procedure to be used at the destination airport, the forecasted visibility for that airport must be at least 3 SM, or 2 SM more than the lowest applicable visibility minimums, whichever is greater.

HELICOPTER IFR ALTERNATES

The pilot must file for an alternate if weather reports and forecasts at the proposed destination do not meet certain minimums. These minimums differ for Part 91 and Part 135 operators.

Alternate landing minimums for flights conducted under Part 135 are described in the OpsSpecs for that operation. All helicopters operated under IFR must carry enough fuel to fly to the intended destination, fly from that airport to the filed alternate, if required, and continue for 30 minutes at normal cruising speed.

HELICOPTER INSTRUMENT APPROACHES

Helicopter instrument flight is relatively new when compared to airplane instrument flight. Many new helicopter instrument approach procedures have been developed to take advantage of advances in both avionics and helicopter technology.

STANDARD INSTRUMENT APPROACH PROCEDURES TO AN AIRPORT

Helicopters flying standard instrument approach procedures (SIAP) must adhere to the MDA or decision altitude for Category A airplanes, and may apply the Part 97.3(d-1) rule to reduce the airplane Category A visibility by half but in no case less than 1/4 SM or 1200 RVR [Figure 7-10 on page 7-11]. The approach can be initiated at any speed up to the highest approach category authorized; however, the speed on the final approach segment must be reduced to the Category A speed of less than 90 KIAS before the MAP in order to apply the visibility reduction. A constant airspeed is recommended on the final approach segment to comply with the stabilized approach concept since a decelerating approach may make early detection of wind shear on the approach path more difficult. [Figure 7-8]

When visibility minimums must be increased for inoperative components or visual aids, use the Inoperative Components and Visual Aids Table (provided in the front cover of the U.S. Terminal Procedures) to derive the Category A minima before applying any visibility reduction. The published visibility may be increased above the standard visibility minima due to penetrations of the 20:1 and 34:1 final approach obstacle identification surfaces (OIS). The minimum visibility required for 34:1 penetrations is 3/4 SM and for 20:1 penetrations 1 SM (see Chapter 5). When there are penetrations of the final approach OIS, a visibility credit for approach lighting systems is not allowed for either airplane or helicopter procedures that would result in values less than the appropriate 3/4 SM or 1 SM visibility requirement. The Part 97.3 visibility reduction rule does not apply, and you must take precautions to avoid any obstacles in the visual segment. Procedures with penetrations of the final approach

OIS will be annotated at the next amendment with **"Visibility Reduction by Helicopters NA."**

Until all the affected SIAPs have been annotated, an understanding of how the standard visibilities are established is the best aid in determining if penetrations of the final approach OIS exists. Some of the variables in determining visibilities are: DA/MDA height above touchdown (HAT), height above airport (HAA), distance of the facility to the MAP (or the runway threshold for non-precision approaches), and approach lighting configurations.

The standard visibility requirement, without any credit for lights, is 1 SM for nonprecision approaches and 3/4 SM for precision approaches. This is based on a Category A airplane 250-320 feet HAT/HAA, and for nonprecision approaches a distance of 10,000 feet or less from the facility to the MAP (or runway threshold). For precision approaches, credit for any approach light configuration, and for non-precision approaches (with a 250 HAT) configured with a MALSR, SSALR, or ALSF-1 normally results in a published visibility of 1/2 SM.

Consequently, if an ILS is configured with approach lights or a nonprecision approach is configured with either MALSR, SSALR, or ALSF-1 lighting configurations and the procedure has a published visibility of 3/4 SM or greater, a penetration of the final approach OIS may exist. Also, pilots will be unable to determine whether there are penetrations of the final approach OIS if a nonprecision procedure does not have approach lights, or is configured with ODALS, MALS, or SSALS/SALS lighting since the minimum published visibility will be 3/4 SM or greater.

As a rule of thumb, approaches with published visibilities of 3/4 SM or more should be regarded as having final approach OIS penetrations and care must be taken to avoid any obstacles in the visual segment. Approaches with published visibilities of 1/2 SM or less are free of OIS penetrations and the visibility reduction in Part 97.3 is authorized.

Helicopter Use of Standard Instrument Approach Procedures

Procedure	Helicopter Visibility Minima	Helicopter MDA/DA	Maximum Speed Limitations
Standard	The greater of: one half the Category A visibility minima, 1/4 statute mile visibility, or 1200 RVR unless annotated (Visibility Reduction by Helicopters NA.)	As published for Category A	The helicopter may initiate the final approach segment at speeds up to the upper limit of the highest Approach Category authorized by the procedure, but must be slowed to no more than 90 KIAS at the MAP in order to apply the visibility reduction.
Copter Procedure	As published	As published	90 KIAS when on a published route/track.
GPS Copter Procedure	As published	As published	90 KIAS when on a published route, track, or holding, 70 KIAS when on the final approach or missed approach segment. Military procedures are limited to 90 KIAS for all segments.

Figure 7-8. Helicopter Use of Standard Instrument Approach Procedures.

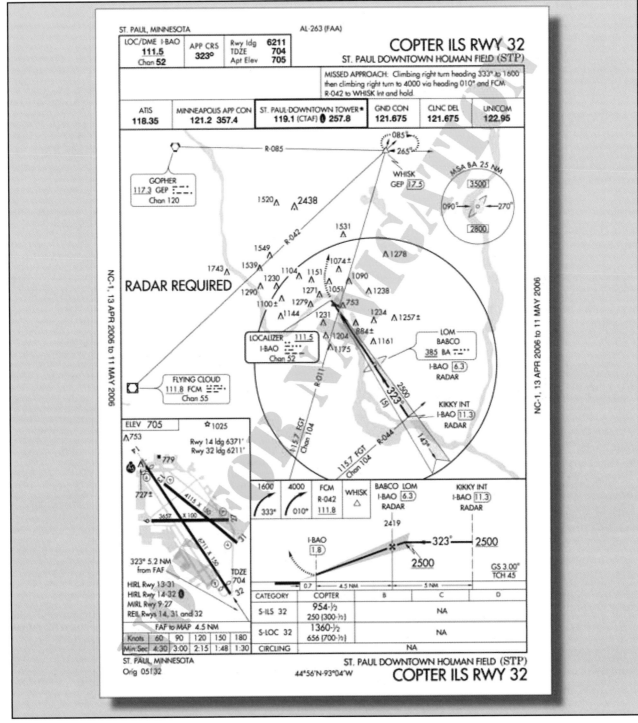

Figure 7-9. KSTP Copter ILS Rwy 32.

COPTER ONLY APPROACHES TO AN AIRPORT OR HELIPORT

Pilots flying Copter standard instrument approach procedures (SIAPs), other than GPS, may use the published minima with no reductions in visibility allowed. The maximum airspeed is 90 KIAS on any segment of the approach or missed approach. Figure 7-9, illustrates a helicopter only ILS runway 32 approach at St. Paul, Minnesota.

Copter ILS approaches to Category (CAT) I facilities with DAs no lower than a 200-foot HAT provide an advantage over a conventional ILS of shorter final segments, and lower minimums (based on the 20:1 missed approach surface). There are also Copter approaches with minimums as low as 100-foot HAT and 1/4 SM visibility. Approaches with a HAT below 200 foot are annotated with the note: "SPECIAL AIRCREW & AIRCRAFT CERTIFICATION REQUIRED" since the FAA must approve the helicopter and its avionics, and the flight crew must have the required experience, training, and checking.

97.3 SYMBOLS AND TERMS USED IN PROCEDURES

(d) (1) "Copter procedures" means helicopter procedures, with applicable minimums as prescribed in §97.35 of this part. Helicopters may also use other procedures prescribed in Subpart C of this part and may use the Category A minimum descent altitude (MDA) or decision height (DH). The required visibility minimum may be reduced to 1/2 the published visibility minimum, but in no case may it be reduced to less than one-quarter mile or 1,200 feet RVR.

Figure 7-10. Part 97 Excerpt.

The ground facilities (approach lighting, signal in space, hold lines, maintenance, etc.) and air traffic infrastructure for CAT II ILS approaches are required to support these procedures. The helicopter must be equipped with an autopilot, flight director or head up guidance system, alternate static source (or heated static source), and radio altimeter. The pilot must have at least a private pilot helicopter certificate, an instrument helicopter rating, and a type rating if the helicopter requires a type rating. Pilot experience requires the following flight times: 250 PIC, 100 helicopter PIC, 50 night PIC, 75 hours of actual or simulated instrument flight time, including at least 25 hours of actual or simulated instrument flight time in a helicopter or a helicopter flight simulator, and the appropriate recent experience, training and check. For "Copter" CAT II ILS operations below 200 feet HAT, approach deviations are limited to 1/4 scale of the localizer or glide slope needle. Deviations beyond that require an immediate missed approach unless the pilot has at least one of the visual references in sight and otherwise meets the requirements of 14 CFR Part 91.175(c). The reward for this effort is the ability to fly "Copter" ILS approaches with minima that are sometimes below the airplane CAT II minima. [Figure 7-11 on page 7-12] The procedure to apply for this certification is available from your local Flight Standards District Office.

COPTER GPS APPROACHES TO AN AIRPORT OR HELIPORT

Pilots flying Copter GPS or WAAS SIAPs must limit the speed to 90 KIAS on the initial and intermediate segment of the approach, and to no more than 70 KIAS on the final and missed approach segments. If annotated, holding may also be limited to 90 KIAS to contain the helicopter within the small airspace provided for helicopter holding patterns. During testing for helicopter holding, the optimum airspeed and leg length combination was determined to be 90 KIAS with a 3 NM outbound leg length. Consideration was

given to the wind drift on the dead reckoning entry leg at slower speeds, the turn radius at faster airspeeds, and the ability of the helicopter in strong wind conditions to intercept the inbound course prior to the holding fix. The published minimums are to be used with no visibility reductions allowed. Figure 7-12 on page 7-13 is an example of a Copter GPS PinS approach that allows the helicopter to fly VFR from the MAP to the heliport.

The final and missed approach protected airspace providing obstacle and terrain avoidance is based on 70 KIAS, with a maximum 10-knot tailwind component. It is absolutely essential that pilots adhere to the 70 KIAS limitation in procedures that include an immediate climbing and turning missed approach. Exceeding the airspeed restriction increases the turning radius significantly, and can cause the helicopter to leave the missed approach protected airspace. This may result in controlled flight into terrain (CFIT) or obstacles.

If a helicopter has a V_{MINI} greater than 70 knots, then it will not be capable of conducting this type of approach. Similarly, if the autopilot in "go-around" mode climbs at a VYI greater than 70 knots, then that mode cannot be used. It is the responsibility of the pilot to determine compliance with missed approach climb gradient requirements when operating at speeds other than V_Y or V_{YI}. Missed approaches that specify an "IMMEDIATE CLIMBING TURN" have no provision for a straight ahead climbing segment before turning. A straight segment will result in exceeding the protected airspace limits.

Protected obstacle clearance areas and surfaces for the missed approach are established on the assumption that the missed approach is initiated at the DA point and for nonprecision approaches no lower than the MDA at the MAP (normally at the threshold of the approach end of the runway). The pilot must begin the missed approach at those points! Flying beyond either point before beginning the missed approach will result in flying below the protected OCS and can result in a collision with an obstacle.

The missed approach segment TERPS criteria for all Copter approaches take advantage of the helicopter's climb capabilities at slow airspeeds, resulting in high climb gradients. [Figure 7-13 on page 7-14] The OCS used to evaluate the missed approach is a 20:1 inclined plane. This surface is twice as steep for the helicopter

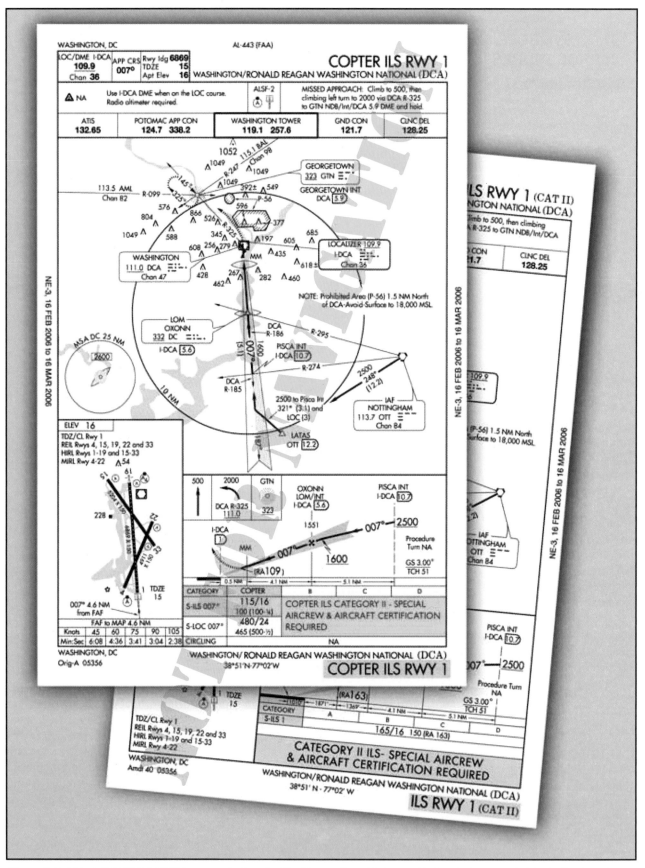

Figure 7-11. This COPTER ILS RWY 1 approach chart for Washington/Ronald Reagan National shows the DA for helicopters is 115 feet. The Category II DA for airplanes is 165 feet. The difference is due to the helicopter missed approach obstacle clearance surface (OCS) of 20:1, compared to the 40:1 OCS for airplanes. In this case, the missed approach must be started no later than the point on the glidepath that the decision height (DH) is reached, in order to miss the Washington Monument.

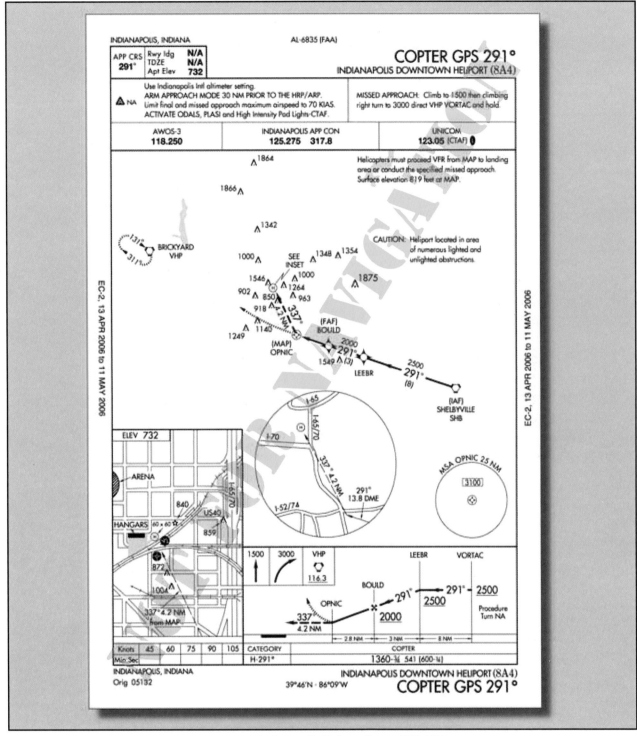

Figure 7-12. Indianapolis Heliport Copter GPS 291°.

as the OCS used to evaluate the airplane missed approach segment. The helicopter climb gradient is therefore required to be double that of the airplane's required missed approach climb gradient.

A minimum climb gradient of at least 400 feet per NM is required unless a higher gradient is published on the approach chart; e.g., a helicopter with a ground speed of 70 knots is required to climb at a rate of 467 feet per minute (FPM)[2]. The advantage of using the 20:1 OCS for the helicopter missed approach segment instead of

the 40:1 OCS used for the airplane is that obstacles that penetrate the 40:1 missed approach segment may not have to be considered. The result is the DA/MDA may be lower for helicopters than for other aircraft. The minimum required climb gradient of 400 feet per NM for the helicopter in a missed approach will provide 96 feet of required obstacle clearance (ROC) for each NM of flight path.

---

[2]467 FPM = 70 KIAS x 400 feet per NM/60 seconds

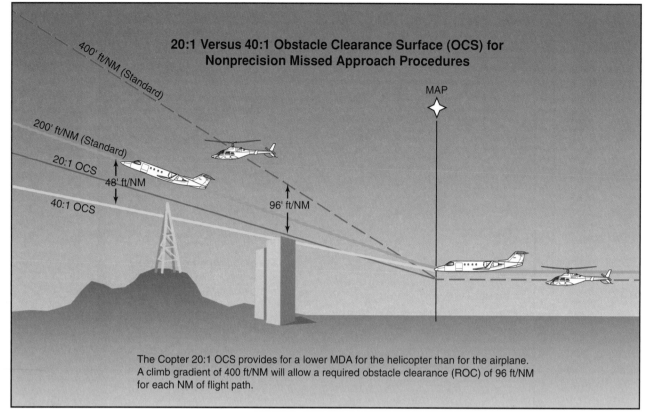

Figure 7-13. Obstacle Clearance Surface.

HELICOPTER APPROACHES TO VFR HELIPORTS

Helicopter approaches to VFR heliports are normally developed either as public procedures to a point-in-space (PinS) that may serve more than one heliport or as a Special procedure to a specific VFR heliport that requires pilot training due to its unique characteristics. These approaches can be developed using VOR or ADF, but RNAV using GPS is the most common system used today. In the future, RNAV using the wide area augmentation system (WAAS) offers the most advantages because it can provide lower approach minimums, narrower route widths to support a network of approaches, and may allow the heliport to be used as an alternate. A majority of the special procedures to a specific VFR heliport are developed in support of helicopter emergency medical services (HEMS) operators and have a "Proceed Visually" segment between the MAP and the heliport. Public procedures are developed as a PinS approach with a "Proceed VFR" segment between the MAP and the landing area. These PinS "Proceed VFR" procedures specify a course and distance from the MAP to the available heliports in the area.

APPROACH TO A POINT-IN-SPACE

The note associated with these procedures is: "PROCEED VFR FROM (NAMED MAP) OR CONDUCT THE SPECIFIED MISSED APPROACH." They may be developed as a special or public procedure where the MAP is located more than 2 SM from the landing site, the turn from the final approach to the visual segment is greater than 30 degrees, or the VFR segment

	Non-Mountainous		Mountainous (14 CFR Part 95)	
Area	Local	Cross Country	Local	Cross Country
Condition	Ceiling-visibility			
Day	500-1	800-2	500-2	800-3
Night – High Lighting Conditions*	500-2	1000-3	500-3	1000-3
Night – Low Lighting Conditions	800-3	1000-5	1000-3	1000-5

Figure 7-14. Weather Minimums and Lighting Conditions for HEMS Operators.

from the MAP to the landing site has obstructions that require pilot actions to avoid them. Figure 7-15 is an example of a public PinS approach that allows the pilot to fly to one of four heliports after reaching the MAP.

For Part 135 operations, pilots may not begin the instrument approach unless the latest weather report indicates that the weather conditions are at or above the authorized IFR or VFR minimums as required by the class of airspace, operating rule and/or OpsSpecs,

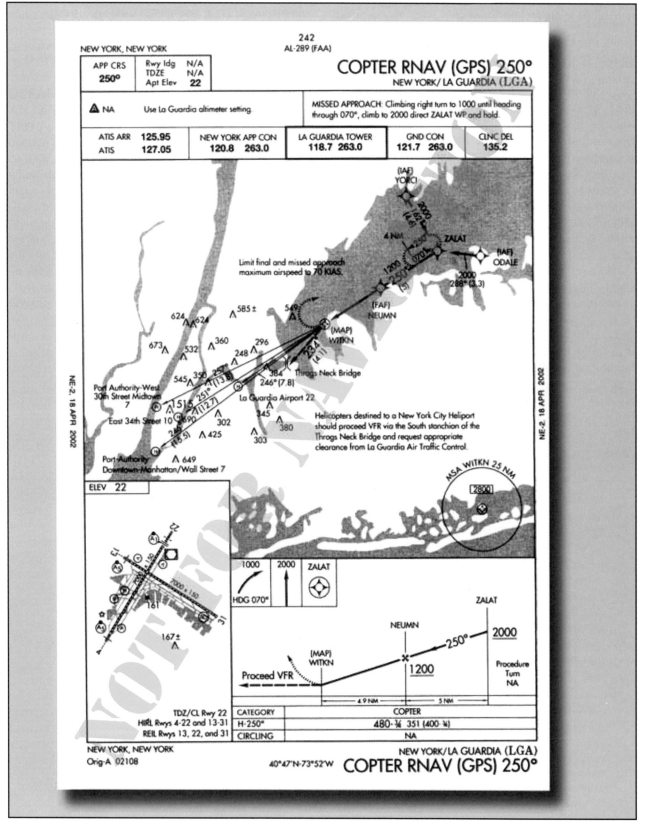

Figure 7-15. KLGA Copter RNAV (GPS) 250°.

whichever is higher. Visual contact with the landing site is not required; however, prior to the MAP, for either Part 91 or 135 operators, the pilot must determine if the flight visibility meets the basic VFR minimums required by the class of airspace, operating rule and/or OpsSpecs (whichever is higher). The visibility is limited to no lower than that published in the procedure until canceling IFR. If VFR minimums do not exist, then the published missed approach procedure must be executed. The pilot must contact air traffic control upon reaching the MAP, or as soon as practical after that, and advise whether executing the missed approach or canceling IFR and proceeding VFR. Figure 7-16 provides examples of the procedures used during a PinS approach for Part 91 and Part 135 operations.

To proceed VFR in uncontrolled airspace, Part 135 operators are required to have at least 1/2 SM visibility and a 300-foot ceiling. Part 135 HEMS operators must have at least 1 SM day or 2 SM night visibility and a 500-foot ceiling provided the heliport is located within 3 NM of the MAP. These minimums apply regardless of whether the approach is located on the plains of Oklahoma or in the Colorado mountains. However, for heliports located farther than 3 NM from the heliport, Part 135 HEMS operators are held to an even higher standard and the minimums and lighting conditions contained in Figure 7-14 apply to the entire route. Mountainous terrain at night with low light conditions requires a ceiling of 1,000 feet and either 3 SM or 5 SM visibility depending on whether it has been determined as part of the operator's local flying area.

In Class B, C, D, and E surface area airspace, a SVFR clearance may be obtained if SVFR minimums exist. On your flight plan, give ATC a heads up about your intentions by entering the following in the remarks section: "Request SVFR clearance after the MAP."

APPROACH TO A SPECIFIC VFR HELIPORT
The note associated with these procedures is: "PROCEED VISUALLY FROM (NAMED MAP) OR CONDUCT THE SPECIFIED MISSED APPROACH." Due to their unique characteristics, these approaches require training. They are developed to hospitals, oilrigs, private heliports, etc. As Specials, they require Flight Standards approval by a Letter of Authorization (LOA) for Part 91 operators or by OpsSpecs for Part 135 operators. The heliport associated with these procedures must be located within 2 SM of the MAP, the visual segment between the MAP and the heliport evaluated for obstacle hazards, and the heliport must meet the appropriate VFR heliport recommendations of Advisory Circular 150/5390-2, Heliport Design.

The visibility minimum is based on the distance from the MAP to the heliport, among other factors, e.g., height above the heliport elevation when at the MAP MDA. The pilot is required to acquire and maintain visual contact with the heliport **final approach and takeoff (FATO)** area at or prior to the MAP. Obstacle or terrain avoidance from the MAP to the heliport is the responsibility of the pilot. If the required weather minimums do not exist, then the published missed approach procedure must be executed at the MAP because IFR obstruction clearance areas are not applied to the visual segment of the approach and a missed

Point-in-Space Approach Examples

Example 1:

Under Part 91 the operator flies the published IFR PinS approach procedure that has a charted MDA of 340 mean sea level (MSL) and visibility of 3/4 SM. When approaching the MAP at an altitude of 340 feet MSL the pilot transitions from Instrument Meteorological Conditions (IMC) to Visual Meteorological Conditions (VMC) and determines that the flight visibility is 1/2 SM. The pilot must determine prior to the MAP whether the applicable basic VFR weather minimums can be maintained from the MAP to the heliport or execute a missed approach. If the pilot determines that the applicable basic VFR weather minimums can be maintained to the heliport the pilot may proceed VFR. If the visual segment is in Class B, C, D, or the surface area of Class E airspace, it may require the pilot to obtain a Special VFR clearance.

Example 2:

For an operator to proceed VFR under Part 135, a minimum visibility of 1/2 SM during the day and 1 SM at night with a minimum ceiling of 300 feet. If prior to commencing the approach the pilot determines the reported visibility is 3/4 SM during the day the pilot descends IMC to an altitude no lower than the MDA and transitions to VMC. If the pilot determines prior to the MAP that the flight visibility is less than 1/2 SM in the visual segment a missed approach must be executed at the MAP.

Figure 7-16. Point-in-Space Approach Examples.

approach segment protection is not provided between the MAP and the heliport. As soon as practicable after reaching the MAP, the pilot advises ATC whether cancelling IFR and proceeding visually, or executing the missed approach.

INADVERTENT IMC

Whether it is a corporate or HEMS operation, helicopter pilots sometimes operate in challenging weather conditions. An encounter with weather that does not permit continued flight under VFR might occur when conditions do not allow for the visual determination of a usable horizon (e.g., fog, snow showers, or night operations over unlit surfaces such as water). Flight in conditions of limited visual contrast should be avoided since this can result in a loss of horizontal or surface reference, and obstacles such as wires become perceptually invisible. To prevent spatial disorientation, loss of control (LOC) or CFIT, pilots should slow the helicopter to a speed that will provide a controlled deceleration in the distance equal to the forward visibility. The pilot should look for terrain that provides sufficient contrast to either continue the flight or to make a precautionary landing. If spatial disorientation occurs, and a climb into instrument meteorological conditions is not feasible due to fuel state, icing conditions, equipment, etc., make every effort to land the helicopter with a slight forward descent to prevent any sideward or rearward motion.

All helicopter pilots should receive training on avoidance and recovery from inadvertent IMC with emphasis on avoidance. An unplanned transition from VFR to IFR flight is an emergency that involves a different set of pilot actions. It requires the use of different navigation and operational procedures, interaction with ATC, and crewmember resource management (CRM). Consideration should be given to the local flying area's terrain, airspace, air traffic facilities, weather (including seasonal affects such as icing and thunderstorms), and available airfield/heliport approaches.

Training should emphasize the identification of circumstances conducive to inadvertent IMC and a strategy to abandon continued VFR flight in deteriorating conditions.[3] This strategy should include a minimum altitude/airspeed combination that provides for an off-airport/heliport landing, diverting to better conditions, or initiating an emergency transition to IFR. Pilots should be able to readily identify the minimum initial altitude and course in order to avoid CFIT. Current IFR en route and approach charts for the route of flight are essential. A GPS navigation receiver with a moving map provides exceptional situational awareness for terrain and obstacle avoidance.

Training for an emergency transition to IFR should include full and partial panel instrument flight, unusual attitude recovery, ATC communications, and instrument approaches. If an ILS is available and the helicopter is equipped, an ILS approach should be made. Otherwise, if the helicopter is equipped with an IFR approach-capable GPS receiver with a current database, a GPS approach should be made. If neither an ILS nor GPS procedure is available use another instrument approach.

Upon entering inadvertent IMC, priority must be given to control of the helicopter. Keep it simple and take one action at a time.

- Control. First use the wings on the attitude indicator to level the helicopter. Maintain heading and increase to climb power. Establish climb airspeed at the best angle of climb but no slower than V_{MINI}.

- Climb. Climb straight ahead until your crosscheck is established. Then make a turn only to avoid terrain or objects. If an altitude has not been previously established with ATC to climb to for inadvertent IMC, then you should climb to an altitude that is at least 1,000 feet above the highest known object, and that allows for contacting ATC.

- Communicate. Attempt to contact ATC as soon as the helicopter is stabilized in the climb and headed away from danger. If the appropriate frequency is not known you should attempt to contact ATC on either VHF 121.5 or UHF 243.0. Initial information provided to ATC should be your approximate location, that inadvertent IMC has been encountered and an emergency climb has been made, your altitude, amount of flight time remaining (fuel state), and number of persons on board. You should then request a vector to either VFR weather conditions or to the nearest suitable airport/heliport that conditions will support a successful approach. If unable to contact ATC and a transponder code has not been previously established with ATC for inadvertent IMC, change the transponder code to 7700.

---

[3] A radio altimeter is a necessity for alerting the pilot when inadvertently going below the minimum altitude. Barometric altimeters are subject to inaccuracies that become important in helicopter IFR operations, especially in cold temperatures. (See Appendix B.)

IFR HELIPORTS

Advisory Circular 150/5390-2, Heliport Design, provides recommendations for heliport design to support non-precision, approach with vertical guidance (APV), and precision approaches to a heliport. When a heliport does not meet the criteria of this AC, FAA Order 8260.42, Helicopter Global Positioning System (GPS) Nonprecision Approach Criteria, requires that an instrument approach be published as a SPECIAL procedure with annotations that special aircrew qualifications are required to fly the procedure. Currently there are no operational civil IFR heliports in the U.S. although the U.S. military has some nonprecision and precision approach procedures to IFR heliports.

EVOLUTION OF AIRBORNE NAVIGATION DATABASES

There are nearly as many different area navigation (RNAV) platforms operating in the National Airspace System (NAS) as there are aircraft types. The range of systems and their capabilities is greater now than at any other time in aviation history. From the simplest panel-mounted LOng RAnge Navigation (LORAN), to the moving-map display global positioning system (GPS) currently popular for general aviation aircraft, to the fully integrated flight management system (FMS) installed in corporate and commercial aircraft, the one common essential element is the database. [Figure A-1]

RNAV systems must not only be capable of determining an aircraft's position over the surface of the earth, but they also must be able to determine the location of other fixes in order to navigate. These systems rely on airborne navigation databases to provide detailed information about these fixed points in the airspace or on the earth's surface. Although, the location of these points is the primary concern for navigation, these databases can also provide many other useful pieces of information about a given location.

HISTORY

In 1973, National Airlines installed the Collins ANS-70 and AINS-70 RNAV systems in their DC-10 fleet; this marked the first commercial use of avionics that required navigation databases. A short time later, Delta Air Lines implemented the use of an ARMA Corporation RNAV system that also used a navigation database. Although the type of data stored in the two systems was basically identical, the designers created the databases to solve the individual problems of each sys-

tem. In other words, the data was not interchangeable. This was not a problem because so few of the systems were in use, but as the implementation of RNAV systems expanded, a world standard for airborne navigation databases had to be created.

Figure A-1. Area Navigation Receivers.

In 1973, Aeronautical Radio, Inc. (ARINC) sponsored the formation of a committee to standardize aeronautical databases. In 1975, this committee published the first standard (ARINC Specification 424), which has remained the worldwide-accepted format for coding airborne navigation databases.

There are many different types of RNAV systems certified for instrument flight rules (IFR) use in the NAS. The two most prevalent types are GPS and the multi-sensor FMS.

Most GPSs operate as stand-alone RNAV systems. A modern GPS unit accurately provides the pilot with the aircraft's present position; however, it must use an airborne navigation database to determine its direction or distance from another location unless a latitude and longitude for that location is manually entered. The database provides the GPS with position information for navigation fixes so it may perform the required geodetic calculations to determine the appropriate tracks, headings, and distances to be flown.

Modern FMSs are capable of a large number of functions including basic en route navigation, complex departure and arrival navigation, fuel planning, and precise vertical navigation. Unlike stand-alone navigation systems, most FMSs use several navigation inputs. Typically, they formulate the aircraft's current position using a combination of conventional distance measuring equipment (DME) signals, inertial navigation systems (INS), GPS receivers, or other RNAV devices. Like stand-alone navigation avionics, they rely heavily on airborne navigation databases to provide the information needed to perform their numerous functions.

DATABASE CAPABILITIES

The capabilities of airborne navigation databases depend largely on the way they are implemented by the avionics manufacturers. They can provide data about a large variety of locations, routes, and airspace segments for use by many different types of RNAV equipment. Databases can provide pilots with information regarding airports, air traffic control frequencies, runways, special use airspace, and much more. Without airborne navigation databases, RNAV would be extremely limited.

PRODUCTION AND DISTRIBUTION

In order to understand the capabilities and limitations of airborne navigation databases, pilots should have a basic understanding of the way databases are compiled and revised by the database provider and processed by the avionics manufacturer.

THE ROLE OF THE DATABASE PROVIDER

Compiling and maintaining a worldwide airborne navigation database is a large and complex job. Within the United States (U.S.), the Federal Aviation Administration (FAA) sources give the database providers information, in many different formats, which must be analyzed,

edited, and processed before it can be coded into the database. In some cases, data from outside the U.S. must be translated into English so it may be analyzed and entered into the database. Once the data is coded following the specifications of ARINC 424 (see ARINC 424 later in this appendix), it must be continually updated and maintained.

Once the FAA notifies the database provider that a change is necessary, the update process begins.[1] The change is incorporated into a 28-day airborne database revision cycle based on its assigned priority. If the information does not reach the coding phase prior to its cutoff date (the date that new aeronautical information can no longer be included in the next update), it is held out of revision until the next cycle. The cutoff date for aeronautical databases is typically 21 days prior to the effective date of the revision.[2]

The integrity of the data is ensured through a process called cyclic redundancy check (CRC). A CRC is an error detection algorithm capable of detecting small bit-level changes in a block of data. The CRC algorithm treats a data block as a single (large) binary value. The data block is divided by a fixed binary number (called a "generator polynomial") whose form and magnitude is determined based on the level of integrity desired. The remainder of the division is the CRC value for the data block. This value is stored and transmitted with the corresponding data block. The integrity of the data is checked by reapplying the CRC algorithm prior to distribution, and later by the avionics equipment onboard the aircraft.

RELATIONSHIP BETWEEN EFB AND FMS DATABASES

The advent of the Electronic Flight Bag (EFB) discussed in Chapter 6 illustrates how the complexity of avionics databases is rapidly accelerating. The respective FMS and EFB databases remain independent of each other even though they may share some of the same data from the database provider's master navigation database. For example, FMS and GPS databases both enable the retrieval of data for the onboard aircraft navigation system.

Additional data types that are not in the FMS database are extracted for the EFB database, allowing replacement of traditional printed instrument charts for the

---

[1] The majority of the volume of official flight navigation data in the U.S. disseminated to database providers is primarily supplied by FAA sources. It is supplemented by airport managers, state civil aviation authorities, Department of Defense (DOD) organizations such as the National Geospatial-Intelligence Agency (NGA), branches of the military service, etc. Outside the U.S., the majority of official data is provided by each country's civil aviation authority, the equivalent of the FAA, and disseminated as an aeronautical information publication (AIP).

[2] The database provider extract occurs at the 21-day point. The edited extract is sent to the avionics manufacturer or prepared with the avionics-packing program. Data not coded by the 21-day point will not be contained in the database extract for the effective cycle. In order for the data to be in the database at this 21-day extract, the actual cutoff is more like 28 days before the effective date.

pilot. The three EFB charting applications include Terminal Charts, En route Moving Map (EMM), and Airport Moving Map (AMM). The Terminal Charts EFB charting application utilizes the same information and layout as the printed chart counterpart. The EMM application uses the same ARINC 424 en route data that is extracted for an FMS database, but adds additional information associated with aeronautical charting needs. The EFB AMM database is a new high-resolution geo-spatial database only for EFB use. The AMM shows aircraft proximity relative to the airport environment. Runways depicted in the AMM correlate to the runway depictions in the FMS navigation database. The other information in the AMM such as ramps, aprons, taxiways, buildings, and hold-short lines are not included in traditional ARINC 424 databases.

THE ROLE OF THE AVIONICS MANUFACTURER
When avionics manufacturers develop a piece of equipment that requires an airborne navigation database, they typically form an agreement with a database provider to supply the database for that new avionics platform. It is up to the manufacturer to determine what information to include in the database for their system. In some cases, the navigation data provider has to significantly reduce the number of records in the database to accommodate the storage capacity of the manufacturer's new product.

The manufacturer must decide how its equipment will handle the records; decisions must be made about each field in the record. Each manufacturer can design their systems to manipulate the data fields in different ways, depending on the needs of the avionics user. Some fields may not be used at all. For instance, the ARINC primary record designed for individual runways may or may not be included in the database for a specific manufacturer's machine. The avionics manufacturer might specify that the database include only runways greater than 4,000 feet. If the record is included in the tailored database, some of the fields in that record may not be used.

Another important fact to remember is that although there are standard naming conventions included in the ARINC 424 specification; each manufacturer determines how the names of fixes and procedures are displayed to the pilot. This means that although the database may

specify the approach identifier field for the VOR/DME Runway 34 approach at Eugene Mahlon Sweet Airport (KEUG) in Eugene, Oregon, as "V34," different avionics platforms may display the identifier in any way the manufacturer deems appropriate. For example, a GPS produced by one manufacturer might display the approach as "VOR 34," whereas another might refer to the approach as "VOR/DME 34," and an FMS produced by another manufacturer may refer to it as "VOR34." [Figure A-2] These differences can cause visual inconsistencies between chart and GPS displays as well as confusion with approach clearances and other ATC instructions for pilots unfamiliar with specific manufacturer's naming conventions.

The manufacturer determines the capabilities and limitations of an RNAV system based on the decisions that it makes regarding that system's processing of the airborne navigation database.

USERS ROLE
Like paper charts, airborne navigation databases are subject to revision. Pilots using the databases are ultimately responsible for ensuring that the database they are operating with is current. This includes checking "NOTAM-type information" concerning errors that may be supplied by the avionics manufacturer or the database supplier. The database user is responsible for learning how the specific navigation equipment handles the navigation database. The manufacturer's documentation is

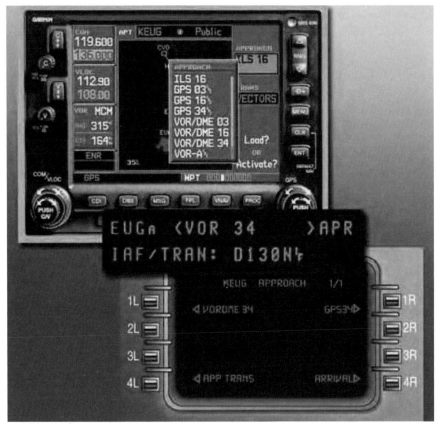

Figure A-2. Naming Conventions of Three Different Systems for the VOR 34 Approach.

the pilot's best source of information regarding the capabilities and limitations of a specific database.
[Figure A-3]

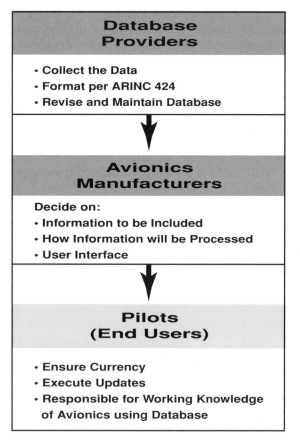

Database Providers

- Collect the Data
- Format per ARINC 424
- Revise and Maintain Database

Avionics Manufacturers

Decide on:
- Information to be Included
- How Information will be Processed
- User Interface

Pilots (End Users)

- Ensure Currency
- Execute Updates
- Responsible for Working Knowledge of Avionics using Database

Figure A-3. Database Roles.

COMPOSITION OF AIRBORNE NAVIGATION DATABASES

The concept of global position is an important concept of RNAV. Whereas short-range navigation deals primarily with azimuth and distance on a relatively small, flat surface, long-range point-to-point navigation must have a method of defining positions on the face of a large and imperfect sphere (or more specifically a mathematical reference surface called a geodetic datum). The latitude-longitude system is currently used to define these positions.

Each location/fix defined in an airborne navigation database is assigned latitude and longitude values in reference to a geodetic datum that can be used by avionics systems in navigation calculations.

THE WGS-84 REFERENCE DATUM

The idea of the earth as a sphere has existed in the scientific community since the early Greeks hypothesized about the shape and size of the earth over 2,000 years ago. This idea has become scientific fact, but it has been modified over time into the current theory of the earth's shape. Since modern avionics rely on databases and

mathematical geodetic computations to determine the distance and direction between points, those avionics systems must have some common frame of reference upon which to base those calculations. Unfortunately, the actual topographic shape of the earth's surface is far too complex to be stored as a reference datum in the memory of today's FMS or GPS data cards. Also, the mathematical calculations required to determine distance and direction using a reference datum of that complexity would be prohibitive. A simplified model of the earth's surface solves both of these problems for today's RNAV systems.

In 1735, the French Academy of Sciences sent an expedition to Peru and another to Lapland to measure the length of a meridian degree at each location. The expeditions determined conclusively that the earth is not a perfect sphere, but a flattened sphere, or what geologists call an **ellipsoid of revolution**. This means that the earth is flattened at the poles and bulges slightly at the equator. The most current measurements show that the polar diameter of the earth is about 7,900 statute miles and the equatorial diameter is 7,926 statute miles. This discovery proved to be very important in the field of geodetic survey because it increased the accuracy obtained when computing long distances using an earth model of this shape. This model of the earth is referred to as the Reference Ellipsoid, and combined with other mathematical parameters, it is used to define the reference for geodetic calculations or what is referred to as the **geodetic datum**.

Historically, each country has developed its own geodetic reference frame. In fact, until 1998 there were more than 160 different worldwide geodetic datums. This complicated accurate navigation between locations of great distance, especially if several reference datums are used along the route. In order to simplify RNAV and facilitate the use of GPS in the NAS, a common reference frame has evolved.

The reference datum currently being used in North America for airborne navigation databases is the North American Datum of 1983 (NAD-83), which for all practical navigation purposes is equivalent to the World Geodetic System of 1984 (WGS-84). Since WGS-84 is the geodetic datum that the constellation of GPS satellites are referenced to, it is the required datum for flight by reference to a GPS navigation receiver certified in accordance with FAA Technical Standard Order (TSO) C129A, Airborne Supplemental Navigation Equipment Using the Global Positioning System (GPS). The World Geodetic Datum was created by the Department of Defense in the 1960s as an earth-centered datum for military purposes, and one iteration of the model was adapted by the Department of Defense as a reference for GPS satellite orbits in 1987. The International Civil

Aviation Organization (ICAO) and the international aviation community recognized the need for a common reference frame and set WGS-84 as the worldwide geodetic standard. All countries were obligated to convert to WGS-84 in January 1998. Many countries have complied with ICAO, but many still have not done so due to the complexity of the transformation and their limited survey resources.

ARINC 424

First published in 1975, the ARINC document, Navigation System Data Base (ARINC 424), sets forth the air transport industry's recommended standards for the preparation of airborne navigation system reference data tapes. This document outlines the information to be included in the database for each specific navigation entity (i.e. airports, navigation aides [NAVAIDs], airways, and approaches), as well as the format in which the data is coded. The ARINC specification determines naming conventions.

RECORDS

The data included in an airborne navigation database is organized into ARINC 424 records. These records are strings of characters that make up complex descriptions of each navigation entity. There are 132 columns or spaces for characters in each record. Not all of the 132 character-positions are used for every record — some of the positions are left blank to permit like information to appear in the same columns of different records, and others are reserved for possible future record expansion. These records are divided into fields that contain specific pieces of information about the subject of the record. For instance, the primary record for an airport, such as KZXY, contains a field that describes the longest runway at that airport. The columns 28 through 30 in the record contain the first three digits in the longest runway's length in feet. If the numbers 0, 6, and 5 were in the number 28, 29, and 30 columns respectively, the longest runway at KZXY would be recorded in the record as 6,500 feet (065). [Figure A-4] Columns 28 through 30, which are designated as "longest runway" in the airport record, would be a different field in the record for a very high frequency omni-directional range (VOR) or an airway. The record type determines what fields are included and how they are organized.

For the purpose of discussion, ARINC records can be sorted into four general groups – fix records, simple route records, complex route records, and miscellaneous records. Although it is not important for pilots to have in-depth knowledge of all the fields contained in the ARINC 424 records, pilots should be aware of the types of records contained in the navigation database and their general content.

Columns—The spaces for data entry on each record. One column can accommodate one character.

Record—A single line of computer data made up of the fields necessary to define fully a single useful piece of data.

Field—The collection of characters needed to define one item of information.

FIX RECORDS

Database records that describe specific locations on the face of the earth can be considered fix records. NAVAIDs, waypoints, intersections, and airports are all examples of this type of record. These records can be used directly by avionics systems and can be included as parts of more complex records like airways or approaches.

Within the 132 characters that make up a fix record, there are several fields that are generally common to all: record type, latitude, longitude, ICAO fix identifier, and ICAO location code. One exception is airports that use FAA identifiers. In addition, fix records contain many fields that are specific to the type of fix they describe. Figure A-5 on page A-6 shows examples of field types for three different fix records.

In each of the above examples, magnetic variation is dealt with in a slightly different manner. Since the locations of these fixes are used to calculate the magnetic courses displayed in the cockpit, their records must include the location's magnetic variation to be used in

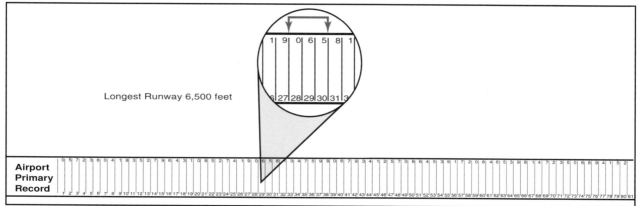

Figure A-4. Longest Runway Field in an Airport Record.

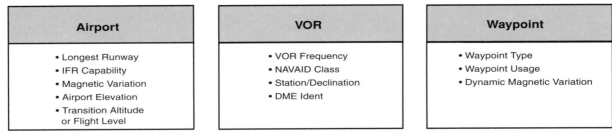

Airport	VOR	Waypoint
• Longest Runway • IFR Capability • Magnetic Variation • Airport Elevation • Transition Altitude or Flight Level	• VOR Frequency • NAVAID Class • Station/Declination • DME Ident	• Waypoint Type • Waypoint Usage • Dynamic Magnetic Variation

Figure A-5. Unique Fields for Three Different Fix Records.

those calculations. In records for airports for instance, the **magnetic variation** is given as the difference in degrees between the measured values of true north and magnetic north at that location. The field labeled "Station Declination" in the record for a VOR differs only slightly in that it is the angular difference between true north and the zero degree radial of the NAVAID the last time the site was checked. The record for a waypoint, on the other hand, contains a field named "**Dynamic Magnetic Variation,**" which is simply a computer model calculated value instead of a measured value.

Another concept pilots should understand relates to how aircraft make turns over navigation fixes. Fixes can be designated as fly-over or fly-by depending on how they are used in a specific route. [Figure A-6] Under certain circumstances, a navigation fix is designated as fly-over. This simply means that the aircraft must actually pass directly over the fix before initiating a turn to a new course. Conversely, a fix may be designated fly-by, allowing an aircraft's navigation system to use its turn anticipation feature, which ensures that the proper radius of turn is commanded to avoid overshooting the new course. Some RNAV systems are not programmed to fully use this feature. It is important to remember a fix can be coded as fly-over in one proce-dure, and fly-by in another, depending on how the fix is used.

SIMPLE ROUTE RECORDS

Route records are those that describe a flight path instead of a fixed position. Simple route records contain strings of fix records and information pertaining to how the fixes should be used by the navigation avionics. A Victor Airway, for example, is described in the database by a series of "en route airway records" that contain the names of fixes in the airway and information about how those fixes make up the airway. These records describe the way the fixes are used in the airway and contain important information including the fix identifier, sequence number, route type, required navigation performance (RNP), outbound and inbound magnetic courses (if appropriate), route distance, and minimum and maximum altitudes for the route.

Sequence number fields are a necessary addition to the navigation database because they allow the avionics system to track the fix order within the route. Most routes can be entered from any point and flown in both directions. The sequence number allows the avionics to keep track of the fixes in order, so that the proper flight path can be followed starting anywhere within the route.

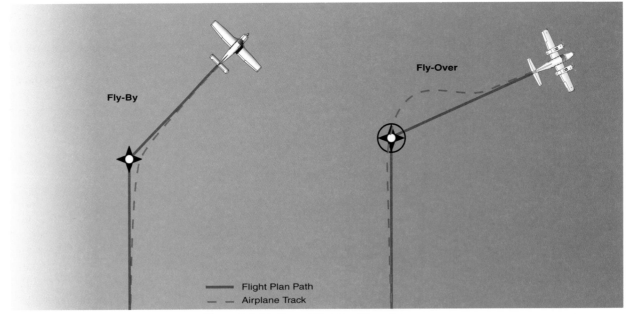

Figure A-6. Fly-By and Fly-Over Waypoints.

COMPLEX ROUTE RECORDS

Complex route records include those strings of fixes that describe complex flight paths like standard instrument departures (SIDs), standard terminal arrival routes (STARs), and instrument approach procedures. Like simple routes, these records contain the names of fixes to be used in the route as well as instructions on how the route will be flown. However, there are several fields included in these records that are unique to this type.

SID procedures are examples of complex routes that are coded in airborne navigation databases. The record for a SID includes many of the same types of information that are found in the en route airway record, and many other pieces of information that pertain only to complex flight paths. Some examples of the fields included in the SID record are the airport identifier, SID identifier, transition identifier, turn direction, recommended NAVAID, magnetic course, and path/terminator.

MISCELLANEOUS RECORDS

There are several other types of records coded into airborne navigation databases, most of which deal with airspace or communications. For example, there are records for restricted airspace, airport minimum safe altitudes, and grid minimum off route altitudes (MORAs). These records have many individual and unique fields that combine to describe the record's subject. Some are used by avionics manufacturers, some are not, depending on the individual capabilities of each RNAV unit.

THE PATH/TERMINATOR CONCEPT

One of the most important concepts for pilots to learn regarding the limitations of RNAV equipment has to do with the way these systems deal with the "Path/Terminator" field included in complex route records.

The first RNAV systems were capable of only one type of navigation: they could fly directly to a fix. This was not a problem when operating in the en route environment in which airways are mostly made up of direct (or very nearly direct) routes between fixes. The instrument approaches that were designed for RNAV also presented no problem for these systems and the databases they used since they consisted mainly of GPS overlay approaches that demanded only direct point-to-point navigation. The desire for RNAV equipment to have the ability to follow more complicated flight paths necessitated the development of the "Path/Terminator" field that is included in complex route records.

There are currently 23 different Path/Terminators in the ARINC 424 standard. They enable RNAV systems to follow the complex paths that make up instrument

departures, arrivals, and approaches. They describe to navigation avionics a path to be followed and the criteria that must be met before the path concludes and the next path begins. One of the simplest and most common Path/Terminators is the track to a fix (TF), which is used to define the great circle route between two known points. [Figure A-7] Additional information on Path/Terminator leg types is contained in Chapter 4.

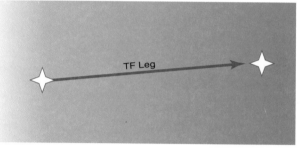

Figure A-7. Path/Terminator. A Path/Terminator value of a TF leg indicates a great circle track directly from one fix to the next.

The GRAND JUNCTION FOUR DEPARTURE for Walker Field in Grand Junction, Colorado, provides a good example of another type of Path/Terminator. [Figure A-8 on page A-8] When this procedure is coded into the navigation database, the person entering the data into the records must identify the individual legs of the flight path and then determine which type of terminator should be used.

The first leg of the departure for Runway 11 is a climb via runway heading to 6,000 feet mean sea level (MSL) and then a climbing right turn direct to a fix. When this is entered into the database, a heading to an altitude (VA) value must be entered into the record's Path/Terminator field for the first leg of the departure route. This Path/Terminator tells the avionics to provide course guidance based on heading, until the aircraft reaches 6,000 feet, and then the system begins providing course guidance for the next leg. After reaching 6,000 feet, the procedure calls for a right turn direct to the Grand Junction (JNC) VORTAC. This leg is coded into the database using the Path/Terminator direct to a fix (DF) value, which defines an unspecified track starting from an undefined position to a specific database fix. After reaching the JNC VORTAC the only Path/Terminator value used in the procedure is a TF leg.

Another commonly used Path/Terminator value is heading to a radial (VR). Figure A-9 on page A-9 shows the CHANNEL ONE DEPARTURE procedure for Santa Ana, California. The first leg of the runway 19L/R procedure requires a climb on runway heading until crossing the I-SNA 1 DME fix or the SLI R-118, this leg must be coded into the database using the VR value in the Path/Terminator field. After crossing the

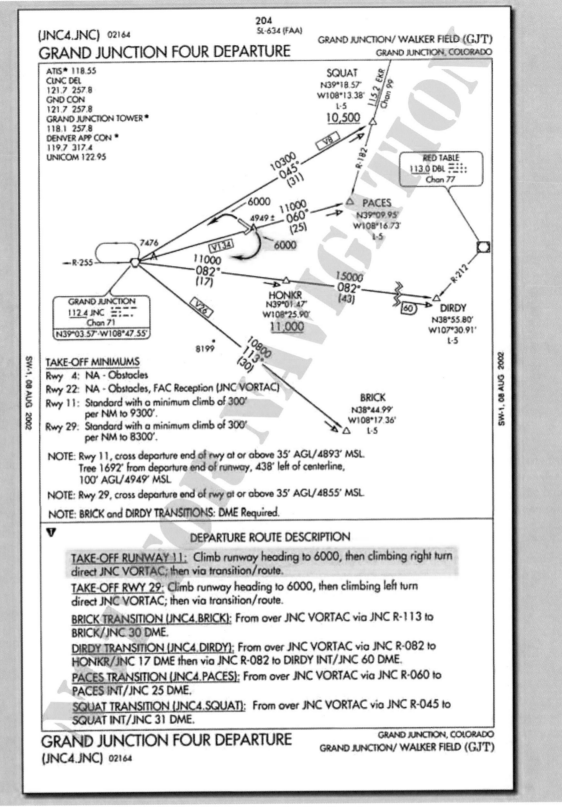

Figure A-8. Grand Junction Four Departure.

I-SNA 1 DME fix or the SLI R-118, the avionics should cycle to the next leg of the procedure that in this case, is a climb on a heading of 175° until crossing SLI R-132. This leg is also coded with a VR Path/Terminator. The next leg of the procedure consists of a heading of 200° until intercepting the SXC R-084. In order for the avionics to correctly process this leg, the database record must include the heading to an intercept (VI) value in the Path/Terminator field. This value directs the avionics to follow a specified heading to intercept the subsequent leg at an unspecified position.

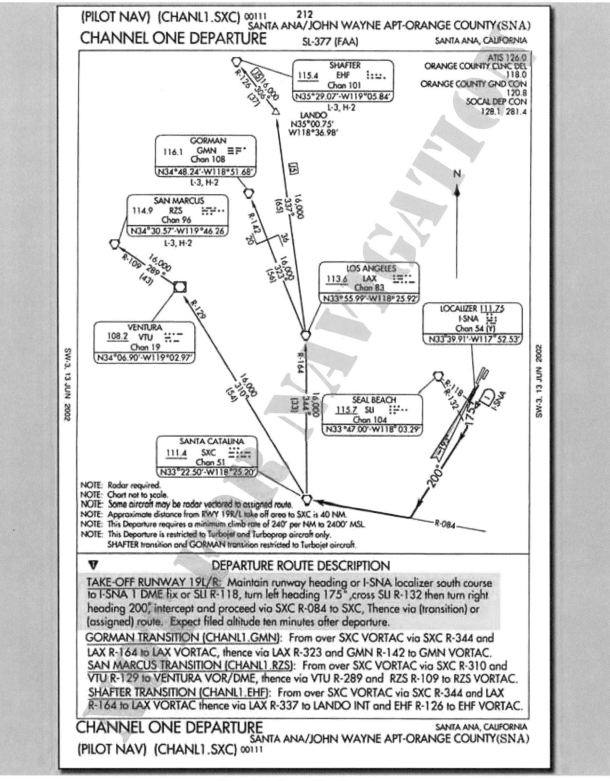

Figure A-9. Channel One Departure.

The Path/Terminator concept is a very important part of airborne navigation database coding. In general, it is not necessary for pilots to have an in-depth knowledge of the ARINC coding standards; however, pilots should be familiar with the concepts related to coding in order to understand the limitations of specific RNAV systems that use databases. For a more detailed discussion of coding standards, refer to ARINC Specification 424-15 Navigation System Data Base.

OPERATIONAL LIMITATIONS OF AIRBORNE NAVIGATION DATABASES

Understanding the capabilities and limitations of the navigation systems installed in an aircraft is one of the pilot's biggest concerns for IFR flight. Considering the vast number of RNAV systems and pilot interfaces available today, it is critical that pilots and flight crews be familiar with the manufacturer's operating manual for each RNAV system they operate and achieve and retain proficiency operating those systems in the IFR environment.

RELIANCE ON NAVIGATION AUTOMATION

Most professional and general aviation pilots are familiar with the possible human factors issues related to cockpit automation. It is particularly important to consider those issues when using airborne navigation databases. Although modern avionics can provide precise guidance throughout all phases of flight including complex departures and arrivals, not all systems have the same capabilities. RNAV equipment installed in some aircraft is limited to direct route point-to-point navigation. Therefore, it is very important for pilots to familiarize themselves with the capabilities of their systems through review of the manufacturer documentation.

Most modern RNAV systems are contained within an integrated avionics system that receives input from several different navigation and aircraft system sensors. These integrated systems provide so much information that pilots may sometimes fail to recognize errors in navigation caused by database discrepancies or misuse. Pilots must constantly ensure that the data they enter into their avionics is accurate and current. Once the transition to RNAV is made during a flight, pilots and flight crews must always be capable and ready to revert to conventional means of navigation if problems arise.

STORAGE LIMITATIONS

As the data in a worldwide database grows more detailed, the required data storage space increases. Over the years that panel-mounted GPS and FMS have developed, the size of the commercially available airborne navigation databases has grown exponentially. Some manufacturer's systems have kept up with this growth and some have not. Many of the limitations of older RNAV systems are a direct result of limited data storage capacity. For this reason, avionics manufacturers must make decisions regarding which types of data records will be extracted from the master database to be included with their system. For instance, older GPS units rarely include all of the waypoints that are coded into master databases. Even some modern FMSs, which typically have much larger storage capacity, do not include all of the data that is available from the database producers. The manufacturers often choose not to include certain types of data that they think is of low importance to the usability of the unit. For example, manufacturers of FMSs used in large airplanes may elect not to include airports where the longest runway is less than 3,000 feet or to include all the procedures for an airport.

Manufacturers of RNAV equipment can reduce the size of the data storage required in their avionics by limiting the geographic area the database covers. Like paper charts, the amount of data that needs to be carried with the aircraft is directly related to the size of the coverage area. Depending on the data storage that is available, this means that the larger the required coverage area, the less detailed the database can be.

Again, due to the wide range of possible storage capacities, and the number of different manufacturers and product lines, the manufacturer's documentation is the pilot's best source of information regarding limitations caused by storage capacity of RNAV avionics.

PATH/TERMINATOR LIMITATIONS

How a specific RNAV system deals with Path/Terminators is of great importance to pilots operating with airborne navigation databases. Some early RNAV systems may ignore this field completely. The ILS/DME RWY 2 approach at Durango, Colorado, provides an example of problems that may arise from the lack of Path/Terminator capability in RNAV systems. Although approaches of this type are authorized only for sufficiently equipped RNAV systems, it is possible that a pilot may elect to fly the approach with conventional navigation, and then re-engage RNAV during a missed approach. If this missed approach is flown using an RNAV system that does not use Path/Terminator values, then the system will most likely ignore the first two legs of the procedure. This will cause the RNAV equipment to direct the pilot to make an immediate turn toward the Durango VOR instead of flying the series of headings that terminate at specific altitudes as dictated by the approach procedure. [Figure A-10] Pilots must be aware of their individual systems Path/Terminator handling characteristics and always review the manufacturer's documentation to familiarize themselves with the capabilities of the RNAV equipment they are operating.

Pilots should be aware that some RNAV equipment was designed without the fly-over capability that was discussed earlier in this appendix. This can cause problems for pilots attempting to use this equipment to fly complex flight paths in the departure, arrival, or approach environments.

CHARTING/DATABASE INCONSISTENCIES

It is important for pilots to remember that many inconsistencies may exist between aeronautical charts and airborne navigation databases. Since there are so many

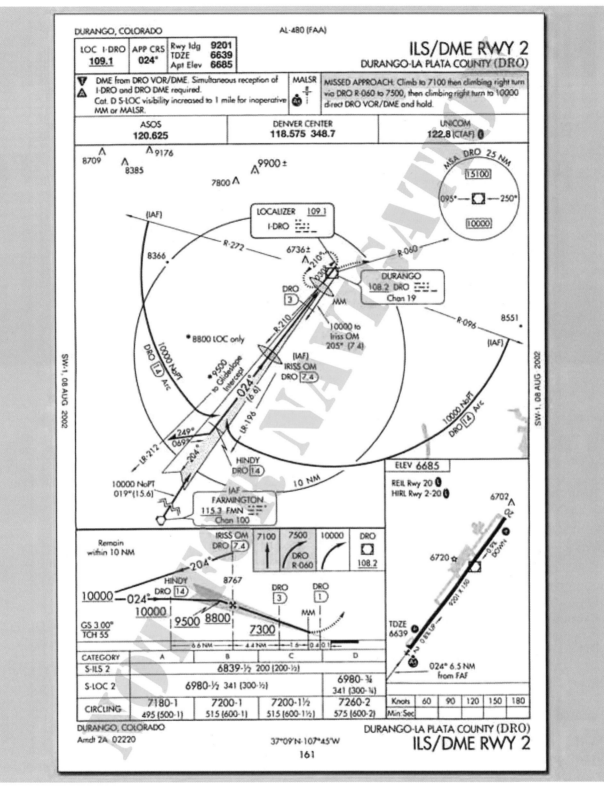

Figure A-10. ILS/DME Runway 2 in Durango, Colorado.

sources of information included in the production of these materials, and the data is manipulated by several different organizations before it eventually is displayed on RNAV equipment, the possibility is high that there will be noticeable differences between the charts and the databases. However, only the inconsistencies that

may be built into the databases are addressed in this discussion.

NAMING CONVENTIONS

As was discussed earlier in this appendix, obvious differences exist between the names of procedures shown on charts and those that appear on the displays of many

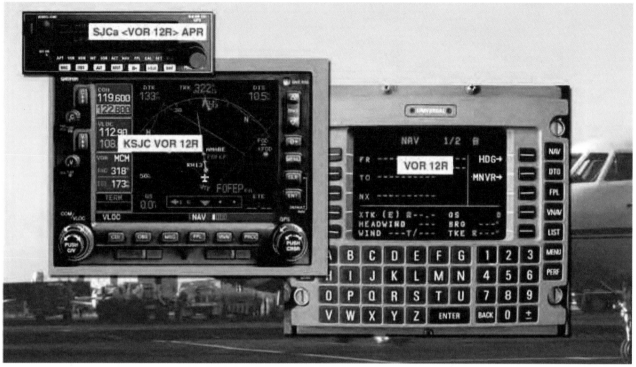

Figure A-11. Three Different Formats for the Same Approach.

RNAV systems. Most of these differences can be accounted for simply by the way the avionics manufacturers elect to display the information to the pilot. It is the avionics manufacturer that creates the interface between the pilot and the database, so the ARINC 424 naming conventions do not really apply. For example, the VOR 12R approach in San Jose, California, might be displayed several different ways depending on how the manufacturer designs the pilot interface. [Figure A-11] Some systems display procedure names exactly as they are charted, but many do not.

Although the three different names shown in Figure A-11 identify the same approach, the navigation system manufacturer has manipulated them into different formats to work within the framework of each specific machine. Of course, the data provided to the manufacturer in ARINC 424 format designates the approach as a 132-character data record that is not appropriate for display, so the manufacturer must create its own naming conventions for each of its systems.

NAVAIDs are subject to naming discrepancies. This problem is complicated by the fact that multiple NAVAIDs can be designated with the same identifier. VOR XYZ may occur several times in a provider's database, so the avionics manufacturer must design a way to identify these fixes by a more specific means than the three-letter identifier. Selection of geographic region is used in most instances to narrow the pilot's selection of NAVAIDs with like identifiers.

Non-directional beacons (NDBs) and locator outer markers (LOMs) can be displayed differently than they are charted. When the first airborne navigation data-

bases were being implemented, NDBs were included in the database as waypoints instead of NAVAIDs. This necessitated the use of five character identifiers for NDBs. Eventually, the NDBs were coded into the database as NAVAIDs, but many of the RNAV systems in use today continue to use the five-character identifier. These systems display the characters "NB" after the charted NDB identifier. Therefore, NDB ABC would be displayed as "ABCNB."

Other systems refer to NDB NAVAIDs using either the NDB's charted name if it is five or fewer letters, or the one to three character identifier. PENDY NDB located in North Carolina, for instance, is displayed on some systems as "PENDY," while other systems might only display the NDBs identifier "ACZ." [Figure A-12]

ISSUES RELATED TO MAGNETIC VARIATION

Magnetic variations for locations coded into airborne navigation databases can be acquired in several ways. In many cases they are supplied by government agencies in the "Epoch Year Variation" format. Theoretically, this value is determined by government sources and published for public use every five years. Providers of airborne navigation databases do not use annual drift values; instead the database uses the "Epoch Year Variation" until it is updated by the appropriate source provider. In the U.S., this is the National Oceanic and Atmospheric Administration (NOAA). In some cases the variation for a given location is a value that has been calculated by the avionics system. These "Dynamic Magnetic Variation" values can be different than those used for locations during aeronautical charting.

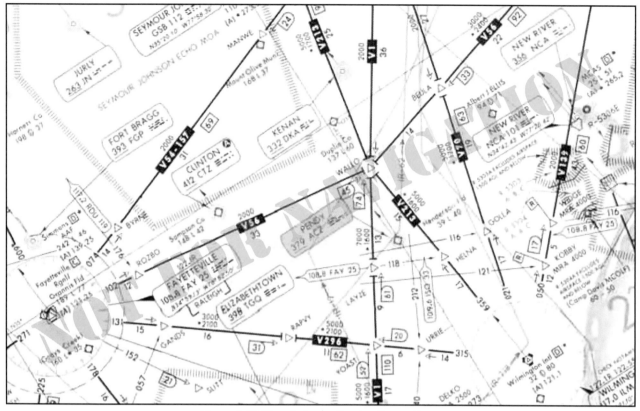

Figure A-12. Manufacturers Naming Conventions.

It is important to remember that even though ARINC standard records for airways and other procedures contain the appropriate magnetic headings and radials for routes, most RNAV systems do not use this information for en route flight. Magnetic courses are computed by airborne avionics using geodesic calculations based on the latitude and longitude of the waypoints along the route. Since all of these calculations are based on true north, the navigation system must have a way to account for magnetic variation. This can cause many discrepancies between the charted values and the values derived by the avionics. Some navigation receivers use the magnetic variation, or station declination, contained in the ARINC data records to make calculations, while other systems have independent ways of determining the magnetic variation in the general area of the VOR or waypoint.

Discrepancies can occur for many reasons. Even when the variation values from the database are used, the resulting calculated course might be different from the course depicted on the charts. Using the magnetic variation for the region, instead of the actual station declination, can result in differences between charted and calculated courses. Station declination is only updated when a NAVAID is "site checked" by the governing authority that controls it, so it is often different than the current magnetic variation for that location. Using an onboard means of determining variation usually entails coding some sort of earth model into the avionics memory. Since magnetic variation for a given location changes predictably over time, this model may only be correct

for one time in the lifecycle of the avionics. This means that if the intended lifecycle of a GPS unit were 20 years, the point at which the variation model might be correct would be when the GPS unit was 10 years old. The discrepancy would be greatest when the unit was new, and again near the end of its life span.

Another issue that can cause slight differences between charted course values and those in the database occurs when a terminal procedure is coded using "Magnetic Variation of Record." When approaches or other procedures are designed, the designers use specific rules to apply variation to a given procedure. Some controlling government agencies may elect to use the Epoch Year Variation of an airport to define entire procedures at that airport. This may cause the course discrepancies between the charted value and the value calculated using the actual variations from the database.

ISSUES RELATED TO REVISION CYCLE

Pilots should be aware that the length of the airborne navigation database revision cycle could cause discrepancies between aeronautical charts and information derived from the database. One important difference between aeronautical charts and databases is the length of cutoff time. Cutoff refers to the length of time between the last day that changes can be made in the revision, and the date the information becomes effective. Aeronautical charts typically have a cutoff date of 10 days prior to the effective date of the charts.

EVOLUTION OF RNAV

The use of RNAV equipment utilizing airborne navigation databases has significantly increased the capabilities of aircraft operating in the NAS. Pilots are now capable of direct flight over long distances with increasing precision. Furthermore, RNAV (RNP) instrument approach procedures are now capable of precision curved flight tracks. [Figure A-13] The availability of RNAV equipment has reached all facets of commercial, corporate, and general aviation. Airborne navigation databases have played a large role in this progress.

Although database providers have implemented a standard for airborne navigation databases, pilots must understand that RNAV is an evolving technology. Information published on current aeronautical charts must be used in cases where discrepancies or uncertainties exist with a navigation database. There are many variables relating to database, manufacturer, and user limitations that must be considered when operating with any RNAV equipment. Manufacturer documentation, aeronautical charts, and FAA publications are the pilot's best source of information regarding these capabilities and limitations.

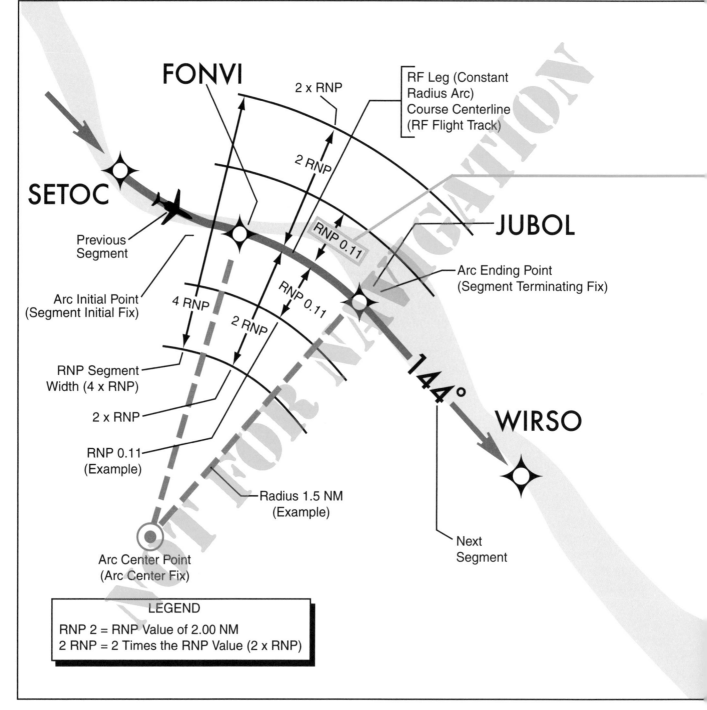

Figure A-13. Example of an RNAV (RNP) RF Leg Segment and Associated FMS Control Display Unit.

```
    ACT  RTE     LEGS      2/3
 -- ROUTE  DISCONTINUITY  -
SETOC    <CTR>----/  1510A
   1.5  ARC  L      GP  3.00°
FONVI            ----/  1218A
   1.5  ARC  R      GP  3.00°
JUBOL            ----/   933A
   146°        1.6NM GP  3.00°
WIRSO            ----/   424A
   0.8  ARC  R      GP  3.00°
FIROP            ----/   234A
   RNP/ACTUAL-------MAP  CTR
   0.11/0.09NM        STEP>
HIXIT
```

RNP navigation enables the geometry of instrument approach procedure design to be very flexible, and allows the incorporation of radius-to-fix (RF) legs enabling the FMS/autopilot to follow curved flight tracks. The constant radius arc RF leg defines a constant radius turn between two database fixes, lines tangent to the arc, and a center fix. While the arc initial point, arc ending point, and arc center point are available as database fixes, implementation of this leg type may not require the arc center point to be available as a fix.

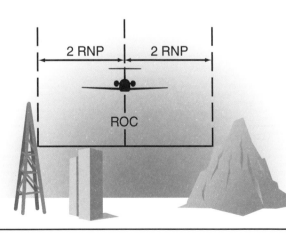

Appendix B
STAYING WITHIN PROTECTED AIRSPACE

At higher altitudes, protected airspace helps to maintain separation between aircraft. At lower altitudes, protected airspace also provides separation from terrain or obstructions. But, what does it mean to be established on course? How wide is the protected airspace of a particular route? How can you tell from the cockpit whether your aircraft is nearing the limits of protected airspace? The intent of this appendix is to answer these questions and explain the general limits of protected airspace by means of typical instrument indications.

Some pilots assume that flying to the tolerances set out in the FAA Instrument Practical Test Standards (PTS) (http://www.faa.gov/education_research/testing/airmen/test_standards/) will keep them within protected airspace. As a result, it is important to observe the last sentence of the following note in the PTS:

"The tolerances stated in this standard are intended to be used as a measurement of the applicant's ability to operate in the instrument environment. They provide guidance for examiners to use in judging the applicant's qualifications. The regulations governing the tolerances for operation under Instrument Flight Rules (IFR) are established in 14 CFR Part 91."

The in-flight presentation of course data can vary widely based upon the selection and distance from a Navigational Aid (NAVAID) or airfield. Consequently, you need to understand that in some cases, flying to the same standards required during your instrument rating flight test does not necessarily ensure that your aircraft will remain within protected airspace during IFR operations or that your aircraft will be in a position from which descent to a landing can be made using normal maneuvers.

For example, the PTS requires tracking a selected course, radial, or bearing within 3/4 of full-scale deflection (FSD) of the course deviation indicator (CDI). Since very high frequency omnidirectional ranges (VORs) use angular cross track deviation, the 3/4 scale deflection equates to 7.5 degrees, and means that the aircraft could be as much as 6.7 NM from the centerline when 51 NM from the VOR station. A VOR receiver is acceptable for IFR use if it indicates within four degrees of the reference when checked at a VOR test facility. If the maximum receiver tolerance is added to the allowable off-course indication, an aircraft could be 11.5 degrees from the centerline, or about 10.4 NM off the course centerline at 51 NM from the station. The primary protected airspace normally extends only 4 NM to each side of the centerline of published airways. (This example does not take into account any misalignment of the signals transmitted by the VOR.) [Figure B-1]

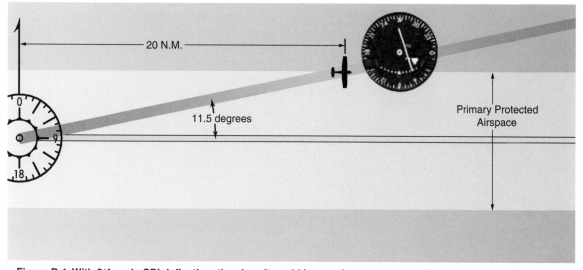

Figure B-1. With 3/4 scale CDI deflection, the aircraft could leave primary protected airspace when 20 NM from the station, assuming the transmitter is accurate and the receiver has a four degree error.

Lateral guidance is more intuitive with Area Navigation (RNAV) systems. For basic GPS, the CDI scale uses linear cross track deviation indications. During approach operations, a Wide Area Augmentation System (WAAS) navigation receiver combines the best of linear and angular deviations resulting in reduced Flight Technical Error (FTE). For departures, en route, and terminal operations, WAAS uses a linear deviation with varying scales. With linear scaling, if the CDI scaling is at 1 NM, a half scale deflection indicates that the aircraft is 1/2 NM off the course centerline, regardless of how far the aircraft is from the waypoints of the route segment. You need to be familiar with the distance and approach parameters that change the CDI scaling, and monitor the navigation unit to be sure the CDI scaling is appropriate for the route segment and phase of flight, e.g., **GPS C129 – Class C1** equipment used with a flight management system (FMS), unlike a C129A receiver, normally remains at the terminal scale of ±1 NM FSD during the approach (instead of ramping down to ±0.3 NM scaling beginning at 2 NM from the FAF). For this class of equipment, if a deviation of ±3/4 FSD is made from centerline during the approach, the aircraft will exceed the primary protected airspace width of ±0.5 NM by 1/4 NM.

Likewise, if a Category (CAT) I ILS is flown with ±3/4 FSD it can preclude an aircraft from safely transitioning to a landing on the runway. At a decision altitude (DA) point located 3,000 feet from the threshold with 3/4 FSD from centerline and above glidepath, the aircraft will be approximately 400 feet from centerline and 36 feet above the glidepath. If the aircraft were operating at 130 knots it would require two track changes within the 14-second transit time from the DA point to the threshold to align the aircraft with the runway. This may not allow landing within the touchdown zone (typically the first 3000 feet of a runway) when combined with strong crosswinds or Category C, D, or E airplane approach speeds.

Staying within protected airspace depends primarily on five factors:

- Accurate flying

- Accurate navigation equipment in the aircraft

- Accurate navigation signals from ground and space-based transmitters

- Accurate direction by air traffic control (ATC)

- Accurate (current) charts and publications

Incorporated within these factors are other related items, for example, flying accurately includes using the navigation equipment correctly, and accurate navigation equipment includes the altimeter.

- It is important for pilots to understand that the altimeter is a barometric device that measures pressure, not altitude. Some pilots may think of the altimeter as a true "altitude indicator," without error. In fact, the pressure altimeter is a barometer that measures changes in atmospheric pressure, and through a series of mechanisms and/or computer algorithms, converts these changes, and displays an altitude. This conversion process assumes standard atmospheric conditions, but since we fly in weather conditions other than standard, errors will result. Also, certain procedures may be annotated "NA" below a given temperature.

- The Instrument Flying Handbook (FAA-H-8083-15), Chapter 3, and the Aeronautical Information Manual (AIM), Chapter 7, include detailed discussions about altimeters and associated errors. Each includes the International Civil Aviation Organization (ICAO) Cold Temperature Error Table for altitude corrections when operating with an outside air temperature (OAT) below +10 degrees C.

The design of protected airspace is a very detailed and complex process, combining the professional skills of many different experts. Terrain elevations and contours, runway configurations, traffic considerations, prevailing winds and weather patterns, and the performance capabilities of the aircraft that will use the procedures must be balanced to create airspace that combines functionality with safety. Although it is not necessary for pilots to have an in-depth knowledge of how airspace is protected, it is useful to understand some of the terms used.

Required Obstacle Clearance (ROC) is the minimum vertical clearance required between the aircraft and ground obstructions over a specific point in an instrument procedure. Procedure designers apply the ROC when designing instrument approach procedures. On the initial segment, the ROC is approximately 1,000 feet, and it is at least 500 feet on the intermediate segment. Obviously, an imaginary surface 1,000 feet above the actual terrain and obstacles would be as rough and irregular as the surface below, so for practical reasons, airspace planners create smooth planes above the highest ground features and obstructions. These are called obstacle clearance surfaces (OCSs). Procedure designers use both level and sloping obstacle clearance surfaces when designing approaches.

Fix Displacement Area (FDA) is an area created by combining the permissible angular errors from the two VOR or nondirectional beacon (NDB) NAVAIDs that define the fix. When the NAVAIDs are close together and the angle that defines the fix is near 90 degrees, the

FDA is relatively small. At greater distances or less favorable angles, the FDA is larger. Airspace planners use the FDA to define the limits of protected airspace. [Figure B-2]

Fix Displacement Tolerance (FDT) is an area that applies to area navigation (RNAV) and equates to a FDA for VOR or NDB NAVAIDs. The FDT has an Along Track (ATRK) tolerance and a Cross Track (XTRK) tolerance.

Flight Technical Error (FTE) is the measure of the pilot or autopilot's ability to control the aircraft so that its indicated position matches the desired position. For example, FTE increases as the CDI swings further from center. If the cockpit instruments show the airplane to be exactly where you want it, the FTE is essentially zero.

Navigation System Error (NSE) is the error attributable to the navigation system in use. It includes the navigation sensor error, receiver error, and path definition error. NSE combines with FTE to produce the Total System Error (TSE). TSE is the difference between true position of the aircraft and the desired position. It combines the flight technical errors and the navigation system tracking errors.

Actual navigation performance (ANP) is an estimate of confidence in the current navigation system's performance. ANP computations consider accuracy, availability, continuity, and integrity of navigation performance at a given moment in time. Required Navigation Performance (RNP) necessitates the aircraft navigation system monitor the ANP and ensures the ANP does not exceed the RNP value required for the operation. The navigation system must also provide the pilot an alert in the primary field of view when ANP exceeds RNP. [Figure B-3 on page B-4]

While you may have thought of protected airspace as static and existing at all times whether aircraft are present or not, protection from conflicts with other aircraft is dynamic and constantly changing as aircraft move through the airspace. With continuous increases in air traffic, some routes have become extremely congested. Fortunately, the accuracy and integrity of aircraft navigation systems has also increased, making it possible to reduce the separation between aircraft routes without compromising safety. RNP is a standard for the navigation performance necessary to accurately keep an aircraft within a specific block of airspace.

Containment is a term central to the basic concept of RNP. This is the idea that the aircraft will remain within a certain distance of its intended position (the stated RNP value) at least 95 percent of the time on any flight.

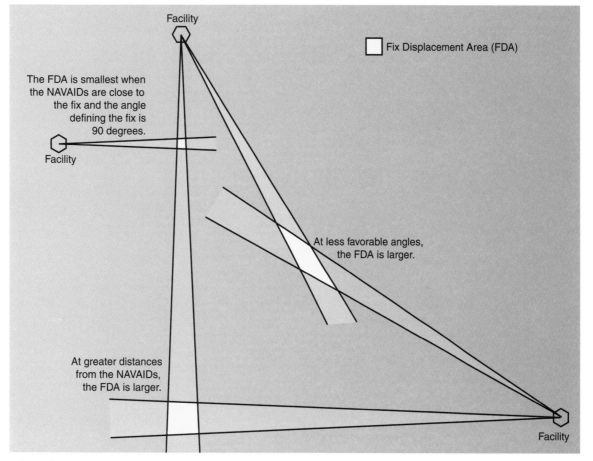

Figure B-2. The size of the protected airspace depends on where the terrestrial NAVAIDs that define it are located.

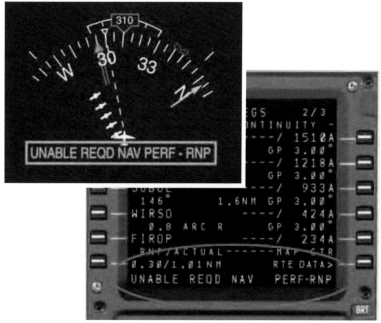

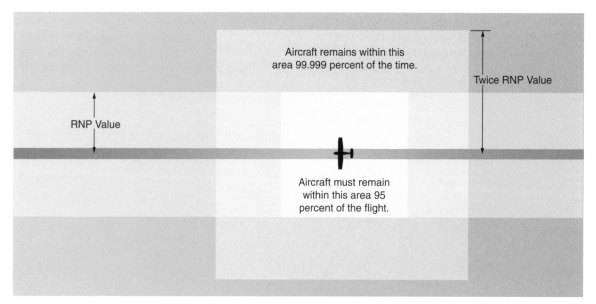

Figure B-3. An alerting system in the pilot's primary view must warn if ANP exceeds RNP. This alerting system is comparable to an "OFF" flag for a VOR or ILS.

This is a very high percentage, but it would not be enough to ensure the required level of safety without another layer of protection outside the basic containment area. This larger area has dimensions that are twice the RNP value, giving the aircraft two times the lateral area of the primary RNP area. Aircraft are expected to be contained within this larger boundary 99.999 percent of the time, which achieves the required level of confidence for safety. [Figure B-4]

Figure B-5 on pages B-6 through B-9 helps explain the cockpit indications and tolerances that will comply with criteria to keep you within protected airspace. The tolerances are predicated on zero instrument error unless noted otherwise. Special Aircraft and

Figure B-4. RNP Containment.

Aircrew Authorization Required (SAAAR) routes are not covered in this table.

For approaches, it is not enough to just stay within protected airspace. For nonprecision approaches, you must also establish a rate of descent and a track that will ensure arrival at the MDA prior to reaching the MAP with the aircraft continuously in a position from which a descent to a landing on the intended runway can be made at a normal rate using normal maneuvers. For precision approaches or approaches with vertical guidance, a transition to a normal landing is made only when the aircraft is in a position from which a descent to a landing on the runway can be made at a normal rate of descent using normal maneuvering.

For a pilot, remaining within protected airspace is largely a matter of staying as close as possible to the centerline of the intended course. There are formal definitions of what it means to be established on course, and these are important in practice as well as theory, since controllers often issue clearances contingent on your being established on a course.

You must be established on course before a descent is started on any route or approach segment. The ICAO Procedures for Air Navigation Services – Aircraft Operations (PANS-OPS) Volume I Flight Procedures, specifies, "Descent shall not be started until the aircraft is established on the inbound track," and that an aircraft is considered established when it is "within half full scale deflection for the ILS and VOR; or within ±5 degrees of the required bearing for the NDB."

In the AIM "established" is defined as "to be stable or fixed on a route, route segment, altitude, heading, etc." The "on course" concept for IFR is spelled out in Part 91.181, which states that the course to be flown on an airway is the centerline of the airway, and on any other route, along the direct course between the NAVAIDS or fixes defining that route.

As new navigational systems are developed with the capability of flying routes and approaches with increased resolution, increased navigation precision and pilot situational awareness is required. For safety, deviations from altitudes or course centerline should be communicated to ATC promptly. This is increasingly important when flights are in close proximity to restricted airspace. Whether you are a high time corporate pilot flying an aircraft that is equipped with state of the art avionics or a relatively new general aviation pilot that ventures into the NAS with only a VOR for navigation, adhering to the tolerances in Figure B-5 will help facilitate your remaining within protected airspace when conducting flights under IFR.

Phase of Flight

NAVAID	DEPARTURE	EN ROUTE	TERMINAL
NDB	RMI ±5 degrees. For departures, the climb area protected airspace initially splays at 15 degees from the ±500-foot width at the departure end of runway (DER) to 2 NM from the DER. The initial climb area width at 2 NM is ±3,756 feet from centerline. After the initial splay, the splay is 4.76 degrees until reaching an en route fix.	RMI ± 5 degrees. Because of angular cross track deviation, the NDB needle becomes less sensitive as you fly away from the NDB and more sensitive as you approach the station. The airway primary width is 4.34 NM either side of centerline to 49.6 NM. From 49.6 NM to the maximum standard service volume of 75 NM, the primary protected airspace splays at 5 degrees.	RMI ± 5 degrees. The maximum standard service volume for a compass locator is 15 NM. The feeder route width is ±4.34 NM.
VOR	CDI ±3/10 FSD (scale ±10 degrees). Same as NDB except after the initial splay, the splay is 2.86 degrees until reaching an en route fix.	CDI ±1/2 FSD up to 51 NM and beyond 51 NM CDI ±2/5 FSD (scale ±10 degrees). Like the NDB, the farther you are from the VOR, the more the signal diverges. The airway primary width is 4 NM either side of centerline to 51 NM. From 51 NM to the maximum standard service volume of 130 NM, the primary protected airspace splays at 4.5 degrees.	CDI ±1/2 FSD (scale ±10 degrees). The maximum standard service volume for a T VOR is 25 NM. The feeder route width is ±4 NM.
ILS	N/A	N/A	CDI ± 3/4 FSD for both lateral and vertical. The standard service volume for a localizer is 18 NM. The localizer total width at 18 NM is ±2.78 NM from centerline and tapers to approximately ±5,000 feet from centerline at the FAF. The standard service volume for the glide slope is 10 NM.

Figure B-5. Cockpit Indications and Tolerances to Keep You Within Protected Airspace. (Continued on Pages B-8 and B-9)

Phase of Flight

FINAL APPROACH	MISSED APPROACH	HOLDING
RMI ±10 degrees. If flown "FROM" the NDB, RMI ±5 degrees at the visual descent point (VDP) or equivalent for a normal landing. The course width for an approach with a FAF may be as small as 2.5 NM at the NDB and as wide as 5 NM at 15 NM from the NDB. For an on-airport facility, no FAF approach, the course width tapers from 6 NM (10 NM from the NDB) to 2.5 NM at the MAP/NDB.	RMI ±10 degrees. The course width widens to ±4 NM at 15 NM from the MAP.	RMI ±5 degrees. Intersections – the size of protected airspace varies with the distance from the NAVAID. See Figure 3-27 on page 3-23.
CDI ±3/4 FSD. If flown "FROM" the VOR, CDI ±1/2 FSD at the VDP or equivalent for a normal landing (scale ±10 degrees). The course width for an approach with a FAF may be as small as 2.0 NM at the VOR and as wide as 5 NM at 30 NM from the VOR. For an on airport facility no FAF approach, the course width tapers from 6 NM (10 NM from the VOR) to 2.0 NM at the MAP/VOR.	CDI ±3/4 FSD (scale ±10 degrees). For both FAF and no FAF approaches, the course width widens to 4 NM at 15 NM from the MAP.	CDI ±1/2 FSD (scale ±10 degrees). Intersections – the size of protected airspace varies with the distance from the NAVAIDs that form the holding fix. See Figure 3-27 on page 3-23.
CAT I CDI ±3/4 FSD for localizer and glidepath at the glide slope intercept. CDI ±1/2 FSD at the DA point for a normal landing. (scale total width may vary from 3 to 6 degrees). The normal length of final is 5 NM from the threshold. The final approach obstacle clearance area width at the FAF is approximately ±5,000 feet from centerline and tapers to as small as ±500 feet from centerline at 200 feet from the runway threshold. The CAT I final approach OCS can be as small as 500 feet below glidepath at the FAF. At a DA point located 3,000 feet from the threshold, the OCS may be as close as 114 feet below the glidepath. Decision range for airplane CAT II CDI ±1/6 FSD for localizer and ±1/4 FSD for glidepath and for helicopter ±1/4 FSD for localizer and glidepath. The tracking performance parameters within the decision range (that portion of the approach between 300 feet AGL and DH) are maximums, with no sustained oscillations about the localizer or glidepath. If the tracking performance is outside of these parameters while within the decision region, execute a go-around since the overall tracking performance is not sufficient to ensure that the aircraft will arrive at the DH on a flight path that permits the landing to be safely completed.	N/A	N/A

Phase of Flight

NAVAID	DEPARTURE	EN ROUTE	TERMINAL
GPS (C-129A)	CDI centered when departing the runway with a maximum of ±3/4 FSD upon reaching the terminal route (scale ± 1NM). For departures, the climb area protected airspace initially splays at 15 degrees from the ±500-foot width either side of centerline at the DER to a nominal distance of 2 NM from the DER. The initial climb area width at 2 NM is ± 3,756 feet from centerline. After the initial splay to the Initial Departure Fix (IDF) a smaller splay continues until reaching a terminal width as small as 2 NM at 10.89 NM from the DER. The horizontal alarm limit (HAL) is ±1 NM within 30 NM of the airport reference point (ARP).	CDI ±1/2 FSD (scale ±5 NM). The airway primary width is ±4 NM from centerline at 30 NM from the airport reference point (ARP). The HAL is ±2 NM for distances greater than 30 NM from the ARP.	CDI ±3/4 FSD (scale ±1 NM within 30 NM of the ARP). For arrivals, the terminal primary width is ±2 NM from centerline at approximately 30 NM from the ARP. The HAL is ±1 NM within 30 NM of the ARP.
WAAS LPV	CDI centered when departing the runway to the IDF with a maximum of ±3/4 FSD upon reaching the terminal route (CDI scale ±1NM). The HAL in the terminal mode is 1 NM.	CDI ±3/4 FSD (scale ±2 NM). The airway primary width is ±4 NM from centerline (equivalent to 2 RNP) at approximately 30 NM from the ARP. The HAL in the en route mode is 2 NM.	CDI ±3/4 FSD (scale ±1 NM within 30 NM of the ARP). For arrivals the terminal primary width is ±2 NM from centerline at approximately 30 NM from the ARP. The HAL in the terminal mode is 1 NM. The terminal mode begins at 30 NM from the ARP or at the initial approach fix (IAF) when more than 30 NM from the ARP.

Figure B-5. Continued

Phase of Flight

FINAL APPROACH	MISSED APPROACH	HOLDING
CDI ±2/3 FSD (scale ±0.3 NM). For conventional GPS approaches the primary width is ±1.0 NM from centerline at the FAF and tapers to ±0.5 NM at the MAP. CDI ±1/3 FSD for "Copter" approaches (scale ±0.3 NM). For "Copter" approaches the primary width is ± 0.55 NM from centerline at the FAF and tapers to ±0.4 NM at the MAP. The HAL is ±0.3 NM on the final approach segment (FAS). NOTE: GPS C129 – Class C1 (FMS equipped) Flight Director/Autopilot required since 1.0 NM scaling on the CDI is used. For Airplane approaches CDI ±1/5 FSD and for Copter approaches CDI ± 1/10 FSD (scale ±1.0 NM).	CDI ±3/4 FSD (scale ±1.0 NM). For missed approaches the primary width at the MAP is 0.5 NM and splays to ±4.0 NM from centerline at 15 NM from the MAP. For Copter approaches CDI ±1/2 FSD (scale ±1.0 NM). For missed approaches the primary width at the MAP is ±0.4 NM and splays to ±1.5 NM from centerline at 7.5 NM from the MAP. The HAL is ±1.0 NM within 30 NM of the ARP.	Terminal (within 30 NM of the ARP). CDI ±3/4 FSD (scale ±1.0 NM). En route (more than 30 NM from the ARP) CDI ±1/2 FSD (scale ±5.0 NM). The HAL for terminal holding is ±1.0 NM within 30 NM of the ARP and ±2 NM when more than 30 NM from the ARP.
CDI ±3/4 FSD lateral and vertical (LPV scale is ±2 degrees or ±0.3 NM FSD at the FAF whichever is less. Nonprecision scale is ±0.3 NM). LPV/LNAV approaches are similar to ILS/LOC approaches. LNAV (nonprecision): The CDI scaling for not vectored to final (VTF) approaches starts out with a linear width of ±1 NM FSD on the intermediate segment. At 2 NM prior to the FAF the scaling begins a change to either an angular ±2 degrees taper or ±0.3 NM FSD whichever is smaller. This change must be completed at the FAF. At the landing threshold point (LTP) the angular scale then becomes linear again with a width of approximately ±350 feet from centerline. For VTF approaches the CDI scaling starts out linear at ±1 NM FSD and changes to a ±2 degrees taper FSD and then becomes linear again with a width of approximately 350 feet from centerline at the LTP. Approaching the runway, a LPV nominal 3 degrees glidepath starts out linear (±150 M FSD) and then approximately 6 NM from the landing threshold becomes angular at a width of ±0.75 degrees and then becomes linear again as early as approximately 1.9 NM from the GPI for a ±45 M FSD or as small as a ±15 M FSD at a distance of approximately 0.6 NM from the landing threshold (depending on the manufacturer). The normal length of final is 5.0 NM from the threshold. The final approach obstacle clearance area width at the FAF is approximately ±4,000 feet from centerline and tapers to ±700 feet from centerline 200 feet from the runway threshold. The final approach OCS can be as small as 500 feet below glidepath at the FAF. At a DA point located 3,000 feet from the threshold, the OCS may be as small as 118 feet below the glidepath. The HAL for LNAV is 0.3 NM. The HAL for LPV is 40 M and the vertical alarm limit (VAL) starts out at 150 M and may be as large as 45 M near the LTP.	CDI ±1/2 FSD (From the LTP to the DER the scale is approximately ±350 feet wide and then changes ±0.3 NM at the DER) The primary width at the DA point for missed approaches (aligned within 3 degrees of the final approach course) is approximately ±1,000 feet from centerline and splays outward for 8,341 feet until reaching a width of ±3,038 feet from centerline. The HAL for missed approaches aligned within 3 degrees of the final approach course is ±0.3 NM at the DER and then changes to a HAL of ±1 NM at the turn initiation point for the first waypoint in the missed approach.	CDI ±3/4 FSD for terminal or en route holding (scale ±1.0 NM terminal and ±2.0 NM en route). The HAL is 1.0 NM when within 30 NM of the ARP and 2.0 NM beyond 30 NM of the ARP.

Appendix C
Acronyms and Glossary

AC — advisory circular

ACARS — aircraft communications addressing and reporting system

ACAS — airborne collision avoidance system

AD — airworthiness directive

ADF — automatic direction finder

ADS — automatic dependent surveillance

ADS-B — automatic dependent surveillance-broadcast

AER — approach end of runway

AFCS — automatic flight control system

A/FD — airport/facility directory

AFM — airplane flight manual or aircraft flight manual

AFSS — Automated Flight Service Station

AGL — above ground level

AIM — aeronautical information manual

AIP — aeronautical information publication

AIS — airmen's information system

ALAR — approach and landing accident reduction

AMASS — airport movement area safety system [delete term]

ANP — actual navigation performance

ANR — advanced navigation route

AOA — airport operating area

AOCC — airline operations control center

AOPA — Aircraft Owners and Pilots Association

AP — autopilot system

APC — auxiliary performance computer

APV — approach with vertical guidance

ARFF — aircraft rescue and fire fighting

ARINC — aeronautical radio incorporated

A-RNAV — advanced area navigation

ARSR — air route surveillance radar

ARTCC — Air Route Traffic Control Center

ARTS — Automated Radar Terminal System

ASDA — accelerate-stop distance available

ASDAR — aircraft to satellite data relay

ASDE-3 — Airport Surface Detection Equipment-3

ASDE-X — Airport Surface Detection Equipment-X

ASOS — automated surface observing system

ASR — airport surveillance radar

ATC — air traffic control

ATCRBS — air traffic control radar beacon system

ATCS — Air Traffic Control Specialist

ATCSCC — Air Traffic Control System Command Center

ATC-TFM — air traffic control-traffic flow management

ATCT — airport traffic control tower

ATD — along-track distance

ATIS — automatic terminal information service

ATM — air traffic management

ATS — air traffic service

ATT — attitude retention system

AVN — Office of Aviation System Standards

AWOS — automated weather observing system

AWSS — automated weather sensor system

Baro-VNAV — barometric vertical navigation

BRITE — bright radar indicator tower equipment

B-RNAV — European Basic RNAV

CAA — Civil Aeronautics Administration

CAASD — Center for Advanced Aviation Systems Development

CARF — central altitude reservation function

CAT — category

CDI — course deviation indicator

CDM — collaborative decision making

CDTI — cockpit display of traffic information

CDU — control display unit

CENRAP — Center Radar ARTS Processing

CFIT — controlled flight into terrain

CFR — Code of Federal Regulations

CGD — combined graphic display

CIP — Capital Investment Plan

CNF — computer navigation fix

CNS — communication, navigation, and surveillance

COP — changeover point

COTS — commercial off the shelf

CPDLC — controller pilot data link communications

CRC — cyclic redundancy check

CRCT — collaborative routing coordination tool

CRM — crewmember resource management

CRT — cathode-ray tube

CTAF — common traffic advisory frequency

CTD — controlled time of departure

CVFP — charted visual flight procedure

DA — density altitude, decision altitude

D-ATIS — digital automatic terminal information service

DACS — digital aeronautical chart supplement

DBRITE — digital bright radar indicator tower equipment

DER — departure end of the runway

DH — decision height

DME — distance measuring equipment

DOD — Department of Defense

DOT — Department of Transportation

DPs — departure procedures

DSR — display system replacement

C-2

DRVSM — domestic reduced vertical separation minimums

DUATS — direct user access terminal system

DVA — diverse vector area

EDCT — expect departure clearance time

EFB — electronic flight bag

EFC — expect further clearance

EFIS — electronic flight information system

EGPWS — enhanced ground proximity warning systems

EICAS — Engine indicating and crew alerting system

EMS — emergency medical service

EPE — estimated position error

ER-OPS — extended range operations

ETA — estimated time of arrival

EWINS — enhanced weather information system

FAA — Federal Aviation Administration

FAF — final approach fix

FAP — final approach point

FATO – Final Approach and Takeoff Area

FB — fly-by

FBWP — fly-by waypoint

FD — winds and temperatures aloft forecast

FD — flight director

FDC NOTAM — Flight Data Center Notice to Airmen

FDP — flight data processing

FIR — flight information region

FIS — flight information system

FIS-B — flight information service-broadcast

FISDL — flight information services data link

FL — flight level

FMC — flight management computer

FMS — flight management system

FO — fly-over

FOM — flight operations manual

FOWP — fly-over waypoint

FPM — feet per minute

FSDO — Flight Standards District Office

FSS — Flight Service Station

FTE — flight technical error

GA — general aviation

GAMA — General Aviation Manufacturer's Association

GBT — ground-based transmitter

GCA — ground controlled approach

GCO — ground communication outlet

GDP — ground delay programs

GDPE — ground delay program enhancements

GLS — Global Navigation Satellite System Landing System

GNE — gross navigation error

GNSS — Global Navigation Satellite System

GPS — Global Positioning System

GPWS — ground proximity warning system

G/S — glide slope

GS — groundspeed

GWS — graphical weather service

HAA — height above airport

HAR — High Altitude Redesign

HAT — height above touchdown

HDD — head-down display

HEMS — helicopter emergency medical service

HF — high frequency

HFDL — high frequency data link

HGS — head-up guidance system

HITS — highway in the sky

HOCSR — host/oceanic computer

system replacement

HSI — horizontal situation indicator

HSAC — Helicopter Safety Advisory Council

HUD — head-up display

IAF — initial approach fix

IAP — instrument approach procedure

IAS — indicated air speed

ICA — initial climb area

ICAO — International Civil Aviation Organization

IF — intermediate fix

IFR — instrument flight rules

ILS — instrument landing system

IMC — instrument meteorological conditions

INS — inertial navigation system

IOC — initial operational capability

IPV — instrument procedure with vertical guidance (this term has been renamed APV)

IRU – Inertial Reference Unit

KIAS — knots indicated airspeed

LAAS — Local Area Augmentation System

LAHSO — land and hold short operations

LDA — localizer type directional aid, landing distance available

LF — low frequency

LNAV — lateral navigation

LOA — letter of agreement/letter of authorization

LOC — localizer

LOM — locator outer marker

LPV — See glossary

LTP — landing threshold point

MAA — maximum authorized altitude

MAHWP — missed approach holding waypoint

MAMS — military airspace management system

MAP — missed approach point

MAP — manifold absolute pressure

MASPS — minimum aviation system performance specification

MAWP — missed approach waypoint

MCA — minimum crossing altitude

McTMA — multi-center traffic management advisor

MDA — minimum descent altitude

MDH — minimum descent height

MEA — minimum en route altitude

MEL — minimum equipment list

METAR — aviation routine weather report

MFD — multifunction display

MIA — minimum IFR altitude

MIT — miles-in-trail [delete term]

MLS — microwave landing system

MNPS — minimum navigation performance specifications

MOA — military operations area

MOCA — minimum obstruction clearance altitude

MOPS — minimum operational performance standards

MORA — minimum off route altitude

MRA — minimum reception altitude

MSA — minimum safe altitude

MSAW — minimum safe altitude warning

MSL — mean sea level

MTA — minimum turning altitude

MVA — minimum vectoring altitude

NA — not authorized

NACO — National Aeronautical Charting Office

NAR — National Airspace Redesign

NAS — National Airspace System

NASA — National Aeronautics and Space Administration

NASSI — National Airspace System status information

NAT — North Atlantic

NATCA — National Air Traffic Controllers Association

NAT/OPS — North Atlantic Operation

NAVAID — navigational aid

NBCAP — National Beacon Code Allocation Plan

ND — navigation displays

NDB — nondirectional beacon

NFDC — National Flight Data Center

NFPO — National Flight Procedures Office

NGA — National Geospatial-Intelligence Agency

NIMA — National Imagery and Mapping Agency

NM — nautical mile

NOAA — National Oceanic and Atmospheric Administration

NOPAC — North Pacific

NOTAM — Notice to Airmen

NOTAM D — Distant NOTAM

NOTAM L — Local NOTAM

NOZ — normal operating zone

NPA — nonprecision approach

NPRM — Notice of Proposed Rulemaking

NRP — national route program

NRR — non-restrictive routing

NRS — National Reference System

NSE — navigation system error

NTAP — Notice to Airmen Publication

NTSB — National Transportation Safety Board

NTZ — no transgression zone

NWS — National Weather Service

OCS — obstacle clearance surface

ODP — obstacle departure procedure

OEP — Operational Evolution Plan

OpsSpecs — operations specifications

OROCA — off-route obstruction clearance altitude

PA — precision approach

PAR — precision approach radar

PARC — performance-based operations aviation rulemaking committee

PCG — positive course guidance

PDC — pre-departure clearance

PDR — preferential departure route

PF — pilot flying

PFD — primary flight display

pFAST — passive final approach spacing tool

PIC — pilot in command

PinS — Point-in-Space

PIREP — pilot weather report

PM — pilot monitoring

POH — pilot's operating handbook

POI — principle operations inspector

PRM — precision runway monitor

P-RNAV — European Precision RNAV

PT — procedure turn

PTP — point-to-point

QFE — transition height

QNE — transition level

QNH — transition altitude

RA — resolution advisory, radio altitude

RAIM — receiver autonomous integrity monitoring

RCO — remote communications outlet

RDOF — radio failure

RJ — regional jet

RNAV — area navigation

RNP — required navigation performance

ROC — required obstacle clearance

RSP — runway safety program

RVR — runway visual range

RVSM — reduced vertical separation minimums

RVV — runway visibility value

RWY — runway

SAAAR — Special Aircraft and Aircrew Authorization Required

SAAR — special aircraft and aircrew requirements

SAMS — special use airspace management system

SAS — stability augmentation system

SATNAV — satellite navigation

SDF — simplified directional facility

SER — start end of runway

SIAP — standard instrument approach procedure

SID — standard instrument departure

SIGMET — significant meteorological information

SM — statute mile

SMA — surface movement advisor

SMGCS — surface movement guidance and control system

SMS — surface management system

SOIA — simultaneous offset instrument approaches

SOP — standard operating procedure

SPECI — non-routine (special) aviation weather report

SSV — standard service volume

STAR — standard terminal arrival

STARS — standard terminal automation replacement system

STC — supplemental type certificate

STMP — special traffic management program

SUA — special use airspace

SUA/ISE — special use airspace/inflight service enhancement

SVFR — special visual flight rules

SWAP — severe weather avoidance plan

TA — traffic advisory

TAA — terminal arrival area

TACAN — tactical air navigation

TAF — terminal aerodrome forecast

TAS — true air speed

TAWS — terrain awareness and warning systems

TCAS — traffic alert and collision avoidance system

TCH — threshold crossing height

TDLS — terminal data link system

TDZ — touchdown zone

TDZE — touchdown zone elevation

TEC — tower en route control

TERPS — U.S. Standard for Terminal Instrument Procedures

TFM — traffic flow management

TIS — traffic information service

TIS-B — traffic information service-broadcast

TLOF – Touchdown and Lift-Off Area

TM — traffic management

TMA — traffic management advisor

TMU — traffic management unit

TOC — top of climb

TOD — top of descent

TODA — takeoff distance available

TOGA — take-off/go around

TORA — takeoff runway available

TPP — terminal procedures publication

TRACAB — see glossary.

TRACON — terminal radar approach control

TSE — total navigation system error

TSO — technical standard order

UAT — universal access transceiver

UHF — ultra high frequency

URET — user request evaluation tool

US — United States

USAF — United States Air Force

VCOA — visual climb over airport

VDP — visual descent point

VFR — visual flight rules

VGSI — visual glide slope indicator

VHF — very high frequency

VLJ — very light jet

VMC — visual meteorological conditions

V$_{MINI}$ — minimum speed–IFR.

VNAV — vertical navigation

V$_{NEI}$ — never exceed speed-IFR.

VOR — very high frequency omnidirectional range

VORTAC — very high frequency omnidirectional range/tactical air navigation

VPA — vertical path angle

V$_{REF}$ — reference landing speed

V$_{SO}$ — stalling speed or the minimum steady flight speed in the landing configuration

WAAS — Wide Area Augmentation System

WAC — World Aeronautical Chart

WP — waypoint

Glossary

Abeam Fix – A fix, NAVAID, point, or object positioned approximately 90 degrees to the right or left of the aircraft track along a route of flight. Abeam indicates a general position rather than a precise point.

Accelerate-Stop Distance Available (ASDA) – The runway plus stopway length declared available and suitable for the acceleration and deceleration of an airplane aborting a takeoff.

Aircraft Approach Category – A grouping of aircraft based on reference landing speed (VREF), if specified, or if VREF is not specified, 1.3 VSO (the stalling speed or minimum steady flight speed in the landing configuration) at the maximum certificated landing weight.

Airport Diagram – A full-page depiction of the airport that includes the same features of the airport sketch plus additional details such as taxiway identifiers, airport latitude and longitude, and building identification. Airport diagrams are located in the U.S. Terminal Procedures booklet following the instrument approach charts for a particular airport.

Airport/Facility Directory (A/FD) – Regional booklets published by the National Aeronautical Charting Office (NACO) that provide textual information about all airports, both VFR and IFR. The A/FD includes runway length and width, runway surface, load bearing capacity, runway slope, airport services, and hazards such as birds and reduced visibility.

Airport Sketch – Depicts the runways and their length, width, and slope, the touchdown zone elevation, the lighting system installed on the end of the runway, and taxiways. Airport sketches are located on the lower left or right portion of the instrument approach chart.

Air Route Traffic Control Center (ARTCC) – A facility established to provide air traffic control service to aircraft operating on IFR flight plans within controlled airspace and principally during the en route phase of flight

Air Traffic Service (ATS) – Air traffic service is an ICAO generic term meaning variously, flight information service, alerting service, air traffic advisory service, air traffic control service (area control service, approach control service, or aerodrome control service).

Approach End of Runway (AER) – The first portion of the runway available for landing. If the runway threshold is displaced, use the displaced threshold latitude/longitude as the AER.

Approach Fix – From a database coding standpoint, an approach fix is considered to be an identifiable point in space from the intermediate fix (IF) inbound. A fix located between the initial approach fix (IAF) and the IF is considered to be associated with the approach transition or feeder route.

Approach Gate –An imaginary point used by ATC to vector aircraft to the final approach course. The approach gate is established along the final approach course 1 NM from the final approach fix (FAF) on the side away from the airport and is located no closer than 5 NM from the landing threshold.

Area Navigation (RNAV) – A method of navigation that permits aircraft operations on any desired course within the coverage of station referenced navigation signals or within the limits of self contained system capability.

Automated Surface Observing System (ASOS)/Automated Weather Sensor System (AWSS) – The ASOS/AWSS is the primary surface weather observing system of the U.S.

Automated Surface Observing System (ASOS) – A weather observing system that provides minute-by-minute weather observations such as temperature, dew point, wind, altimeter setting, visibility, sky condition, and precipitation. Some ASOS stations include a precipitation discriminator which can differentiate between liquid and frozen precipitation.

Automated Weather Observing System (AWOS) – A suite of sensors which measure, collect, and disseminate weather data. AWOS stations provide a minute-by-minute update of weather parameters such as wind speed and direction, temperature and dew point, visibility, cloud heights and types, precipitation, and barometric pressure. A variety of AWOS system types are available (from AWOS 1 to AWOS 3), each of which includes a different sensor array.

Automated Weather Sensor System (AWSS) – The AWSS is part of the Aviation Surface Weather Observation Network suite of programs and provides pilots and other users with weather information through the Automated Surface Observing System. The AWSS sensor suite automatically collects, measures, processes, and broadcasts surface weather data.

Automated Weather System – Any of the automated weather sensor platforms that collect weather data at airports and disseminate the weather information via radio and/or landline. The systems currently consist of the Automated Surface Observing System (ASOS), Automated Weather Sensor System (AWSS) and Automated Weather Observation System (AWOS).

Automatic Dependent Surveillance-Broadcast (ADS-B) – A surveillance system that continuously broadcasts GPS position information, aircraft identification, altitude, velocity vector, and direction to all other aircraft and air traffic control facilities within a specific area. Automatic dependent surveillance-broadcast (ADS-B) information will be displayed in the cockpit via a cockpit display of traffic information (CDTI) unit, providing the pilot with greater situational awareness. ADS-B transmissions will also provide controllers with a more complete picture of traffic and will update that information more frequently than other surveillance equipment.

Automatic Terminal Information Service (ATIS) – A recorded broadcast available at most airports with an operating control tower that includes crucial information about runways and instrument approaches in use, specific outages, and current weather conditions, including visibility.

Center Radar ARTS Presentation/Processing (CENRAP) – CENRAP was developed to provide an alternative to a non-radar environment at terminal facilities should an ASR fail or malfunction. CENRAP sends aircraft radar beacon target information to the ASR terminal facility equipped with ARTS.

Changeover Point (COP) – A COP indicates the point where a frequency change is necessary between navigation aids when other than the midpoint on an airway, to receive course guidance from the facility ahead of the aircraft instead of the one behind. These COPs divide an airway or route segment and ensure continuous reception of navigational signals at the prescribed minimum en route IFR altitude.

Charted Visual Flight Procedure (CVFP) – A CVFP may be established at some towered airports for environmental or noise considerations, as well as when necessary for the safety and efficiency of air traffic operations. Designed primarily for turbojet aircraft, CVFPs depict prominent landmarks, courses, and recommended altitudes to specific runways.

Cockpit display of traffic information (CDTI) – The display and user interface for information about air traffic within approximately 80 miles. It will typically combine and show traffic data from TCAS, TIS-B, and ADS-B. Depending on features, the display may also show terrain, weather, and navigation information.

Collision Hazard – A condition, event, or circumstance that could induce an occurrence of a collision or surface accident or incident.

Columns - See Database Columns

Contact Approach – An approach where an aircraft on an IFR flight plan, having an air traffic control authorization, operating clear of clouds with at least one mile flight visibility, and a reasonable expectation of contin-

uing to the destination airport in those conditions, may deviate from the instrument approach procedure and proceed to the destination airport by visual reference to the surface. This approach will only be authorized when requested by the pilot and the reported ground visibility at the destination airport is at least one statute mile.

Controlled Flight Into Terrain (CFIT) – A situation where a mechanically normally functioning airplane is inadvertently flown into the ground, water, or an obstacle. There are two basic causes of CFIT accidents; both involve flight crew situational awareness. One definition of situational awareness is an accurate perception by pilots of the factors and conditions currently affecting the safe operation of the aircraft and the crew. The causes of CFIT are the flight crews' lack of vertical position awareness or their lack of horizontal position awareness in relation to terrain and obstacles.

Database Columns – The spaces for data entry on each record. One column can accommodate one character.

Database Field – The collection of characters needed to define one item of information.

Database Identifier – A specific geographic point in space identified on an aeronautical chart and in a naviation database, officially designated by the controlling state authority or derived by Jeppesen. It has no ATC function and should not be used in filing flight plans nor used when communicating with ATC.

Database Record – A single line of computer data made up of the fields necessary to define fully a single useful piece of data.

Decision Altitude (DA) –A specified altitude in the precision approach at which a missed approach must be initiated if the required visual reference to continue the approach has not been established. The term "Decision Altitude (DA)" is referenced to mean sea level and the term "Decision Height (DH)" is referenced to the threshold elevation. Even though DH is charted as an altitude above MSL, the U.S. has adopted the term "DA" as a step toward harmonization of the United States and international terminology. At some point, DA will be published for all future instrument approach procedures with vertical guidance.

Decision Height (DH) – See Decision Altitude

Departure End of Runway (DER) – The end of runway available for the ground run of an aircraft departure. The end of the runway that is opposite the landing threshold, sometimes referred to as the stop end of the runway.

Descend Via – A descend via clearance instructs you to follow the altitudes published on a STAR. You are not authorized to leave your last assigned altitude unless specifically cleared to do so. If ATC amends the altitude or route to one that is different from the published procedure, the rest of the charted descent procedure is canceled. ATC will assign you any further route, altitude, or airspeed clearances, as necessary.

Digital ATIS (D-ATIS) – An alternative method of receiving ATIS reports by aircraft equipped with datalink services capable of receiving information in the cockpit over their Aircraft Communications Addressing and Reporting System (ACARS) unit.

Diverse Vector Area (DVA) – An airport may establish a diverse vector area if it is necessary to vector aircraft below the minimum vectoring altitude to assist in the efficient flow of departing traffic. DVA design requirements are outlined in TERPS and allow for the vectoring of aircraft immediately off the departure end of the runway below the MVA.

Dynamic Magnetic Variation – A field which is simply a computer model calculated value instead of a measured value contained in the record for a waypoint.

Electronic Flight Bag (EFB) – An electronic display system intended primarily for cockpit or cabin use. EFB devices can display a variety of aviation data or perform basic calculations (e.g., performance data, fuel calculations, etc.). In the past, some of these functions were traditionally accomplished using paper references or were based on data provided to the flight crew by an airline's "flight dispatch" function. The scope of the EFB system functionality may also include various other hosted databases and applications. Physical EFB displays may use various technologies, formats, and forms of communication. These devices are sometimes referred to as auxiliary performance computers (APC) or laptop auxiliary performance computers (LAPC).

Ellipsoid of Revolution – The surface that results when an ellipse is rotated about one of its axes.

En Route Obstacle Clearance Areas – Obstacle clearance areas for en route planning are identified as primary, secondary, and turning areas, and they are designed to provide obstacle clearance route protection width for airways and routes.

Expanded Service Volume – When ATC or a procedures specialist requires the use of a NAVAID beyond the limitations specified for standard service volume, an expanded service volume (ESV) may be established. See standard service volume.

Feeder Route – A feeder route is a route depicted on IAP charts to designate courses for aircraft to proceed from the en route structure to the IAF. Feeder routes, also referred to as approach transitions, technically are not considered approach segments but are an integral part of many IAPs.

Field - See Database Field

Final Approach and Takeoff Area (FATO) – The FATO is a defined heliport area over which the final approach to a hover or a departure is made. The touchdown and lift-off area (TLOF) where the helicopter is permitted to land is normally centered in the FATO. A safety area is provided around the FATO.

Fix – A geographical position determined by visual reference to the surface, by reference to one or more radio NAVAIDs, by celestial plotting, or by another navigational device. Note: Fix is a generic name for a geographical position and is referred to as a fix, waypoint, intersection, reporting point, etc.

Flight Information Region (FIR) – A FIR is an airspace of defined dimensions within which Flight Information Service and Alerting Service are provided. Flight Information Service (FIS) is a service provided for the purpose of giving advice and information useful for the safe and efficient conduct of flights. Alerting Service is a service provided to notify appropriate organizations regarding aircraft in need of search and rescue aid, and assist such organizations as required.

Flight Level (FL) – A flight level is a level of constant atmospheric pressure related to a reference datum of 29.92 in.Hg. Each flight level is stated in three digits that represents hundreds of feet. For example, FL 250 represents an altimeter indication of 25,000 feet.

Floating Waypoints – Floating waypoints represent airspace fixes at a point in space not directly associated with a conventional airway. In many cases they may be established for such purposes as ATC metering fixes, holding points, RNAV-direct routing, gateway waypoints, STAR origination points leaving the en route structure, and SID terminating points joining the en route structure.

Fly-By (FB) Waypoint – A waypoint that requires the use of turn anticipation to avoid overshooting the next flight segment.

Fly-Over (FO) Waypoint – A waypoint that precludes any turn until the waypoint is overflown, and is followed by either an intercept maneuver of the next flight segment or direct flight to the next waypoint.

Four Corner Post Configuration – An arrangement of air traffic pathways in a terminal area that brings incoming flights over fixes at four corners of the traffic area, while outbound flights depart between the fixes, thus minimizing conflicts between arriving and departing traffic.

Gateway Fix – A navigational aid or fix where an aircraft transitions between the domestic route structure and the oceanic route airspace.

Geodetic Datum – The reference plane from which geodetic calculations are made. Or, according to ICAO Annex 15, the numerical or geometrical quantity or set of such quantities (mathematical model) that serves as a reference for computing other quantities in a specific geographic region such as the latitude and longitude of a point.

Glidepath Angle (GPA) – The angular displacement of the vertical guidance path from a horizontal plane that passes through the reference datum point (RDP). This angle is published on approach charts (e.g., 3.00°, 3.20°, etc.). GPA is sometimes referred to as vertical path angle (VPA).

Global Navigation Satellite System (GNSS) – An umbrella term adopted by the International Civil Aviation Organization (ICAO) to encompass any independent satellite navigation system used by a pilot to perform onboard position determinations from the satellite data.

Gross Navigation Error (GNE) – In the North Atlantic area of operations, a gross navigation error is a lateral separation of more than 25 NM from the centerline of an aircraft's cleared route, which generates an Oceanic Navigation Error Report. This report is also generated by a vertical separation if you are more than 300 feet off your assigned flight level.

Ground Communication Outlet (GCO) – An unstaffed, remotely controlled ground/ground communications facility. Pilots at uncontrolled airports may contact ATC and AFSS via Very High Frequency (VHF) radio to a telephone connection. This lets pilots obtain an instrument clearance or close a VFR/IFR flight plan.

Head-Up Display (HUD) – See Head-Up Guidance System (HGS)

Head-Up Guidance System (HGS) – A system which projects critical flight data on a display positioned between the pilot and the windscreen. In addition to showing primary flight information, the HUD computes an extremely accurate instrument approach and landing guidance solution, and displays the result as a guidance cue for head-up viewing by the pilot.

Height Above Touchdown (HAT) – The height of the DA above touchdown zone elevation (TDZE).

Highway in the Sky (HITS) – A graphically intuitive pilot interface system that provides an aircraft operator with all of the attitude and guidance inputs required to safely fly an aircraft in close conformance to air traffic procedures.

Initial Climb Area (ICA) – An area beginning at the departure end of runway (DER) to provide unrestricted climb to at least 400 feet above DER elevation.

Instrument Approach Waypoint – Fixes used in defining RNAV IAPs, including the feeder waypoint (FWP), the initial approach waypoint (IAWP), the intermediate waypoint (IWP), the final approach waypoint (FAWP), the RWY WP, and the APT WP, when required.

Instrument Landing System (ILS) – A precision instrument approach system that normally consists of the following electronic components and visual aids; localizer, glide slope, outer marker, middle marker, and approach lights.

Instrument Procedure with Vertical Guidance (IPV) – Satellite or Flight Management System (FMS) lateral navigation (LNAV) with computed positive vertical guidance based on barometric or satellite elevation. This term has been renamed APV.

International Civil Aviation Organization (ICAO) – ICAO is a specialized agency of the United Nations whose objective is to develop standard principles and techniques of international air navigation and to promote development of civil aviation.

Intersection – Typically, the point at which two VOR radial position lines cross on a route, usually intersecting at a good angle for positive indication of position, resulting in a VOR/VOR fix.

Landing Distance Available (LDA) – ICAO defines LDA as the length of runway, which is declared available and suitable for the ground run of an aeroplane landing.

Lateral Navigation (LNAV) – Azimuth navigation, without positive vertical guidance. This type of navigation is associated with nonprecision approach procedures or en route.

Local Area Augmentation System (LAAS) – LAAS further increases the accuracy of GPS and improves signal integrity warnings.

Localizer Performance with Vertical Guidance (LPV) – LPV is one of the four lines of approach minimums found on an RNAV (GPS) approach chart. Lateral guidance accuracy is equivalent to a localizer. The HAT is published as a DA since it uses an electronic glide path that is not dependent on any ground equipment or barometric aiding and may be as low as 200 feet and $1/2$ SM visibility depending on the airport terrain and infrastructure. WAAS avionics approved for LPV is required. Baro-VNAV is not authorized to fly the LPV line of minimums on a RNAV (GPS) procedure since it uses an internally generated descent path that is subject to cold temperature effects and incorrect altimeter settings.

Loss of Separation – An occurrence or operation that results in less than prescribed separation between aircraft, or between an aircraft and a vehicle, pedestrian, or object.

LPV – See Localizer Performance with Vertical Guidance

Magnetic Variation – The difference in degrees between the measured values of true north and magnetic north at that location.

Maximum Authorized Altitude (MAA) – An MAA is a published altitude representing the maximum usable altitude or flight level for an airspace structure or route segment. It is the highest altitude on a Federal airway, jet route, RNAV low or high route, or other direct route for which an MEA is designated at which adequate reception of navigation signals is assured.

Metering Fix – A fix along an established route over which aircraft will be metered prior to entering terminal airspace. Normally, this fix should be established at a distance from the airport which will facilitate a profile descent 10,000 feet above airport elevation (AAE) or above.

Mid-RVR – The RVR readout values obtained from sensors located midfield of the runway.

Mileage Break – A point on a route where the leg segment mileage ends, and a new leg segment mileage begins, often at a route turning point.

Military Airspace Management System (MAMS) – A Department of Defense system to collect and disseminate information on the current status of special use airspace. This information is provided to the Special Use Airspace Management System (SAMS). The electronic interface also provides SUA schedules and historical activation and utilization data.

Minimum Crossing Altitude (MCA) – An MCA is the lowest altitude at certain fixes at which the aircraft must cross when proceeding in the direction of a higher minimum en route IFR altitude. MCAs are established in all cases where obstacles intervene to prevent pilots from maintaining obstacle clearance during a normal climb to a higher MEA after passing a point beyond which the higher MEA applies.

Minimum Descent Altitude (MDA) – The lowest altitude, expressed in feet above mean sea level, to which descent is authorized on final approach or during circle-to-land maneuvering in execution of a standard instrument approach procedure where no electronic glide slope is provided.

Minimum En Route Altitude (MEA) – The MEA is the lowest published altitude between radio fixes that assures acceptable navigational signal coverage and meets obstacle clearance requirements between those fixes. The MEA prescribed for a Federal airway or seg-ment, RNAV low or high route, or other direct route applies to the entire width of the airway, segment, or route between the radio fixes defining the airway, segment, or route.

Minimum IFR Altitude (MIA) – Minimum altitudes for IFR operations are prescribed in Part 91. These MIAs are published on NACO charts and prescribed in Part 95 for airways and routes, and in Part 97 for standard instrument approach procedures.

Minimum Navigation Performance Specifications (MNPS) – A set of standards which require aircraft to have a minimum navigation performance capability in order to operate in MNPS designated airspace. In addition, aircraft must be certified by their State of Registry for MNPS operation. Under certain conditions, non-MNPS aircraft can operate in MNPS airspace, however, standard oceanic separation minima is provided between the non-MNPS aircraft and other traffic.

Minimum Obstruction Clearance Altitude (MOCA) – The MOCA is the lowest published altitude in effect between radio fixes on VOR airways, off-airway routes, or route segments that meets obstacle clearance requirements for the entire route segment. This altitude also assures acceptable navigational signal coverage only within 22 NM of a VOR.

Minimum Reception Altitude (MRA) – An MRA is determined by FAA flight inspection traversing an entire route of flight to establish the minimum altitude the navigation signal can be received for the route and for off-course NAVAID facilities that determine a fix. When the MRA at the fix is higher than the MEA, an MRA is established for the fix, and is the lowest altitude at which an intersection can be determined.

Minimum Safe Altitudes (MSA) – MSAs are published for emergency use on IAP charts. For conventional navigation systems, the MSA is normally based on the primary omnidirectional facility on which the IAP is predicated. For RNAV approaches, the MSA is based on the runway waypoint (RWY WP) for straight-in approaches, or the airport waypoint (APT WP) for circling approaches. For GPS approaches, the MSA center will be the Missed Approach Waypoint (MAWP).

Minimum Vectoring Altitude (MVA) – Minimum vectoring altitude charts are developed for areas where there are numerous minimum vectoring altitudes due to variable terrain features or man-made obstacles. MVAs are established for use by ATC when radar ATC is exercised.

Missed Approach Holding Waypoint (MAHWP) – An approach waypoint sequenced during the holding portion of the missed approach procedure that is usually a fly-over waypoint, rather than a fly-by waypoint.

Missed Approach Waypoint (MAWP) – An approach waypoint sequenced during the missed approach procedure that is usually a fly-over waypoint, rather than a fly-by waypoint.

National Airspace System (NAS) – Consists of a complex collection of facilities, systems, equipment, procedures, and airports operated by thousands of people to provide a safe and efficient flying environment.

Navigational Gap – A navigational course guidance gap, referred to as an MEA gap, describes a distance along an airway or route segment where a gap in navigational signal coverage exists. The navigational gap may not exceed a specific distance that varies directly with altitude.

Nondirectional Radio Beacon (NDB) – An L/MF or UHF radio beacon transmitting nondirectional signals whereby the pilot of an aircraft equipped with direction finding equipment can determine bearing to or from the radio beacon and "home" on or track to or from the station. When the radio beacon is installed in conjunction with the ILS marker, it is normally called a compass locator.

Non-RNAV DP – A DP whose ground track is based on ground-based NAVAIDS and/or dead reckoning navigation.

Obstacle Clearance Surface (OCS) – An inclined or level surface associated with a defined area for obstruction evaluation.

Obstacle Departure Procedure (ODP) – A procedure that provides obstacle clearance. ODPs do not include ATC related climb requirements. In fact, the primary emphasis of ODP design is to use the least onerous route of flight to the en route structure while attempting to accommodate typical departure routes.

Obstacle Identification Surface (OIS) – The design of a departure procedure is based on TERPS, a living document that is updated frequently. Departure design criteria assumes an initial climb of 200 feet per NM after crossing the departure end of the runway (DER) at a height of at least 35 feet above the ground. Assuming a 200 feet per NM climb, the departure is structured to provide at least 48 feet per NM of clearance above objects that do not penetrate the obstacle slope. The slope, known as the obstacle identification slope (OIS), is based on a 40 to 1 ratio, which is the equivalent of a 152-foot per NM slope.

Off-Airway Routes – The FAA prescribes altitudes governing the operation of aircraft under IFR for off-airway routes in a similar manner to those on federal airways, jet routes, area navigation low or high altitude routes, and other direct routes for which an MEA is designated.

Off-Route Obstruction Clearance Altitude (OROCA) – An off-route altitude that provides obstruction clearance with a 1,000 foot buffer in non-mountainous terrain areas and a 2,000 foot buffer in designated mountainous areas within the U.S. This altitude may not provide signal coverage from ground-based navigational aids, air traffic control radar, or communications coverage.

Operations Specifications (OpsSpecs) – A published document providing the conditions under which an air carrier and operator for compensation or hire must operate in order to retain approval from the FAA.

Pilot Briefing Information – The current format for charted IAPs issued by NACO. The information is presented in a logical order facilitating pilot briefing of the procedures. Charts include formatted information required for quick pilot or flight crew reference located at the top of the chart.

Point-in-Space (PinS) Approach – An approach normally developed to heliports that do not meet the IFR heliport design standards but meet the standards for a VFR heliport. A helicopter PinS approach can be developed using conventional NAVAIDs or RNAV systems. These procedures have either a VFR or visual segment between the MAP and the landing area. The procedure will specify a course and distance from the MAP to the heliport(s) and include a note to proceed VFR or visually from the MAP to the heliport, or conduct the missed approach.

Positive Course Guidance (PCG) – A continuous display of navigational data that enables an aircraft to be flown along a specific course line, e.g., radar vector, RNAV, ground-based NAVAID.

Precision Runway Monitor (PRM) – Provides air traffic controllers with high precision secondary surveillance data for aircraft on final approach to parallel runways that have extended centerlines separated by less than 4,300 feet. High resolution color monitoring displays (FMA) are required to present surveillance track data to controllers along with detailed maps depicting approaches and a no transgression zone.

Preferential Departure Route (PDR) – A specific departure route from an airport or terminal area to an en route point where there is no further need for flow control. It may be included in an instrument Departure Procedure (DP) or a Preferred IFR Route.

Preferred IFR Routes – A system of preferred IFR routes guides you in planning your route of flight to minimize route changes during the operational phase of flight, and to aid in the efficient orderly management of air traffic using federal airways.

Principal Operations Inspector (POI) – Scheduled air carriers and operators for compensation or hire are assigned a principal operations inspector (POI) who works directly with the company and coordinates FAA operating approval.

Record – See Database Record

Reduced Vertical Separation Minimums (RVSM) – RVSM airspace is where air traffic control separates aircraft by a minimum of 1,000 feet vertically between flight level (FL) 290 and FL 410 inclusive. RVSM airspace is special qualification airspace; the operator and the aircraft used by the operator must be approved by the Administrator. Air traffic control notifies operators of RVSM by providing route planing information.

Reference Landing Speed (VREF) – The speed of the airplane, in a specified landing configuration, at the point where it descends through the 50-foot height in the determination of the landing distance.

Remote Communications Outlet (RCO) – An unmanned communications facility remotely controlled by air traffic personnel. RCOs serve FSSs and may be UHF or VHF. RCOs extend the communication range of the air traffic facility. RCOs were established to provide ground-to-ground communications between air traffic control specialists and pilots located at a satellite airport for delivering en route clearances, issuing departure authorizations, and acknowledging IFR cancellations or departure/landing times.

Reporting Point – A geographical location in relation to which the position of an aircraft is reported. (See Compulsory Reporting Points)

Required Navigation Performance (RNP) – RNP is a statement of the navigation performance necessary for operation within a defined airspace. On-board monitoring and alerting is required.

RNAV DP – A DP developed for RNAV-equipped aircraft whose ground track is based on satellite or DME/DME navigation systems.

Roll-out RVR – The RVR readout values obtained from sensors located nearest the rollout end of the runway.

Runway Heading – The magnetic direction that corresponds with the runway centerline extended, not the painted runway numbers on the runway. Pilots cleared to "fly or maintain runway heading" are expected to fly or maintain the published heading that corresponds with the extended centerline of the departure runway (until otherwise instructed by ATC), and are not to apply drift correction; e.g., RWY 4, actual magnetic heading of the runway centerline 044.22°, fly 044°.

Runway Hotspots – Locations on a particular airport that historically have hazardous intersections. Hot spots alert pilots to the fact that there may be a lack of visibility at certain points or the tower may be unable to see that particular intersection. Whatever the reason, pilots need to be aware that these hazardous intersections exist and they should be increasingly vigilant when approaching and taxiing through these intersections. Pilots are typically notified of these areas by a Letter to Airmen or by accessing the FAA Office of Runway Safety.

Runway Incursion – an occurrence at an airport involving an aircraft, vehicle, person, or object on the ground that creates a collision hazard or results in a loss of separation with an aircraft that is taking off, intending to take off, landing, or intending to land.

Runway Safety Program (RSP) – Designed to create and execute a plan of action that reduces the number of runway incursions at the nation's airports.

Runway Visual Range (RVR) – An estimate of the maximum distance at which the runway, or the specified lights or markers delineating it, can be seen from a position above a specific point on the runway centerline. RVR is normally determined by visibility sensors or transmissometers located alongside and higher than the centerline of the runway. RVR is reported in hundreds of feet.

Runway Visibility Value (RVV) – The visibility determined for a particular runway by a transmissometer. A meter provides a continuous indication of the visibility (reported in miles or fractions of miles) for the runway. RVV is used in lieu of prevailing visibility in determining minimums for a particular runway.

Significant Point – [ICAO Annex 11] A specified geographical location used in defining an ATS route or the flight path of an aircraft and for other navigation and ATS purposes.

Special Instrument Approach Procedure – A procedure approved by the FAA for individual operators, but not published in FAR 97 for public use.

Special Use Airspace Management System (SAMS) – A joint FAA and military program designed to improve civilian access to special use airspace by providing information on whether the airspace is active or scheduled to be active. The information is available to authorized users via an Internet website.

Standard Instrument Departure (SID) – An ATC requested and developed departure route designed to increase capacity of terminal airspace, effectively control the flow of traffic with minimal communication, and reduce environmental impact through noise abatement procedures.

Standard Service Volume – Most air navigation radio aids which provide positive course guidance have a designated standard service volume (SSV). The SSV defines the reception limits of unrestricted NAVAIDS which are usable for random/unpublished route navigation. Standard service volume limitations do not apply to published IFR routes or procedures. See the AIM for the SSV for specific NAVAID types.

Standard Terminal Arrival (STAR) – Provides a common method for departing the en route structure and navigating to your destination. A STAR is a pre-planned instrument flight rule ATC arrival procedure published for pilot use in graphic and textual form to simplify clearance delivery procedures. STARs provide you with a transition from the en route structure to an outer fix or an instrument approach fix or arrival way-point in the terminal area, and they usually terminate with an instrument or visual approach procedure.

Standardized Taxi Routes – Coded taxi routes that follow typical taxiway traffic patterns to move aircraft between gates and runways. ATC issues clearances using these coded routes to reduce radio communication and eliminate taxi instruction misinterpretation.

STAR Transition – A published segment used to connect one or more en route airways, jet routes, or RNAV routes to the basic STAR procedure. It is one of several routes that bring traffic from different directions into one STAR. NACO publishes STARs for airports with procedures authorized by the FAA, and these STARs are included at the front of each Terminal Procedures Publication regional booklet.

Start End of Runway (SER) – The beginning of the takeoff runway available.

Station Declination – The angular difference between true north and the zero radial of a VOR at the time the VOR was last site checked.

Surface Incident – An event during which authorized or unauthorized/unapproved movement occurs in the movement area or an occurrence in the movement area associated with the operation of an aircraft that affects or could affect the safety of flight.

Surface Movement Guidance Control System (SMGCS) – Facilitates the safe movement of aircraft and vehicles at airports where scheduled air carriers are conducting authorized operations. The SMGCS low visibility taxi plan includes the improvement of taxiway and runway signs, markings, and lighting, as well as the creation of SMGCS low visibility taxi route charts.

Synthetic Vision – A visual display of terrain, obstructions, runways, and other surface features that creates a virtual view of what the pilot would see out the window. This tool could be used to supplement normal vision in low visibility conditions, as well as to increase situational awareness in IMC.

Takeoff Distance Available (TODA) – ICAO defines TODA as the length of the takeoff runway available plus the length of the clearway, if provided.

Takeoff Runway Available (TORA) – ICAO defines TORA as the length of runway declared available and suitable for the ground run of an aeroplane takeoff.

Tangent Point (TP) –The point on the VOR/DME RNAV route centerline from which a line perpendicular to the route centerline would pass through the reference facility.

Terminal Arrival Area (TAA) – TAAs are the method by which aircraft are transitioned from the RNAV en route structure to the terminal area with minimal ATC interaction. The TAA consists of a designated volume of airspace designed to allow aircraft to enter a pro-tected area, offering guaranteed obstacle clearance where the initial approach course is intercepted based on the location of the aircraft relative to the airport.

Threshold – The beginning of the part of the runway usable for landing.

Top of Climb (TOC) – An identifiable waypoint repre-senting the point at which cruise altitude is first reached. TOC is calculated based on your current air-craft altitude, climb speed, and cruise altitude. There can only be one TOC waypoint at a time.

Top of Descent (TOD) – Generally utilized in flight management systems, top of descent is an identifiable waypoint representing the point at which descent is first initiated from cruise altitude. TOD is generally calcu-lated using the destination elevation (if available) and the descent speed schedule.

Touchdown and Lift-Off Area (TLOF) – The TLOF is a load bearing, usually paved area at a heliport where the helicopter is permitted to land. The TLOF can be located at ground or rooftop level, or on an elevated structure. The TLOF is normally centered in the FATO.

Touchdown RVR – The RVR visibility readout values obtained from sensors serving the runway touchdown zone.

Touchdown Zone Elevation (TDZE) – The highest elevation in the first 3,000 feet of the landing surface.

Tower En Route Control (TEC) – The control of IFR en route traffic within delegated airspace between two or more adjacent approach control facilities. This serv-ice is designed to expedite air traffic and reduces air traffic control and pilot communication requirements.

TRACAB – A new type of air traffic facility that con-sists of a radar approach control facility located in the tower cab of the primary airport, as opposed to a sepa-rate room.

Traffic Information Service-Broadcast (TIS-B) – An air traffic surveillance system that combines all available traffic information on a single display.

Traffic Management Advisor (TMA) – A software suite that helps air traffic controllers to sequence arriving air traffic.

Transition Altitude (QNH) – The altitude in the vicinity of an airport at or below which the vertical position of an aircraft is controlled by reference to altitudes (MSL).

Transition Height (QFE) – Transition height is the height in the vicinity of an airport at or below which the vertical position of an aircraft is expressed in height above the airport reference datum.

Transition Layer – Transition layer is the airspace between the transition altitude and the transition level. Aircraft descending through the transition layer will set altimeters to local station pressure, while departing aircraft climbing through the transition layer will be using standard altimeter setting (QNE) of 29.92 inches of Mercury, 1013.2 millibars, or 1013.2 hectopascals.

Transition Level (QNE) – The lowest flight level available for use above the transition altitude.

Turn Anticipation – The capability of RNAV systems to determine the point along a course, prior to a turn WP, where a turn should be initiated to provide a smooth path to intercept the succeeding course, and to enunciate the information to the pilot.

Turn WP [Turning Point] –A WP which identifies a change from one course to another.

User-defined Waypoint – User-defined waypoints typically are created by pilots for use in their own random RNAV direct navigation. They are newly established, unpublished airspace fixes that are designated geographic locations/positions that help provide positive course guidance for navigation and a means of checking progress on a flight. They may or may not be actually plotted by the pilot on enroute charts, but would normally be communicated to ATC in terms of bearing and distance or latitude/longitude. An example of user-defined waypoints typically includes those derived from database-driven area navigation (RNAV) systems whereby latitude/longitude coordinate-based waypoints are generated by various means including keyboard input, and even electronic map mode functions used to establish waypoints with a cursor on the display. Another example is an offset phantom waypoint, which is a point in space formed by a bearing and distance from NAVAIDs such as VORs, VORTACs, and TACANs, using a variety of navigation systems.

User Request Evaluation Tool (URET) – The URET helps provide enhanced, automated flight data management. URET is an automated tool provided at each radar position in selected en route facilities. It uses flight and radar data to determine present and future trajectories for all active and proposed aircraft flights. A graphic plan display depicts aircraft, traffic, and notification of predicted conflicts. Graphic routes for current plans and trial plans are displayed upon controller request. URET can generate a predicted conflict of two aircraft, or between aircraft and airspace.

Vertical Navigation (VNAV) – Traditionally, the only way to get glidepath information during an approach was to use a ground-based NAVAID, but modern area navigation systems allow flight crews to display an internally generated descent path that allows a constant rate descent to minimums during approaches that would otherwise include multiple level-offs.

Vertical Navigation Planning – Included within certain STARs is information provided to help you reduce the amount of low altitude flying time for high performance aircraft, like jets and turboprops. An expected altitude is given for a key fix along the route. By knowing an intermediate altitude in advance when flying a high performance aircraft, you can plan the power or thrust settings and aircraft configurations that result in the most efficient descent, in terms of time, fuel requirements, and engine wear.

Visual Approach – A visual approach is an ATC authorization for an aircraft on an IFR flight plan to proceed visually to the airport of intended landing; it is not an IAP. Also, there is no missed approach segment. When it is operationally beneficial, ATC may authorize pilots to conduct a visual approach to the airport in lieu of the published IAP. A visual approach can be initiated by a pilot or the controller.

Visual Climb Over the Airport (VCOA) – An option to allow an aircraft to climb over the airport with visual reference to obstacles to attain a suitable altitude from which to proceed with an IFR departure.

Waypoints – Area navigation waypoints are specified geographical locations, or fixes, used to define an area navigation route or the flight path of an aircraft employing area navigation. Waypoints may be any of the following types: predefined, published, floating, user-defined, fly-by, or fly-over.

Waypoint (WP) – A predetermined geographical position used for route/instrument approach definition, progress reports, published VFR routes, visual reporting points or points for transitioning and/or circumnavigating controlled and/or special use airspace, that is defined relative to a VORTAC station or in terms of latitude/longitude coordinates.

Wide Area Augmentation System (WAAS) – A method of navigation based on GPS. Ground correction stations transmit position corrections that enhance system accuracy and add vertical navigation (VNAV) features.

Index

A

B

C

R

Radar Approaches, 5-61
Radar Departure, 2-34
Radar DP, 2-18
Radar Required, 4-8
Radar Systems, 1-16, 1-18, 1-19
Radar Vectors to Final Approach Course, 4-6
Random RNAV Routes, 3-28
Receiver Autonomous Integrity Monitoring (RAIM), 5-11
Recommended Altitudes, 5-18
Reduced Vertical Separation Minimum (RVSM), 1-16, 3-27, 3-40, 6-2, 6-10
Reference Landing Speed (V_{REF}), 5-7
Regional Jet (RJ), 1-9, 6-1
Release Time, 1-11
Remote Communications Outlet (RCO), 2-27, 5-15
Required Climb Gradient, 2-15
Required Navigation Performance (RNP), 1-13, 1-14, 1-15, 2-28, 3-26, 3-37, 3-38, 5-23, 5-26, 6-2, 6-8, B-3
Required Obstacle Clearance (ROC), 2-12, 3-13, 5-38, B-2
RNAV, see Area Navigation
RNP, see Required Navigation Performance
Rollout RVR, 2-9
Runway Guard Lights, 2-2
Runway Hotspots, 2-4
Runway Incursion, 1-6, 2-3, 2-6
Runway Safety Program (RSP), 2-6
Runway Template Action Plan, 6-5
Runway Visibility Value (RVV), 2-9, 5-5
Runway Visual Range (RVR) 2-1, 2-8, 2-9, 5-5
RVSM, 1-16, 3-27, 3-40, 6-2, 6-10

S

Satellite-Based Navigation, 3-10
Satellite Navigation (SATNAV), 1-15, 1-16, 1-22
Secondary Obstacle Clearance Area, 3-8
Separation Standard, 6-6
Simplified Directional Facility (SDF), 5-66, 5-68
Simultaneous Close Parallel (Independent) ILS, 5-50
Simultaneous Independent Approaches, 1-17
Simultaneous Offset Instrument Approach (SOIA)/PRM, 1-17, 5-53, 5-56
Simultaneous Parallel (Independent) ILS, 5-50
SPECI, 5-4
Special Aircraft and Aircrew Authorization Required (SAAAR), 5-23, B-4
Special Airport Qualification, 4-26
Special Instrument Approach Procedure (SIAP), 5-44

Special Navigation Qualifications, 4-26
Special Pilot Qualifications, 4-26
Special Traffic Management Program (SMTP), 1-32
Special Use Airspace, 6-12
Speed Adjustments, 3-23
Speed Restrictions, 4-11
Stability Augmentation System (SAS), Helicopter, 7-2
Stabilization, Helicopter, 7-2
Stabilized Approach, 5-30, 5-31
Stabilized Descent, 4-4
Stand-Alone GPS Procedures, 5-11
Standard Alternate Minimums, 2-12
Standard Instrument Approach, Helicopter, 7-9
Standard Instrument Approach Procedure (SIAP), 1-26
Standard Instrument Departure (SID), 1-28, 2-17, 2-19 to 2-36, 4-15
Standard Parallels, 3-27
Standard Service Volume (SSV), 3-3
Standard Taxi Routes, 2-4, 2-5
Standard Terminal Arrival Route (STAR), 1-26, 1-28, 4-1, 4-2, 4-15 to 4-23, 5-28
   STAR Transition, 4-15
Standard Terminal Automation Replacement System (STARS), 1-18, 1-19, 1-28
Station Declination, A-6
Staying Within Protected Airspace, B-2
Stepdown Fixes, 4-17
Sterile Cockpit Rules, 4-13
Stop Bar Lights, 2-2
Straight-In Approaches, 5-8
Substitute Airway and Route Segments, 3-3
Suitability of a Specific IAP, 5-1
Supplement Alaska, 1-29
Surface Incident, 2-6
Surface Management System, 6-5
Surface Movement Advisor (SMA), 6-5
Surface Movement Guidance and Control System (SMGCS), 1-25, 1-26, 2-2, 2-3
Surface Movement Safety, 2-1
Surveillance Systems, 1-23
Synthetic Vision, 6-14
System Capacity, 1-6
System Safety, 1-5

T

TAA, 4-9, 5-23, 5-45
Tailwind Adjustment, 4-2
Takeoff Alternate, 2-12
Takeoff Distance Available (TODA), 2-13
Takeoff Minimums, 2-6, 2-8
   Takeoff Minimums and Obstacle Departure